Solar Cells: From Materials to Device Technology

Solar Cells: From Materials to Device Technology

Editor: Anderson Sheilds

Callisto Reference
www.callistoreference.com

Callisto Reference,
118-35 Queens Blvd., Suite 400,
Forest Hills, NY 11375, USA

Visit us on the World Wide Web at:
www.callistoreference.com

ISBN: 978-1-64116-826-7 (Hardback)

Cataloging-in-Publication Data

Solar cells : from materials to device technology / edited by Anderson Sheilds.
p. cm.
Includes bibliographical references and index.
ISBN 978-1-64116-826-7
1. Solar cells. 2. Solar cells--Materials. 3. Solar cells--Technological innovations. 4. Solar energy.
5. Renewable energy sources. I. Sheilds, Anderson.

TK2960 .S65 2023
621.312 44--dc23

Table of Contents

Preface

A solar cell refers to a type of device that transforms light into electrical energy by directly using the photovoltaic effect technology. Majority of the solar cells are made using silicon, which is the most common semiconductor material. Solar cells are usually made from n-type and p-type silicon semiconductors. The cells can be organized as arrays, which are made up of thousands of individual cells. These arrays can serve as an electric power plant, which generates electrical energy using sunlight for distribution in residential, industrial and commercial areas. Solar cell panels are also utilized to generate electricity in numerous distant terrestrial regions, where conventional electric power sources are either unavailable or impractical to install. Solar cells are used frequently in power consumer devices such as portable radios, electronic toys and pocket calculators. This book unfolds the concepts and innovative models around prospective developments with respect to solar cells materials and device technology. It is a resource guide for experts as well as students.

Significant researches are present in this book. Intensive efforts have been employed by authors to make this book an outstanding discourse. This book contains the enlightening chapters which have been written on the basis of significant researches done by the experts.

Finally, I would also like to thank all the members involved in this book for being a team and meeting all the deadlines for the submission of their respective works. I would also like to thank my friends and family for being supportive in my efforts.

Editor

1

High-Dose Electron Radiation and Unexpected Room-Temperature Self-Healing of Epitaxial SiC Schottky Barrier Diodes

Guixia Yang [1], Yuanlong Pang [1], Yuqing Yang [1], Jianyong Liu [1], Shuming Peng [1], Gang Chen [2], Ming Jiang [3], Xiaotao Zu [3], Xuan Fang [4], Hongbin Zhao [5], Liang Qiao [3,*] and Haiyan Xiao [3,*]

1 Institute of Nuclear Physics and Chemistry, China Academy of Engineering Physics, P.O. Box 919-220, Mianyang 621900, China; biansechong@163.com (G.Y.); pyl641122@163.com (Y.P.); yangsapphire@aliyun.com (Y.Y.); m13568266352@163.com (J.L.); pengshuming@caep.cn (S.P.)

2 State Key Laboratory of Wide-Bandgap Semiconductor Power Electronics, Nanjing 210000, China; steelchg@163.com

3 School of Physical, University of Electronic Science and Technology of China, Chengdu 610054, China; mjianglw@gmail.com (M.J.); xtzu@uestc.edu.cn (X.Z.)

4 School of Science, Changchun University of Science and Technology, 7089 Wei-Xing Road, Changchun 130022, China; fangxuan110@126.com

5 State Key Laboratory of Advanced Materials for Smart Sensing, General Research Institute for Nonferrous Metals, Beijing 100088, China; zhaohongbin@grinm.com

* Correspondence: liang.qiao@uestc.edu.cn (L.Q.); hyxiao@uestc.edu.cn (H.X.)

Abstract: Silicon carbide (SiC) has been widely used for electronic radiation detectors and atomic battery sensors. However, the physical properties of SiC exposure to high-dose irradiation as well as its related electrical responses are not yet well understood. Meanwhile, the current research in this field are generally focused on electrical properties and defects formation, which are not suitable to explain the intrinsic response of irradiation effect since defect itself is not easy to characterize, and it is complex to determine whether it comes from the raw material or exists only upon irradiation. Therefore, a more straightforward quantification of irradiation effect is needed to establish the direct correlation between irradiation-induced current and the radiation fluence. This work reports the on-line electrical properties of 4H-SiC Schottky barrier diodes (SBDs) under high-dose electron irradiation and employs in situ noise diagnostic analysis to demonstrate the correlation of irradiation-induced defects and microscopic electronic properties. It is found that the electron beam has a strong radiation destructive effect on 4H-SiC SBDs. The on-line electron-induced current and noise information reveal a self-healing like procedure, in which the internal defects of the devices are likely to be annealed at room temperature and devices' performance is restored to some extent.

Keywords: electron irradiation; room temperature self-healing; noise; electron-induced current; I–V curve

---

1. Introduction

Silicon carbide (SiC) is an ideal material for high-frequency, high-power, and high-temperature devices due to its strong hardness, excellent thermal/mechanical stability, high-thermal conductivity, and controlled electrical properties [1,2]. With the advance of state-of-the-art growth and epitaxy techniques in the past decade, a variety of high-quality SiC-based structures and devices have been fabricated [1], making them very competitive toward third-generation wide-gap semiconductors. Thus,

SiC-based structures have recently regained significant attention in advanced device applications. Furthermore, owing to its large intrinsic fundamental band gap (up to 3.26 eV [3]), SiC is not suitable to visible and infrared light stimulation or irradiation, and their radiation resistance is stronger than that of the conventional semiconductors (such as Si-based materials); therefore, SiC-based diodes are very promising candidates to replace silicon devices for atomic battery light sensors, nuclear cell photosensors [4,5], or particle detectors [6,7].

Required by long-term exposure applications, the radiation resistance and durability of SiC diodes (as radiation detectors and atomic battery light sensors) are very critical and have been subjected to extensive research. For particle detector applications, the detectivity and radiation resistance of SiC diodes were studied under gamma ray [7,8], proton [7,8], neutron [9], electron [8,10], and X-ray [6] irradiations. The results indicated that the electrical properties of radiated SiC diodes and their radiation resistance depended on types, energies, flux, fluence, and the absorbed dose of irradiated particles. It was found that when the total dose of 32-MeV proton was 8.5×10^{12} cm^{-2} and the total absorbed dose of 1.25-MeV γ-ray was 1000 kGy (Air), the electrical properties of 6H-SiC diodes were basically unchanged [7]. Within the range of 1–10 Gy absorbed dose, the radiation damage to 4H-SiC diodes caused by 22-MeV electron and 6-MeV X-ray photon beams hardly affected the radiation-induced current [6]. When the neutron fluence exceeded 5.7×10^{16} cm^{-2}, the charge collection efficiency of the 4H-SiC diode started to drop [9]. For atomic battery light sensor applications, electrons were an important radiation source [4]; thus, many studies had focused on the ability of 4H-SiC diodes' resistance to electron irradiation. For example, Eiting et al. studied the radiation resistance of 4H-SiC p-i-n junction betavoltaic irradiated with 8.5 GBq ^{33}P source with an average beta decay energy of 77 keV, and found that the devices did not experience degradation under the irradiation of ^{33}P source for four half-lives of ^{33}P source, thus demonstrating a good electron radiation resistance [4]. Chandrashekhar et al. experimentally studied the devices' radiation resistance performance under the 1 mCi ^{63}Ni β-radiation source for ten days, and the results suggested such irradiation conditions did not cause the degradation of 4H-SiC diodes [5].

In practical applications for both radiation detectors and atomic battery light sensors in many extreme conditions, such as nuclear, aerospace, military, and astrophysics, the total expected irradiation dose is normally enormous. However, the most common research of experimental or theoretical study on SiC electron radiation resistance is still relatively low; for example, the largest total 8.2-MeV electron fluence ever reported is 9.48×10^{14} cm^{-2} [8,10]. Large fluence of electron irradiation is not only common in the radiation detection experiments, but also critical in the integrated design of the next-generation atomic battery, and thus deserves further study.

Moreover, the general principle of particle detector and atomic battery light sensor is based on induced-current generation (by radiation of electrons, neutrons, photons, and heavy ions) and detection (of the induced-voltage signals in diodes). The previous research were mostly focused on the correlation between the internal defects of SiC materials (or devices) and their electrical parameters under electron irradiation, such as relationship between defect levels and I–V as well as C–V characteristic curves [5,8]. However, this apparent relationship is not suitable to explain the intrinsic response of irradiation effect, since defects are not easy to characterize and it is complex to distinguish whether the defects come from the raw material or exists only upon irradiation. Therefore, a more straightforward quantification of irradiation effect is greatly needed to establish the direct correlation between irradiation-induced current and the radiation fluence.

For this purpose, low-frequency noise information, especially 1/f noise, can be used as a characterization means for the sensitive detection of internal defects in electronic materials [11] and devices [12,13]. This method in principle can also be utilized to characterize radiation damage to electronic devices [12]. Under the beam radiation, the internal defects and the structural damage will lead to the increase of low-frequency noise power spectral density S_V [14,15]. For example, Babcock et al. [16] studied the radiation resistance of Ultra High Vacuum/Chemical Vapor Deposition SiGe heterojunction bipolar transistor (HBT) irradiated by ^{60}Co γ-ray and found that the increase of the total absorbed dose resulted in the performance degradation of the SiGe HBT. The current gain β was decreased with the increase of absorbed dose, and the internal defects were increased, while the noise information S_V increased in the low-frequency range, as compared with that before irradiation. Therefore, the 1/f noise can be used as a tool to characterize the defects in materials and devices, and correlate the defects with devices performance.

Here, we reported the study of electron-induced current and radiation resistance of SiC Schottky barrier diodes (SBDs) under high-fluence electron irradiation as well as an in situ noise diagnostics for defect–electrical property analysis. Through on-line electron-induced current, I–V curve, noise information, and SBDs' radiation resistance to the environment of electron irradiation had been analyzed, and a self-driven healing process was observed at room temperature, which led to some extent of electrical performance recovery.

2. Materials and Methods

2.1. 4H-SiC SBDs Samples and Irradiation Experimental Conditions

The epitaxial 4H-SiC SBDs used in this experiment were provided by State Key Laboratory of Wide-Band Gap Semiconductor Power Electronics located at Nanjing, China. Its structure is shown in Figure 1a, where the photosensitive area of SBDs is 3 mm $\times$ 3 mm and the voltage-withstand range is -100 to 100 V. The 4H-SiC SBDs were 4H-SiC epilayers grown by chemical vapor deposition on SiC substrates of 360 mm thickness, with nitrogen doped with a net doping density of 1×10^{18} cm^{-3} and micropipe density of 1 micropipe cm^{-2}. The buffer was n-type, 1 μm thick, with a net free carrier concentration of 1×10^{18} cm^{-3}. The epilayer was n-type, 12 μm thick, with a net free carrier concentration of 3×10^{15} cm^{-3}. The ohmic contact was obtained by deposition of a 1000-Å-thick layer of Ni and a 3-μm-thick layer of Au. The Schottky contact was obtained by radio frequency magneton sputtering of a 1000-Å-thick layer of Ni at room temperature. In order to reduce the influence of environment on the device, SiO_2 of 1000 Å thickness and Si_3N_4 of 1000 Å thickness were grown on Ni by sputtering method.

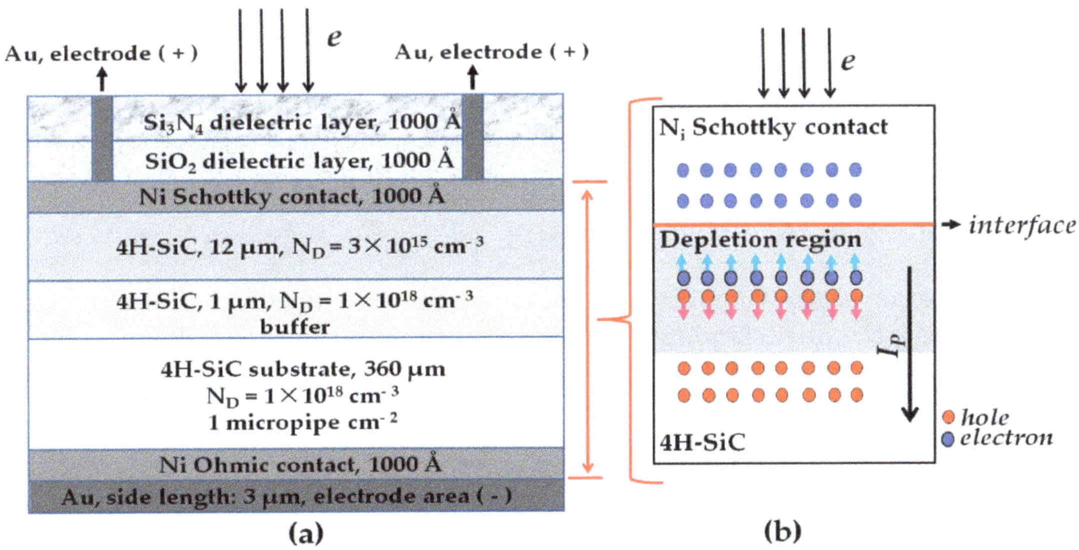

Figure 1. (**a**) The schematic diagram of Silicon carbide (SiC) Schottky barrier diodes (SBDs) and (**b**) the principle of electron-induced current generation under irradiation.

Electron beam irradiation experiments were performed on the electron accelerator of Sichuan Forever Holding Co., Ltd, located at Mianyang, China. The electron energy was 1.8 MeV, the electron flux was 9.62×10^{12} cm^{-2} s^{-1}, and the total electron fluence was 9.05×10^{17} cm^{-2}. In order to avoid overheating of SiC SBDs during electron beam irradiation, the bottom of the irradiated metal platform was continuously cooled with water, and meanwhile, the temperature of the SiC SBDs was monitored by a thermocouple and controlled at a constant temperature of 25 °C. The home-made real-time on-line current test system was used to monitor and record the changes of the electron-induced current of SiC SBDs during irradiation and 30 min after electron irradiation. Then the devices were kept at room temperature (25 °C) for 72 h without heating. The devices performance tended to decrease with the increasing electron fluence, but it was not certain whether the devices were disabled or not. In the case of reverse breakdown, the devices would be disabled completely, and the reverse breakdown state can be regarded as the worst-case state of devices performance. The devices were reverse biased to breakdown at 200 V voltage by Keithley 6517B high impedance/electrometer located at Mianyang, China toward the end of the experiment, and the damage of the device after irradiation was evaluated when the devices performance at reverse breakdown state was considered as the worst-case state. The I–V curves and noise information of SiC SBDs before and after electron beam irradiation, room-temperature self-healing, and reverse breakdown were measured by Keithley 2635 sourcemeter and the self-developed noise parameter test system located at Mianyang, China.

2.2. The Electron-Induced Current Test

When subjected to radiations of electron beam, γ-ray, and neutron, current can be induced in SBDs due to the ionization effect within Schottky barrier junctions, which is shown in Figure 1b. The principle of the electron-induced current can be understood by taking the electrons as an example—1.8 MeV electrons penetrate the SBDs completely. The n-type 4H-SiC material of the Schottky barrier junction was irradiated by incident electrons with an energy larger than the material's fundamental electronic band gap; thus electron–hole pairs would be generated in n-type 4H-SiC material. The Fermi lever in the Ni Schottky contact was less than the Fermi lever in the n-type 4H-SiC material; therefore, the average energy of electrons in the n-type 4H-SiC material was greater than the average energy of those in the Ni Schottky contact. The difference in the average electron energy can be expected to transfer electrons from the n-type 4H-SiC material to the Ni Schottky contact until the average electron energies were equal. While the holes can only stay in n-type 4H-SiC material, the electrons were pulled toward the Ni Schottky contact. As a result, the width of the depletion region became narrower, and the contact potential difference decreased. The incident electron energy was converted into electrical energy. Once the external circuit was short-circuited, current could flow through the Schottky barrier junction. In this way, the separation of the electron–hole pairs can be achieved, and the induced current I_P, whose direction was from the Ni Schottky contact to the n-type 4H-SiC material through the Schottky barrier junction, can be produced. The current equation of Schottky barrier junction under electron beam irradiation is

$$I = I_0(e^{eU/kT} - 1) - I_P \tag{1}$$

where I and I_0 are the current flowing through the Schottky barrier junction and the reverse saturation current, respectively; e, U, and T stand for the electron charge, the applied voltage, and temperature, respectively; and k is Boltzmann constant ($k = 1.38 \times 10^{-23}$ J K^{-1}).

When the applied voltage U is 0 V, the Formula (1) becomes

$$I = -I_P. \tag{2}$$

In this study, the open-circuit induced current of SiC SBDs has been tested by a real-time on-line current test system, as shown in Figure 2a. This test system consisted of the fixtures, low-loss cables, 10-channel scanning card, data interface, Keithley 6517B high impedance/electrometer, and self-compiled control software (installed in the control computer). The fixtures were used to fix SiC SBDs and transmit current signals to low-loss cables. The 10-channel scanning card was installed at the rear panel of Keithley 6517B high impedance/electrometer and they were used to capture multiple electron-induced currents. The self-compiled control software was used to control and record current data detected by Keithley 6517B high impedance/electrometer in real time. The system can achieve a 10-channel real-time signal acquisition with the current and voltage accuracies of each signal acquisition reaching way up to 1 fA and 1 nV, respectively.

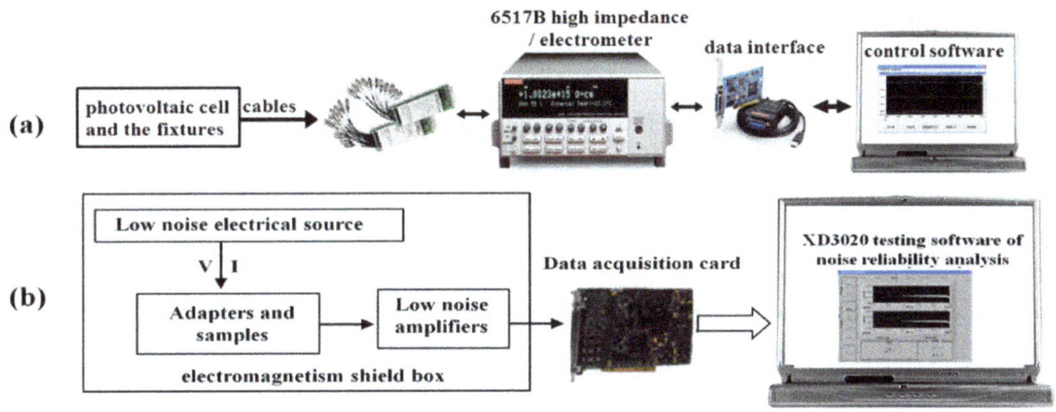

Figure 2. The illustration for (**a**) real-time on-line current test system and (**b**) test system for noise parameters.

2.3. Noise Information Test

In this study, the changes of noise information before and after irradiation of SiC SBDs have been tested by using the noise parameter test system. The noise parameter test system consists of low-noise bias, adapter, Stanford Research Systems SR560 voltage amplifier, acquisition card, and XD3020 [17] noise analysis software with the system bandwidth being 0–10^6 Hz, the background noise being <4 nV/$\sqrt{}$Hz (@1 kHz), and the signal magnification capability being 10^0–10^5, as shown in Figure 2b.

The noise of the diode was typically composed of two or three components of shot noise, 1/f noise, and generation-recombination noise (G-R noise). The noise power spectral density (S) can be written as:

$$S(f) = A + \frac{B}{f^{\gamma}} \tag{3}$$

where A is the shot noise amplitude, B, the 1/f noise amplitude, and the exponents of γ is usually taken to be unity.

In this work, based on the noise spectrum information of SiC SBDs, the electron irradiation effects of SiC SBDs were studied by using the parameter information in Equation (3). The damage degree of the device had been determined according to the change of noise power spectral density before and after irradiation and room-temperature self-healing.

2.4. I–V Curve Test

The current–voltage (I–V) curve of a semiconductor device contained the electrical information of dark current as well as break-over voltage. The real-time on-line electron-induced current and I–V characteristics can characterize the device to some extent. The I–V curve with voltage range of −100 to 2.5 V before and after irradiation of SiC SBDs had been measured with Keithley 2635 sourcemeter in this work. The damage degree of the device had been determined according to the change of I–V curve before and after irradiation and room-temperature self-healing.

3. Results and Discussion

3.1. High-Dose Electron Irradiation

Figure 3 shows the real-time curves of the open-circuit electron-induced current in SiC SBDs. It can be seen that when the energy of electrons was 1.8 MeV, and the electron flux was 9.62×10^{12} cm^{-2} s^{-1}, a current of 1.27×10^{-5} A was induced due to the ionization effect. However, as the electron fluence further increased, the electron-induced current of the device decreased continuously. When the electron fluence reached 1×10^{15} cm^{-2}, the current decreased to 1.24×10^{-5} A, accounting for only a decrease of 2.36%, which was consistent with the results reported by Nava et al. with the total electron fluence being 9.48×10^{14} cm^{-2} [8,10]. SiC SBDs can be considered as sufficient to resist the irradiation of electrons with a fluence under 1×10^{15} cm^{-2}. However, with a further increase of electron fluence, the electron-induced current of SiC SBDs began to decrease sharply, particularly when the electron fluence was $\leq 3 \times 10^{17}$ cm^{-2}. When the electron fluence was 1×10^{16} cm^{-2}, the current decreased to 1.03×10^{-5} A, showing a decrease of 18.90%. When the electron fluence was 1×10^{17} cm^{-2}, the current decreased to 5.00×10^{-6} A, accounting for a decrease of 60.63%. At the end of irradiation, the total electron fluence reached 9.05×10^{17} cm^{-2} and the current became 1.82×10^{-6} A, showing a decrease of 85.70%. It was noteworthy that when the electron fluence was $>3.87 \times 10^{17}$ cm^{-2}, the electron-induced current fluctuation of SiC SBDs increased sharply and the current fluctuation could reach up to 86%.

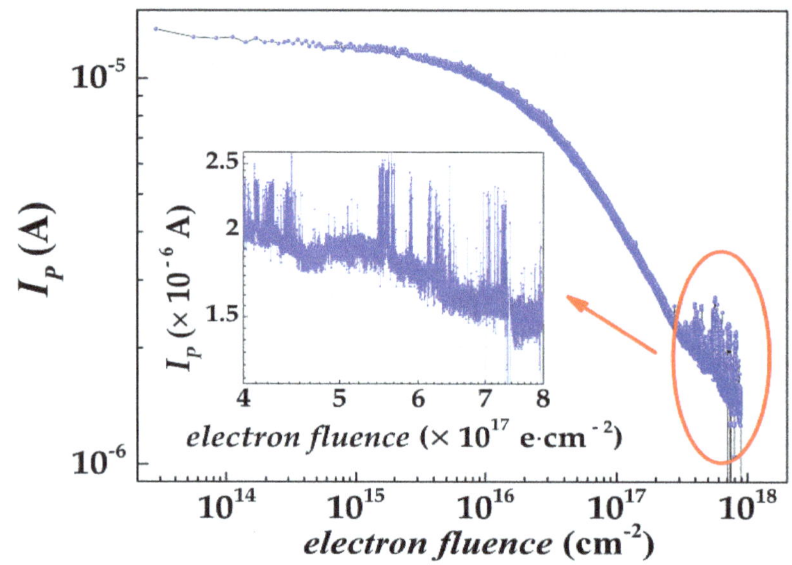

Figure 3. The real-time changing profiles of the induced current with electron fluence.

When SiC SBDs were irradiated by high dose of electrons, a large number of defects would be generated inside the silica layer [18], silicon carbid layer [19–22], and the interfaces [23,24] between the metal and silicon carbide. These defects would produce a removal effect [25,26] on charge carriers at 1.8 MeV electron irradiation, leading to the decrease of current.

Hemmingsson et al. studied the deep level defects in electron-irradiated 4H-SiC epitaxial layers grown by chemical vapor deposition using deep level transient spectroscopy (DLTS). The measurements performed on electron-irradiated p-n junctions in the temperature range 100–750 K revealed both electron and hole traps with thermal ionization energies ranging from 0.35 to 1.65 eV [27], which led to the deterioration of device performance. Iwamoto et al. investigated the formation and evolution of defects in 4H-SiC Schottky barrier diodes and correlated with the SBDs' performances [28]. The SBDs were irradiated with 1 MeV electrons to a fluence of 1.00×10^{15} cm^{-2}. Current–voltage, capacitance–voltage, and DLTS measurements were used to study the effect of defects on the SBDs performance. It was found that the DLTS defect levels (EH_1, EH_3, and $Z_{1/2}$) were very likely to be partly responsible for the charge collection efficiency reduction after electron irradiation. EH_1 and EH_3 were related to carbon interstitials and $Z_{1/2}$ was related to carbon vacancies. DLTS study on n-type 4H-SiC (0001) epilayers also showed that the carrier lifetime in an n-type 4H-SiC epilayer, measured by differential microwave photoconductance decay, had been significantly improved from 0.73 μs (as-grown) to 1.62 μs (after oxidation, 1300 °C) as the $Z_{1/2}$ and $EH_{6/7}$ centers had been reduced from $(0.3–2) \times 10^{13}$ cm^{-3} to below the detection limit (1×10^{11} cm^{-3}) by thermal oxidation of epilayers at

1150–1300 °C [29]. Based on the above results, we suggested that the current fluctuation might due to the increasing defects in the devices, such as interstitials and vacancies. These defects led to a slower and less stable carrier motion, which resulted in current fluctuation.

Figure 4 shows the I–V curves (including reverse breakdown region) of SiC SBDs before and after irradiation. From Figure 4b, it can be seen that the reserve current of SiC SBDs before irradiation increases with the increase of reserve bias, showing typical diode characteristics. For the 2-h irradiation sample, the forward current of the device dropped sharply from 2.39×10^{-5} A (before irradiation) to 2.06×10^{-8} A at a voltage of 2.5 V, while the absolute value of the reverse dark current further increased (Figure 2b). When the voltage became -100 V, the current was -8.76×10^{-7} A, showing a serious degradation of the device characteristics. At the reverse breakdown of the device, the I–V curve of the device had been tested. It was found that the forward current was further reduced compared to the case of 2 h after irradiation. When the voltage became 2.5 V, the current dropped to 1.42×10^{-9} A, while the voltage became -100 V, the current was -1.3×10^{-9} A, which was similar to the I–V characteristics of the resistance. The electron capture levels had been proved to be induced by the electron irradiation, and it would drastically influence the resistance in the bulk crystal and the Schottky barrier of SiC SBDs [30]. Under 2 MeV electron irradiation, the Schottky barrier was found to decrease and the resistance of the bulk crystal was found to increase with the electron fluence, leading to the decrease of the forward current. The Schottky barrier of SiC SBDs decreased from 1.25 eV (pre-irradiation) to 1.17 eV (post-irradiation, 1.00×10^{17} cm^{-2}). The decrease of the barrier height was thus responsible for the increase of the reverse current. On the other hand, the increase of electron-induced defects in passivation layers was expected to result in the increase of the reverse leakage [31,32]. In this work, the I–V curves (2 h after electron irradiation) exhibited similar behavior. The SiC SBDs could be considered as the resistance since the barrier height disappeared and the bulk crystal resistance increased further after the SiC SBDs were reverse biased to breakdown. The I–V curves (reverse breakdown) of SiC SBDs were similar to the I–V characteristics of the resistance.

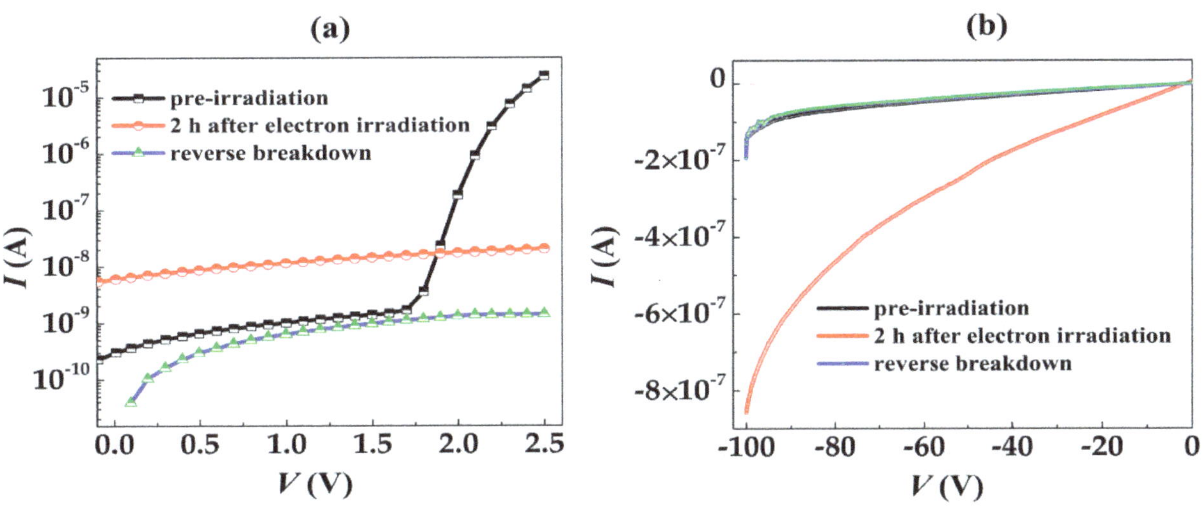

Figure 4. The I–V curves for SiC SBDs before and after electron beam irradiation. (**a**) The foward bias region and (**b**) the reverse bias region.

Besides the electrical properties (electron-induced current and I–V curves), the noise information of SiC SBDs also reflects the internal defect states before and after irradiation and the reverse breakdown, as shown in Figure 5a. It can be seen that in the frequency range of 10^2 Hz to 10^5 Hz, the noise power spectral density, S_V, exhibited a clear dependence on the electron irridation, with S_V (reverse breakdown) > S_V (2 h after irradiation) > S_V (pre-irradiation). Ziel et al. [33] proposed that under small fluence irradiation, induced current was mainly contributed by two components: (1) the current $I_0 e^{qU/kT}$ due to the injection of electrons from the semiconductor into the metal and (2) $-I_0$ due to the injection of electrons from the metal into the semiconductor. Both currents contained carriers that pass independently and randomly through the junction barrier, thereby showing as pure intermediate frequency shot noise. In this work, the shot noise of current SiC SBDs was in the frequency range of 10^2 to 10^5 Hz.

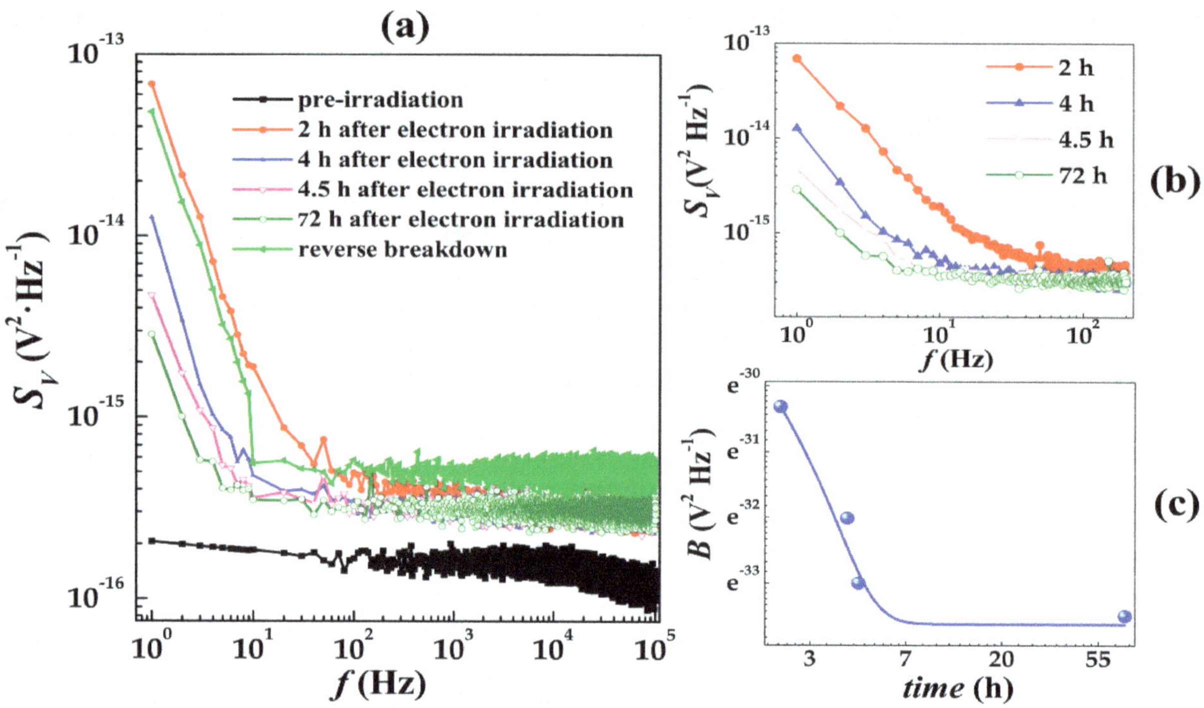

Figure 5. The noise power spectral densities of SiC SBDs. (**a**) Before and after electron beam irradiation and in a reverse breakdown state; (**b**) the low-frequency noise curve within 72 h after electron irradiation; and (**c**) the changing profiles of the low-frequency noise amplitude within 72 h after electron irradiation.

The current noise power spectral density (S_I) of SBDs' shot noise is [34]

$$S_I = 2e(I + 2I_0) \tag{4}$$

The SBDs resistance can be obtained by Equation (1)

$$R = \frac{dU}{dI} \approx \frac{kT}{e(I + I_0)} \tag{5}$$

The voltage noise power spectral density S_V is

$$A = S_V = S_I \times R^2 = \frac{2(kT)^2(I + 2I_0)}{e(I + I_0)^2} \tag{6}$$

The current I is

$$I = \frac{2ne}{t} \tag{7}$$

where R stands for the diode resistance, n and t are the carrier concentration and time for drifting through the barrier, respectively.

When $I \gg I_0$, Equation (6) can be simplified as

$$S_V \approx \frac{2(kT)^2}{eI} = \frac{t(kT)^2}{e^2 n} \tag{8}$$

As the current decreased, n decreased, which made S_V to increase according to Equation (8), thus leading to S_V (2 h after irradiation) > S_V (pre-irradiation). When SiC SBDs first experienced the electron beam irradiation and then subjected to reverse breakdown, a large number of defects would be produced inside SiC SBDs, which made the noise power spectral density of SiC SBDs to further increase compared with SBDs that only irradiated by electrons, therefore showing S_V (reverse breakdown) > S_V (2 h after irradiation).

3.2. Room-Temperature Self-Healing

As shown in Figure 5a, the noise spectra of SiC SBDs before and after irradiation as well as in reverse breakdown state are 1/f noise in the frequency range of 10^0 to 10^2 Hz. 1/f noise had an extrinsic origin that arose from defects [35]. The defects of electronic origin lying in the gap acted as electron traps leading to both mobility and carrier density fluctuations. In the frequency range of 10^0 to 10^2 Hz, S_V reached the minimum before irradiation, while reached the maximum at 2 h after electron irradiation, and S_V even was greater than that of the reverse breakdown. However, if the SiC SBDs were left alone for 72 h after electron irradiation at room temperature, their S_V could become lower than that of S_V at reverse breakdown state, showing a significant decrease. The observed S_V (2 h after irradiation) > S_V (72 h after irradiation) was partially due to the fact that part of the defects were rapidly annealed at room temperature; thus, the effect of carrier removal was weakened, leading to the increase of carrier concentration. According to Hooge's equation [36], the relationship between S_V and carrier concentration n are given by Equation (9).

$$S_V(f) = \frac{B}{f^\gamma} \propto \frac{\alpha_H}{n} \quad (9)$$

where α_H stands for the Hooge parameter. The S_V corresponding to 1/f noise was inversely proportional to the carrier concentration inside the device. As carrier concentration increased, S_V decreased.

The noise power spectra at 2 h and 72 h after irradiation (shown in Figure 5a) cannot clearly reflect the annealing effect of the internal defects in SiC SBDs at room temperature. Therefore, at 72 h after irradiation exposure, we have carried out the tracking measurement on the low-frequency noise of SiC SBDs irradiated by electron beam at room temperature, which is shown in Figure 5b. It can be clearly seen that in the frequency range of 10^0~10^2 Hz, the noise power spectral density of SiC SBDs decreased with time, and the observable frequency range of 1/f noise gradually decreased. At 2 h after irradiation, 1/f noise can be observed in the frequency range of 1–100 Hz; at 4 h after irradiation exposure, 1/f noise can be observed in the frequency range of 1–30 Hz; and at 72 h after irradiation exposure, 1/f noise frequency further decreased into 1–6 Hz, while 1/f noise will be annihilated by shot noise above these frequency ranges.

The temporal variation of 1/f noise amplitude of the SiC SBDs irradiated by electrons can be calculated by Equation (9), and the results are shown in Figure 5c, where the 1/f noise amplitude B decreases exponentially with time t, and this relationship is further well fitted by Equation (10).

$$B = 7.81 \times 10^{-13} e^{-t/0.81} + 2.50 \times 10^{-15} \quad (10)$$

where the unit of B is $V^2 \cdot Hz^{-1}$, the unit of t is s.

Then the declining rate is

$$\frac{dB}{dt} = -9.64 \times 10^{-13} e^{-t/0.81} \quad (11)$$

At 4.5 h after irradiation, B declined sharply, while the declining rate decreased with the increase of time, and the declining rate followed Equation (11). At 72 h after irradiation, the declining rate tended to be constant. The value of B decreased from 6.75×10^{-14} $V^2 \cdot Hz^{-1}$ (2 h after irradiation) to 2.05×10^{-15} $V^2 \cdot Hz^{-1}$ (72 h after irradiation), accounting for 96.96% decrease.

Figure 5b,c shows that the internal defects of the SiC SBDs continuously decrease at room temperature, weakening the removal effects of the defects on carriers. Back in 1966, Fischerrr et al. [37] had systematically investigated the temperature annealing of traps produced by 6–88 MeV electron irradiation in n-type Ge. In their experiment, the electron irradiation temperature was 85 K, and it was found that some traps disappeared when the temperature increased to near 200 K, thereby, leading to an increase of the carrier concentration. In 2009, Messina et al. [38] found that a significant portion of E'_γ centers, which induced in amorphous silica at room temperature by γ-irradiation up to 79 kGy, spontaneously decayed after the end of irradiation. In 1984, Yamaguchi et al. [39] found that effective room-temperature self-healing of radiation-induced defects in both p-type and n-type InP after electron irradiation leading to the recovery of InP solar cell properties. Besides Ge, amorphous silica, and InP solar cell, room-temperature self-healing of radiation-induced defects were also shown in Si-based devices [25,40–43], as reported by Pease et al. Pease et al. [25] found the normalized change in reciprocal resistivity in the drain to source ($\overline{\Delta 1/R_{DS}}$) of Si-based power Metal-Oxide-Semiconductor Field-Effect Transistors degraded between 2 and 10% (most less than 5%) over a period of 24 h after room-temperature proton and neutron irradiation. These phenomena suggested that the internal defects generated inside the semiconductor material under irradiation can be annealed and annihilated below room temperature. However, as far as the room-temperature annealing of SiC materials is concerned, currently there is still not enough experimental evidences to demonstrate that some defects of SiC materials can be annealed after electron irradiation at room temperature. In this work, we used low-frequency noise to show that electron irradiation can produce the defects in SiC devices and these induced defects can be further annealed at room temperature. Assuming only electron irradiation and α_H remained constant, regardless of the amount of electron irradiation, part of internal defects in SiC material at 72 h after irradiation can be calculated to decrease by 3.04% as compared with the same internal defects at 2 h after irradiation. Since $1/f$ noise cannot be directly used to differentiate the types of defects, it is impossible to know what kinds of defects are annealed at room temperature. However, this property can still be used to characterize the declining trend of internal defects with time after electron irradiation via the noise information, and this method is also applicable for the defects characterization in other semiconductor materials and devices.

Furthermore, noise characterization were used to demonstrate that the SiC SBDs after electron irradiation have undergone an unexpected self-annealing or relaxation process at room temperature, thereby, reducing the induced defects while increasing the carrier concentration. Such room-temperature annealing was consistent with the on-line electron-induced current curve after irradiation and vice versa. As shown in Figure 6, the data collected by the on-line current–voltage test system shows that the electron-induced current of the device does not decrease to the background level within 19 min after the exposure to electron irradiation, but instead presents a linear increasing profile with the current rising from 10^{-10} to 10^{-7} A. This phenomenon suggested that within 19 min after the irradiation, the device was subjected to a room-temperature annealing or relaxation, curing some of the internal defects, and recovering the electrical performance to a certain level. This phenomenon was also consistent with the I–V characteristics of the SiC SBDs, demonstrating that although the performance of the SiC SBDs was greatly lowered after the exposure to 9.05×10^{17} cm^{-2} electron irradiation, this type of semiconductor device can still recover its electrical properties to some extent under this unexpected self-annealing mechanism.

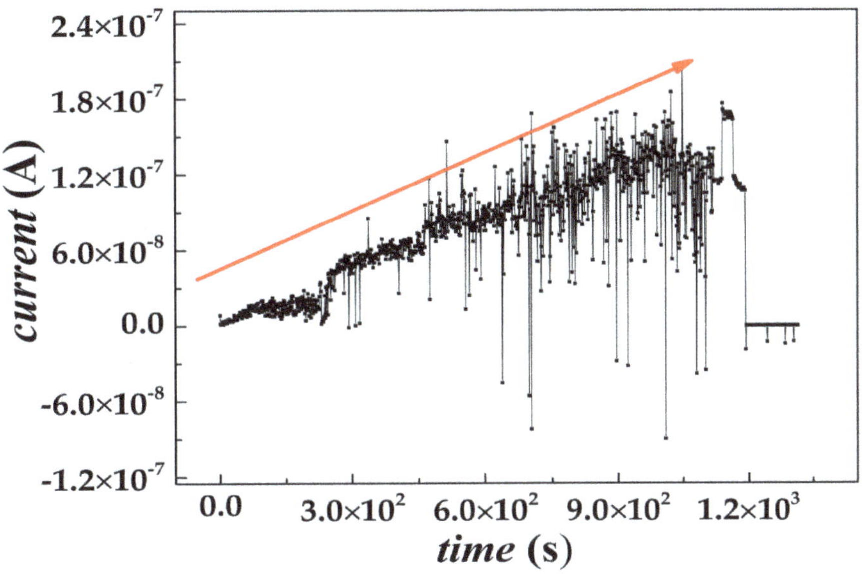

Figure 6. The real-time electron-induced current curve of SiC SBDs after the end of exposure to electron beam irradiation.

4. Conclusions

The electrical performance of SiC SBDs irradiated by high-dose, high-energy (1.8 MeV) electron beam had been investigated in this work. It was found that electron beam had a strong radiation destructive effect on 4H-SiC SBDs. Their electrical performance was greatly reduced when exposed to high-energy electron beam with a fluence as high as 9.05×10^{17} cm^{-2}, the electron-induced current reduced by 85.70%, and the device characteristics degraded seriously. The on-line electron-induced current and noise information revealed a self-healing like procedure, in which the internal defects of the devices were likely to be annealed at room temperature and devices performance was restored to some extent. Although the mechanism of this self-healing is under investigation, the high-dose irradiation and noise diagnostics study reported here can provide useful information for designing next-generation atomic battery and detectors for extreme environment applications.

5. Patents

Guixia Yang, Yuanlong Pang, Fansong Zeng, et al. Low-noise bias device, the Chinese patent of invention, patent number: ZL 201510729381.1.

Author Contributions: All the authors contributed to the conception, design, and performance of the experiment, the analysis of the data, and the writing of the paper. G.Y., H.X., and L.Q. initiated and discussed the research problem; Y.P., Y.Y., and M.J. performed the irradiation and on-line current experiments; G.Y., S.P., and J.L. performed the I–V and noise experiments; X.F. and H.Z. performed defect analysis; L.Q., X.Z., and G.Y. made the figures and analyzed the data; and G.Y., H.X., and L.Q. wrote the paper.

Acknowledgments: This research was funded by NSAF Joint Foundation of China (Grant No. U1530129), the Institution of Nuclear Physics and Chemistry (INPC), China. Project (Grant No. 2015CX03), National key laboratory of Materials Behavior and Evaluation Technology in Space Environment (Grant No. 6142910010403).

References

1. Morkoç, H.; Strite, S.; Gao, G.B.; Lin, M.E.; Sverdlov, B.; Burns, M. Large-band-gap sic, iii-v nitride, and ii-vi znse-based semiconductor device technologies. *J. Appl. Phys.* **1994**, *76*, 1363–1398. [CrossRef]
2. Spiazzi, G.; Buso, S.; Citron, M.; Corradin, M. Performance evaluation of a schottky sic power diode in a boost pfc application. *IEEE Trans. Power Electron.* **2003**, *18*, 1249–1253. [CrossRef]

3. Persson, C.; Lindefelt, U. Detailed band structure for 3c-, 2h-, 4h-, 6h-sic, and si around the fundamental band gap. *Phys. Rev. B* **1996**, *54*, 10257–10260. [CrossRef]
4. Eiting, C.J.; Krishnamoorthy, V.; Rodgers, S.; George, T. Demonstration of a radiation resistant, high efficiency sic betavoltaic. *Appl. Phys. Lett.* **2006**, *88*, 064101. [CrossRef]
5. Chandrashekhar, M.V.S.; Thomas, C.I.; Li, H.; Spencer, M.G.; Lal, A. Demonstration of a 4h sic betavoltaic cell. *Appl. Phys. Lett.* **2006**, *88*, 1009. [CrossRef]
6. Bruzzi, M.; Nava, F.; Russo, S.; Sciortino, S.; Vanni, P. Characterisation of silicon carbide detectors response to electron and photon irradiation. *Diam. Relat. Mat.* **2001**, *10*, 657–661. [CrossRef]
7. Metzger, S.; Henschel, H.; Kohn, O.; Lennartz, W. Silicon carbide radiation detector for harsh environments. *IEEE Trans. Nucl. Sci.* **2002**, *49*, 1351–1355. [CrossRef]
8. Nava, F.; Vittone, E.; Vanni, P.; Verzellesi, G.; Fuochi, P.G.; Lanzieri, C.; Glaser, M. Radiation tolerance of epitaxial silicon carbide detectors for electrons, protons and gamma-rays. *Nucl. Instrum. Methods Phys. Res. Sect. A* **2003**, *505*, 645–655. [CrossRef]
9. Seshadri, S.; Dulloo, A.R.; Ruddy, F.H.; Seidel, J.G.; Rowland, L.B. Demonstration of an sic neutron detector for high-radiation environments. *IEEE Trans. Electron Devices* **1999**, *46*, 567–571. [CrossRef]
10. Castaldini, A.; Cavallini, A.; Rigutti, L.; Nava, F. Low temperature annealing of electron irradiation induced defects in 4h-sic. *Appl. Phys. Lett.* **2004**, *85*, 3780–3782. [CrossRef]
11. Celik-Butler, Z.; Hsiang, T.Y. Determination of si-sio 2 interface trap density by 1/f noise measurements. *IEEE Trans. Electron. Dev.* **2015**, *35*, 1651–1655. [CrossRef]
12. Adak, O.; Rosenthal, E.; Meisner, J.; Andrade, E.F.; Pasupathy, A.N.; Nuckolls, C.; Hybertsen, M.S.; Venkataraman, L. Flicker noise as a probe of electronic interaction at metal-single molecule interfaces. *Nano Lett.* **2015**, *15*, 4143–4149. [CrossRef] [PubMed]
13. Folkes, P.A. Fluctuating deep-level trap occupancy model for hooge's 1/f noise parameter for semiconductor resistors. *J. Appl. Phys.* **1994**, *64*, 487–489.
14. Fleetwood, D.M.; Meisenheimer, T.L.; Scofield, J.H. 1/f noise and radiation effects in mos devices. *IEEE Trans. Electron. Dev.* **1994**, *41*, 1953–1964. [CrossRef]
15. Arora, S.K.; Singh, R.; Kumar, R.; Kanjilal, D.; Mehta, G.K. In situ 1/f noise studies on swift heavy ion irradiated p-type silicon. *Nucl. Instrum. Methods Phys. Res. Sect. B* **1999**, *156*, 265–269. [CrossRef]
16. Babcock, J.A.; Cressler, J.D.; Vempati, L.S.; Clark, S.D.; Jaeger, R.C.; Harame, D.L. Ionizing radiation tolerance of high-performance sige hbt's grown by uhv/cvd. *IEEE Trans. Nucl. Sci.* **1995**, *42*, 1558–1566. [CrossRef]
17. Li, Y.; Zhou, Q. The Noise Measurement and Analysis System of Optoelectronic Coupled Devices Based on Virtual Instrument. In Proceedings of the SPIE 2008 International Conference on Optical Instruments and Technology, Beijing, China, 16–19 November 2008.
18. Aitken, J.M.; Young, D.R.; Pan, K. Electron trapping in electron-beam irradiated SiO_2. *J. Appl. Phys.* **1978**, *49*, 3386–3391. [CrossRef]
19. David, M.L.; Alfieri, G.; Monakhov, E.M.; Hallén, A.; Blanchard, C.; Svensson, B.G.; Barbot, J.F. Electrically active defects in irradiated 4h-sic. *J. Appl. Phys.* **2004**, *95*, 4728–4733. [CrossRef]
20. Barry, A.L.; Lehmann, B.; Fritsch, D.; Bräunig, D. Energy dependence of electron damage and displacement threshold energy in 6h silicon carbide. *IEEE Trans. Nucl. Sci.* **1991**, *38*, 1111–1115. [CrossRef]
21. Storasta, L.; Bergman, J.P.; Janzén, E.; Henry, A.; Lu, J. Deep levels created by low energy electron irradiation in 4h-sic. *J. Appl. Phys.* **2004**, *96*, 4909–4915. [CrossRef]
22. Bardeleben, H.J.V.; Cantin, J.L.; Henry, L.; Barthe, M.F. Vacancy defects in p-type 6h−sic created by low-energy electron irradiation. *Phys. Rev. B* **2000**, *62*, 10841–10846. [CrossRef]
23. Sheridan, D.C.; Chung, G.; Clark, S.; Cressler, J.D. The effects of high-dose gamma irradiation on high-voltage 4h-sic schottky diodes and the sic-SiO_2 interface. *IEEE Trans. Nucl. Sci.* **2001**, *48*, 2229–2232. [CrossRef]
24. Çınar, K.; Coşkun, C.; Aydoğan, Ş.; Asıl, H.; Gür, E. The effect of the electron irradiation on the series resistance of au/ni/6h-sic and au/ni/4h-sic schottky contacts. *Nucl. Instrum. Methods Phys. Res. Sect. B* **2010**, *268*, 616–621. [CrossRef]
25. Pease, R.L.; Enlow, E.W.; Dinger, G.L.; Marshall, P. Comparison of proton and neutron carrier removal rates. *IEEE Trans. Nucl. Sci.* **1987**, *34*, 1140–1146. [CrossRef]
26. McGarrity, J.M.; McLean, F.B.; DeLancey, W.M.; Palmour, J.; Carter, C.; Edmond, J.; Oakley, R.E. Silicon carbide jfet radiation response. *IEEE Trans. Nucl. Sci.* **1992**, *39*, 1974–1981. [CrossRef]

27. Hemmingsson, C.; Son, N.T.; Kordina, O.; Bergman, J.P.; Janzén, E.; Lindström, J.L.; Savage, S.; Nordell, N. Deep level defects in electron-irradiated 4h sic epitaxial layers. *J. Appl. Phys.* **1997**, *81*, 6155–6159. [CrossRef]
28. Iwamoto, N.; Johnson, B.C.; Hoshino, N.; Ito, M.; Tsuchida, H.; Kojima, K.; Ohshima, T. Defect-induced performance degradation of 4h-sic schottky barrier diode particle detectors. *J. Appl. Phys.* **2013**, *113*, 1065. [CrossRef]
29. Toru, H.; Tsunenobu, K. Reduction of deep levels and improvement of carrier lifetime in n-type 4h-sic by thermal oxidation. *Appl. Phys. Express* **2009**, *2*, 041101.
30. Ohyama, H.; Takakura, K.; Watanabe, T.; Nishiyama, K.; Shigaki, K.; Kudou, T.; Nakabayashi, M.; Kuboyama, S.; Matsuda, S.; Kamezawa, C.; et al. Radiation damage of sic schottky diodes by electron irradiation. *J. Mater. Sci. Mater. Electron.* **2005**, *16*, 455–458. [CrossRef]
31. Muzykov, P.G.; Bolotnikov, A.V.; Sudarshan, T.S. Study of leakage current and breakdown issues in 4h–sic unterminated schottky diodes. *Solid-State Electron.* **2009**, *53*, 14–17. [CrossRef]
32. Luo, Z.; Chen, T.; Ahyi, A.C.; Sutton, A.K.; Haugerud, B.M.; Cressler, J.D.; Sheridan, D.C.; Williams, J.R.; Marshall, P.W.; Reed, R.A. Proton radiation effects in 4h-sic diodes and mos capacitors. *IEEE Trans. Nucl. Sci.* **2004**, *51*, 3748–3752.
33. Van Der Ziel, A. Theory of shot noise in junction diodes and junction transistors. *Proc. IRE* **1955**, *43*, 1639–1646. [CrossRef]
34. Van Der Ziel, A. Noise in solid-state devices and lasers. *Proc. IEEE* **1970**, *58*, 1178–1206. [CrossRef]
35. Kar, S.; Raychaudhuri, A.K. Temperature and frequency dependence of flicker noise in degenerately doped si single crystals. *J. Phys. D Appl. Phys.* **2001**, *34*, 3197–3202. [CrossRef]
36. Hooge, F.N. The relation between $1/f$ noise and number of electrons. *Physica B* **1990**, *162*, 344–352. [CrossRef]
37. Fischer, J.; Corelli, J. Production and annealing of defects in 6–88 mev electron-irradiated n-type germanium. *J. Appl. Phys.* **1966**, *37*, 3287–3297. [CrossRef]
38. Messina, F.; Agnello, S.; Cannas, M.; Parlato, A. Room temperature instability of $e'\gamma$ centers induced by γ irradiation in amorphous SiO_2. *J. Phys. Chem. A* **2009**, *113*, 1026–1032. [CrossRef]
39. Yamaguchi, M.; Itoh, Y.; Ando, K. Room-temperature annealing of radiation-induced defects in inp solar cells. *Appl. Phys. Lett.* **1984**, *45*, 1206–1208. [CrossRef]
40. Zhou, X.J.; Fleetwood, D.M.; Schrimpf, R.D.; Faccio, F.; Gonella, L. Radiation effects on the $1/f$ noise of field-oxide field effect transistors. *IEEE Trans. Nucl. Sci.* **2008**, *55*, 2975–2980. [CrossRef]
41. Haohao, Z.; Jinshun, B.; Yuan, D.; Yannan, X.; Ming, L. Proton irradiation effects and annealing behaviors of 16mb magneto-resistive random access memory(mram). In Proceedings of the 2016 13th IEEE International Conference on Solid-State and Integrated Circuit Technology (ICSICT), Hangzhou, China, 25–28 October 2016; pp. 1194–1196.
42. Sato, S.i.; Beernink, K.; Ohshima, T. Charged particle radiation effects on flexible a-si/a-sige/a-sige triple junction solar cells for space use. In Proceedings of the 2013 IEEE 39th Photovoltaic Specialists Conference (PVSC) PART 2, Tampa, FL, USA, 16–21 June 2013; pp. 76–82.
43. Simons, M.; Monteith, L.K.; Hauser, J.R. Some observations on charge buildup and release in silicon dioxide irradiated with low energy electrons. *IEEE Trans. Electron. Dev.* **1968**, *15*, 966–973. [CrossRef]

2

Dopant-Free Hole Transport Materials with a Long Alkyl Chain for Stable Perovskite Solar Cells

Kai Wang [1,†], Haoran Chen [1,†], Tingting Niu [1], Shan Wang [1], Xiao Guo [1] and Hong Wang [2,*]

1 Key Laboratory of Flexible Electronics (KLOFE) & Institute of Advanced Materials (IAM), Jiangsu National Synergetic Innovation Center for Advanced Materials (SICAM), Nanjing Tech University (Nanjing Tech), Nanjing 211816, China

2 Zhongshan Institute of Modern industrial Technology, South China University of Technology, Zhongshan 528437, China

* Correspondence: phhwang@scut.edu.cn

† These authors contributed equally to this work.

Abstract: Hole transport materials are indispensable to high efficiency perovskite solar cells. Two new hole transporting materials (HTMs), named 4,4′-(9-nonyl-9H-carbazole-3,6-diyl)bis (*N*,*N*-bis(4-methoxyphenyl)aniline) (CZTPA-1) and 4,4′-(9-methyl-9H-carbazole-3,6-diyl)bis (*N*,*N*-bis(4-methoxyphenyl)aniline)(CZTPA-2), were developed by different alkyl substitution methods. The two compounds, containing a carbazole core and triphenylamine (TPA) groups with different lengths of the alkyl chain, were designed and synthesized through a two-step synthesis approach. The power conversion efficiency (PCE) was found to be affected by the length of the alkyl chain, reaching 7% for CZTPA-1 and 11% for CZTPA-2. Furthermore, the CZTPA-2 still maintained 89.7% of its original performance after 400 h. The proposed results demonstrate the effect of carbon chain substituents on the efficiency of perovskite solar cells (PSCs).

Keywords: perovskite solar cell; alkyl chain; hole transporting materials; stable

---

1. Introduction

Perovskite solar cells (PSCs) have attracted much attention in recent years owing to their excellent optoelectronic properties, high absorption coefficients, easy solution processability, high carrier mobility, and so on [1–5]. Many efforts have been made to increase the efficiency of PSCs. Surprisingly, it has increased dramatically from 3.8% [1] to 24.2% [6] in a short period of time.

As is well-known, 2,2′,7,7′-Tetrakis[*N*,*N*-di(4-methoxyphenyl)amino]-9,9′-spirobifluorene (Spiro-OMeTAD) [7,8] is a common hole transporting material (HTM) with good performance. However, it is still a formidable challenge to develop this material due to various drawbacks, such as a complicated synthesis process and high cost. Therefore, new HTMs with a simple synthesis process and low cost need urgently to be developed.

HTMs construct the hole transport layer (HTL) that is needed to block electron transport, enhance hole transport, and prevent direct contact between the perovskite layer and the electrode, which causes annihilation. An ideal HTM should have such characteristics as good hole mobility, good hydrophobicity, suitable energy levels, and the ability to be prepared in solution [9–11]. Currently, common HTMs can be classified as inorganic substances, organic polymers, or organic small molecules depending on the type of material. Low cost, high stability, and hole mobility are among the many advantages that HTMs based on inorganic materials have. Kamat and co-workers first introduced copper iodide as an HTM and achieved an average PCE of 6% [12]. Then, inorganic semiconductors (CuSCN, NiO, CuI) as HTMs were developed [13,14]. In addition to inorganic HTMs, polymer-based HTMs have been explored in PSCs. Poly-[bis(4-phenyl)(2,4,6-trimethylphenyl)amine] (PTAA) was the first conjugated polymer to be used as an HTM in PSCs [15] and maintained the highest PCE of any reported polymeric HTM [16]. Accompanied by the development of conjugated polymer HTMs, Poly(3-hexylthiophene-2,5-diyl (P3HT) [17] was originally used in organic solar cells (OSCs) as the active layer was introduced into the hole transport layer. Compared to polymeric HTMs, small molecule HTMs have the advantages of a determined molecular weight and simple purification for PSCs. As a group with good stability and solubility, triphenylamine (TPA) is widely used in organic small molecule HTMs [18–20]. The TPA group in particular can influence the optoelectrical properties of HTMs due to its non-planar geometry. Furthermore, the carbazole group has a high carrier mobility in OSCs and as good a performance as HTMs; thus, it is currently regarded as a core in PSCs [21–27].

In 2014, Sun and co-workers [27] designed a series of HTMs based on carbazole, named X19 and X51. The X51 realized a PCE of 9.8% due to a higher charge-carrier mobility and conductivity than X19. Subsequently, Nazeeruddin and co-workers [28] studied bridged carbazole with biphenyl, by using silolothiophene as the bridge, which obtained a PCE of 13.1%. They found that compared to the spirofluorene-linked triphenylamine HTMs, novel silolothiophene-linked methoxy triphenylamines (Si-OMeTPAs) enable more stable PSCs. In the same year, Tang and co-workers [29] used carbazole as a core with four TPAs as the side groups to achieve a PCE of up to 18.32%.

Recently, Khaja Nazeeruddin and co-workers [30] used anthra[1,2-b:4,3-b′:5,6-b′:8,7-b‴]tetrathiophene as the core, by changing the alkyl chain length of the methoxy groups on the triarylamine sites, to develop a series of materials. The device based on a methyl substitute named ATT-OMe was found to have the best PCE of 18.13%, which was better than that of the device based on other longer alkyl chain HTMs. They believed that the presence of alkyl chains decreased the hole-transport properties. However, the performance of PSCs based on carbazole is far from that of the classical Spiro-OMeTAD. Therefore, more efforts are needed to develop new small molecule HTMs that match with perovskite instead of the Spiro-OMeTAD.

In this work, we developed two HTMs—4,4′-(9-nonyl-9H-carbazole-3,6-diyl)bis (*N*,*N*-bis(4-methoxyphenyl)aniline) (CZTPA-1) and 4,4′-(9-methyl-9H-carbazole-3,6-diyl)bis (*N*,*N*-bis(4-methoxyphenyl)aniline) (CZTPA-2)—based on TPA as the end group. They both have the advantage of simple synthesis steps and low cost. Both HTMs are synthesized by one-step Suzuki coupling. The cost of the raw material 3,6-Dibromocarbazole ($0.3/g) plus 4-methoxy-*N*-(4-methoxyphenyl)-*N*-(4-(4,4,5,5-tetramethyl-1,3,2-dioxaborolan-2-yl)phenyl)aniline ($15/g) is clearly lower than that of Spiro-OMeTAD ($220/g). Notably, the CZTPA-2 with the longer alkyl chain achieved a better PCE of 11.79% with a short current density (J_{sc}) of 21.80 mA/cm^2, an open circuit voltage (V_{oc}) of 0.99 V, and a fill factor (FF) of 54.59%. This is attributed to the significantly improved hole mobility of CZTPA-2, resulting in a significant increase in device efficiency.

2. Materials and Methods

2.1. Materials

Unless otherwise noted, all reagents used in the experiments were purchased from commercial sources and used without further purification. 3,6-Dibromo-9H-carbazole, *N*,*N*-bis(4-Methoxyphenyl)-4-(4,4,5,5-tetraMethyl-1,3,2-dioxaborolan-2-yl)-BenzenaMine, lead iodide (PbI_2), methylammonium iodide (MAI), acetonitrile (99.8%), chlorobenzene (99.9%), and dimethylformamide (DMF) (99%) were purchased from Sigma-Aldrich. 4-tert-butylpyridine (TBP) and Li-bis-(trifluoromethanesulfonyl) imide (Li-TFSI) were purchased from TCI. 2,2′,7,7′-tetrakis-(*N*,*N*-di-p-methoxyphenylamine)-9,9′-spirobifluorene (Spiro-OMeTAD) (99.0%) was purchased from Xi'an Polymer Light Technology Co., Ltd.

Perovskite precursor: The perovskite precursor was obtained by mixing PbI_2 and MACl (in a molar ratio of 1:1) in DMF with a concentration of 350 mg/mL, and was then stirred at 60 °C overnight in a glovebox.

Spiro-OMeTAD: The 2,2′,7,7′-Tetrakis(*N*,*N*′-di-p-methoxyphenylamine)-9,9′-spirobifluorene (Spiro-OMeTAD) was doped with TBP and Li-TFSI. A total of 73.2 mg of Spiro-OMeTAD (Xi'an Polymer Light Technology Co., Ltd., Xi'an, China) was dissolved in 1 mL of chlorobenzene (CB) with 28.8 μL of 4-tert-butylpyridine (TBP) and 17.6 μL of Li-bis-(trifluoromethanesulfonyl) imide (Li-TFSI).

2.2. Device Fabrication

The SnO_2 layer was spin-coated on an ITO substrate at 3000 rpm for 30 s, which was cleaned in UV–ozone and then annealed at 150 °C for 30 min. The perovskite layer was spin-coated at 4000 rpm for 30 s by an anti-solvent method. The details of the operation are as follows: 100 μL CB was rapidly added after 5 s of spin-coating with perovskite solution; and the perovskite films were annealed at 100 °C for 5 min. All of the above processes were performed in the nitrogen glovebox. Then, Spiro-OMeTAD and two HTMs were dissolved in the CB (10 mg/mL), and then spin-coated upon the perovskite layer at 3000 rpm for 30 s. Then, molybdenum trioxide (MoO_3) and gold (Au) were thermally evaporated on the hole transporting layer. The effective area of the cell is 0.05 cm^2.

2.3. Device Characterization

The cross-sectional images of PSC were taken by scanning electron microscopy (SEM) (ZEISS Merlin, Carl Zeiss Microscopy, Jena, Germany). The UV–visible absorption spectra were measured using a UV Spectrophotometer (SHIMADZU UV-1750, East Test Technology Co., Ltd, Shenzhen, China). Photoluminescence (PL) spectra were obtained using a spectrofluorometer (HitachiF-7000, Hitachi High-Technologies Corporation, Shenzhen, China). Thermogravimetric (TGA) analysis was performed on a Mettler Toledo TGA2.

The device was measured under AM 1.5 G solar irradiation with an intensity of 100 mW/cm^2 through an Enlitech SS-F5-3A solar simulator. The instrument was calibrated on standard solar cells. *J*–*V* properties were measured by the method of Enlitech Ltd. (Kaohsiung, Taiwan) and a Keithley (Cleveland, OH, USA) 2400 source meter under dark conditions. The external quantum efficiency (EQE) spectra were measured using a solar cell IPCE test system (CROWNTECH Inc., model QTEST HIFINITY (Macungie, PA, USA)).

The synthesis method and details of the experimental procedure are shown in the supporting information (SI).

3. Results

The details of the experimental procedure are shown in the Supplementary Information (SI). The design principle was to improve planarity as well as increase solubility. The carbazole group with simple structures has good hole transporting ability. By inserting N atoms with different alkyl chain lengths, the optoelectronic properties and solubility can be adjusted. Herein, we chose 4-methoxy-*N*-(4-methoxyphenyl)-*N*-(4-(4,4,5,5-tetramethyl-1,3,2-dioxaborolan-2-yl)phenyl)aniline units as the raw material, used Suzuki coupling, and substituted the carbazole core with different lengths of alkyl chains. The aimed-for materials were obtained by simple column chromatography separation and recrystallization.

Figure 1a shows the absorption spectra of CZTPA-1 and CZTPA-2 in chloroform and as a coating on quartz substrates. Absorption peaks at 335 nm for CZTPA-1 and 327 nm for CZTPA-2 in the solution were observed. Relative to the solution, the CZTPA-1 film exhibits a redshift of 2 nm with an onset of 409 nm, corresponding to an optical bandgap of 3.03 eV, whereas CZTPA-2 aligns to a narrower bandgap of 2.98 eV (onset of 416 nm). Both HTMs have a slight redshift in the film compared with the solution, suggesting that aggregation exists in the films.

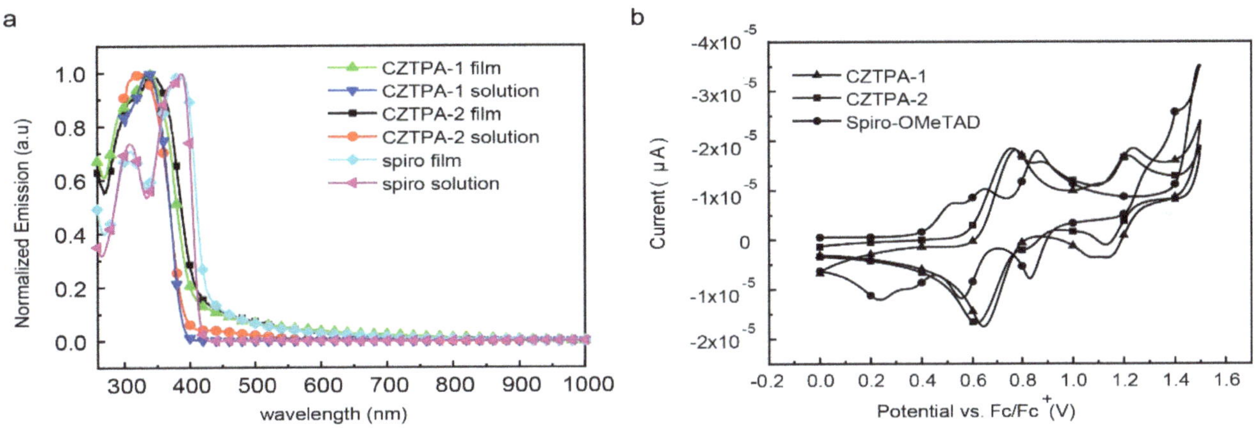

Figure 1. (**a**) Optical absorption spectra of 4,4′-(9-nonyl-9H-carbazole-3,6-diyl)bis(*N*,*N*-bis(4-methoxyphenyl)aniline) (CZTPA-1) and 4,4′-(9-methyl-9H-carbazole-3,6-diyl)bis(*N*,*N*-bis(4-methoxyphenyl)aniline) (CZTPA-2) in chloroform and thin films spin-coated from chloroform. (**b**) Cyclic voltammetry of CZTPA-1 and CZTPA-2.

From Figure 1b, the highest occupied molecular orbital (HOMO) level of CZTPA-1 is −4.89 eV, while in the CZTPA-2 there is a higher HOMO of −4.83 eV (the energy level of Ferroceneis is −0.54 eV by the cyclic voltammetry method). Compared to the two HTMs, the Spiro-OMeTAD has the higher energy level as the HOMO is −4.64 eV. Based on the relationship of $E_{LUMO} = E_{HOMO} + E_g$, the lowest unoccupied molecular orbital (LUMO) level value was calculated to be −1.86 eV for CZTPA-1, −1.85 eV for CZTPA-2, and −1.74 eV for Spiro-OMeTAD (Table 1). CZTPA-2 has similar electrochemical properties compared with CZTPA-1, which indicates that the electrochemical properties of the molecule remain stable with a change in the alkyl chain length.

Table 1. Optical and electrochemical properties of CZTPA-1 and CZTPA-2.

	λ_{max} Sol (a) (nm)	λ_{max} Film (b) (nm)	λ_{onset} (nm)	E_g^{opt} (c) (eV)	E_{HOMO} (eV)	E_{LUMO} (d) (eV)
CZTPA-1	335	337	409	3.03	−4.89	−1.86
CZTPA-2	327	339	416	2.98	−4.83	−1.85
Spiro-OMeTAD	386	390	428	2.90	−4.64	−1.74

(a) Maximum absorption peak in CH_2Cl_3 solution; (b) Maximum absorption peak of films on quartz glass; (c) Optical bandgap calculated from the absorption onset of films: $E_g^{opt} = 1240/\lambda_{onset}$ eV; (d) $E_{LUMO} = E_{HOMO} + E_g^{opt}$.

We then found that CZTPA-1 and CZTPA-2 match well with the energy level of the perovskite from Figure 2a. Furthermore, the thermal stability of the two HTMs was measured by thermogravimetric analysis (TGA) (Figure 2b). Both CZTPA-1 and CZTPA-2 exhibit good thermal stability with decomposition temperatures (T_d, 5% weight loss) at 391.8 °C and 384.6 °C, respectively. As we expected, increasing the length of the alkyl chains reduces the thermal stability.

The photoluminescence (PL) spectra in Figure 2c show the maximum emission peak at 431 nm for CZTPA-1 and at 425 nm for CZTPA-2 in film. CZTPA-2 was slightly more blue-shifted than CZTPA-1, caused by the increase in the length of the alkyl chains.

Figure 2d shows the PL spectra of the perovskite, perovskite with Spiro-OMeTAD, perovskite with CZTPA-1, and perovskite with CZTPA-2. Strong PL quenching was observed after the HTMs were coated on perovskite films. Respectively, compared with the original perovskite film, the PL intensity was reduced to 7%, 22%, and 18% after coating with Spiro-OMeTAD, CZTPA-1, and CZTPA-2. Thus, we think that CZTPA-2 has better charge separation than CZTPA-1 and a smaller J_{sc} and FF in the PSCs compared to Spiro-OMeTAD devices. In short, this means that the hole transfer capabilities of CZTPA-2 are superior because of their better charge transfer capability.

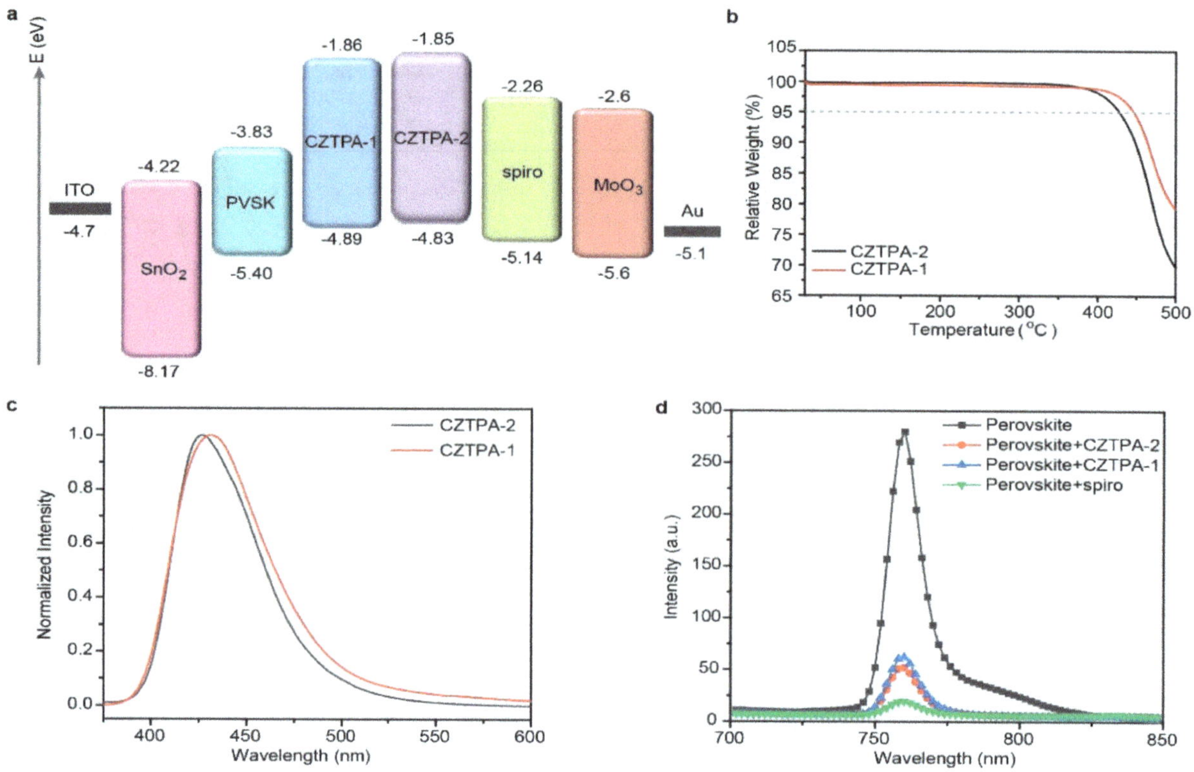

Figure 2. (**a**) Energy level diagram of a perovskite solar cell (PSC) with CZTPA-1, CZTPA-2, and 2,2′,7,7′-Tetrakis[*N*,*N*-di(4-methoxyphenyl)amino]-9,9′-spirobifluorene (Spiro-OMeTAD). (**b**) Thermogravimetric analysis (TGA) diagrams of CZTPA-1 and CZTPA-2. (**c**) Photoluminescence (PL) spectra of CZTPA-1 and CZTPA-2 thin film, excitation at 350 nm. (**d**) Photoluminescence spectra of perovskite, perovskite with Spiro-OMeTAD, perovskite with CZTPA-1, and perovskite with CZTPA-2, excitation at 500 nm.

Figure 3 shows two cross-sectional scanning electron microscopy (SEM) images of the PSC with the structure of ITO/SnO_2/perovskite/HTMs/Au. The PSC includes an ≈460 nm perovskite capping layer and an ≈35 nm HTM layer (CZTPA-1 or CZTPA-2). From the SEM image, we found that the HTM layer deposited on the perovskite layer and the boundaries of each layer are clear. Atomic force microscopy (AFM) images of the two materials spin-coated onto perovskite films are shown in Figure S7, which represent the perovskite/CZTPA-2 (a) and the perovskite/CZTPA-1 (b) films, respectively. The roughness of both samples is slightly high, which may be due to the lower thickness of the HTMs. After a comparison, it was found that the roughness of the film in which the HTM is CZTPA-2 (RMS = 20.968 nm) is significantly lower than that of CZTPA-1 (RMS = 28.662 nm); this is attributed to the solubility of CZTPA-2 being higher such that it could better cover the film and further improve the carrier transport of the device.

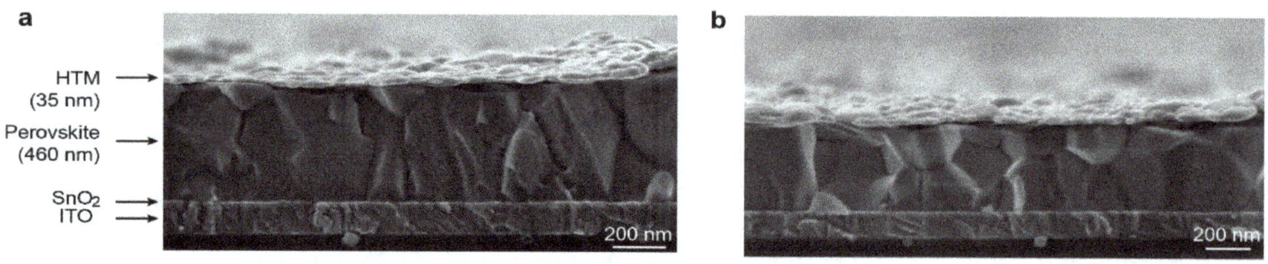

Figure 3. (**a**) A cross-sectional SEM image of the CZTPA-1 PSC. The scale bar is 200 nm. (**b**) A cross-sectional SEM image of the CZTPA-2 PSC. The scale bar is 200 nm.

In order to compare the properties of the two HTMs, we used them as the hole transport layer of perovskite solar cells to compare device performance. Between them, the effective area of the cells is 0.05 cm^2, and the scanning rate is 0.02 V/s. As shown in Figure 4a,b, the PSCs with CZTPA-2 achieve a PCE of 11.79% with an open-circuit voltage (V_{oc}) of 0.99 V, a short-circuit current density (J_{sc}) of 21.8 mA cm^{-2}, and a fill factor (FF) of 54.59%, while the PSCs with CZTPA-1 achieve a lower PCE of 6.05% under the condition of no doping. In contrast, the best cell is based on Spiro-OMeTAD, as the HTM achieves the PCE of 16.77%. We also compared the Spiro-OMeTAD PSC without any doping, which achieved a PCE of 11.74%. The photovoltaic performance of the PSCs based on the two HTMs (dopant-free) and the Spiro-OMeTAD were investigated under AM 1.5 illumination (100 mW cm^{-2}). The champion device performance plots of CZTPA-1 and CZTPA-2 are shown in Table 2. The device fabricated with CZTPA-1 as the HTM yielded a promising PCE of 6.05% with a J_{sc} of 20.58 mA cm^{-2}, a V_{oc} of 0.77 V, and an FF of 38.01%. We calculated the average PCE of the CZTPA-2 devices to be 10.15% ± 0.90. The CZTPA-1 devices own an average PCE of 5.27% ± 0.57. The average PCE of the Spiro-OMeTAD (dopant-free) is 10.02% ± 0.98, and the corresponding PCE of Spiro-OMeTAD is 15.65% ± 0.71. All data are based on the values obtained from 20 devices. From these, we can speculate that CZTPA-2 exhibits better performance compared with CZTPA-1. The FF and V_{oc} of CZTPA-1 are obviously lower than those of CZTPA-2 in the PSC devices. Furthermore, CZTPA-2 obtains a slightly higher PCE compared with the dopant-free Spiro-OMeTAD, which is attributed to the higher J_{sc}.

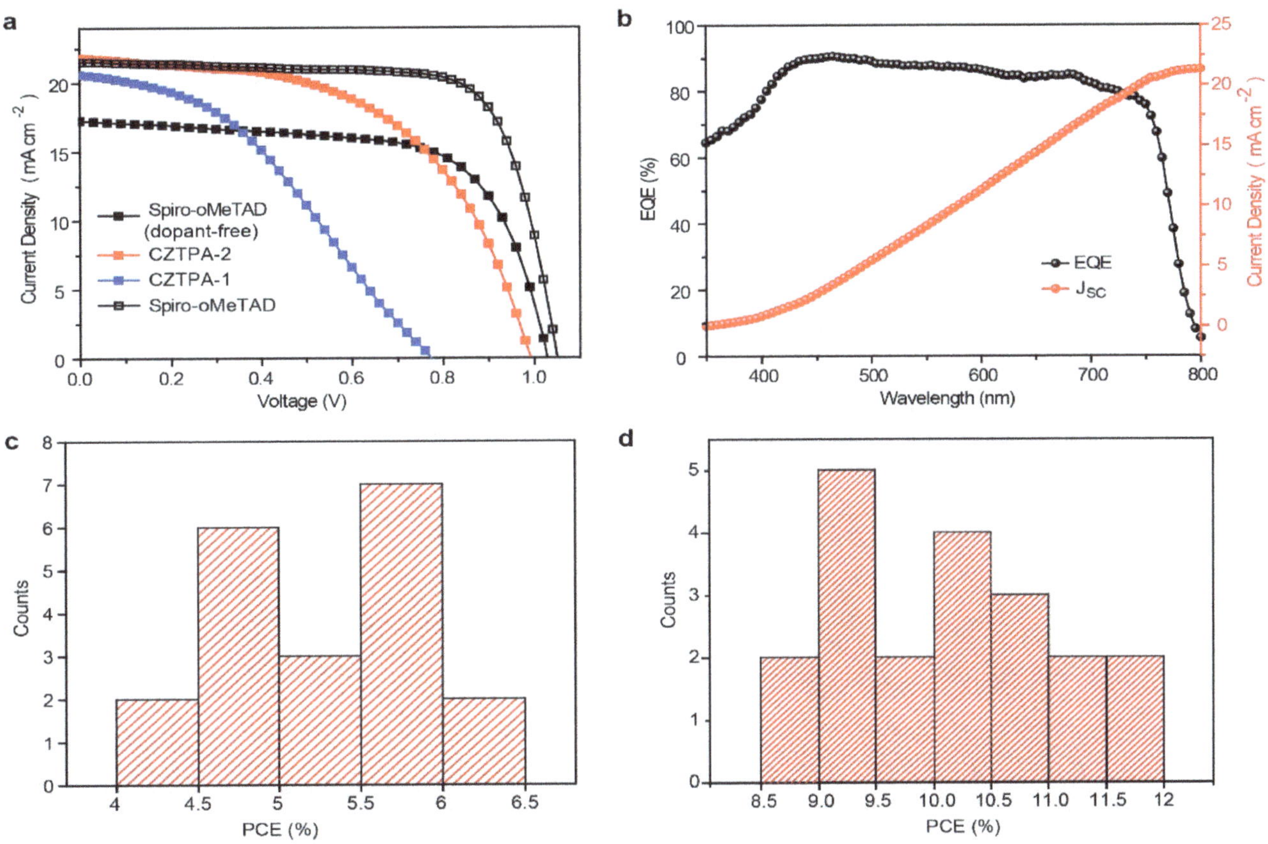

Figure 4. (**a**) *J*–*V* characteristics of PSCs based on the two hole transport materials (HTMs), the Spiro-OMeTAD, and the Spiro-OMeTAD (dopant-free). (**b**) External quantum efficiency (EQE) and J_{sc} spectra of PSCs with CZTPA-2. (**c**) Histograms of PCEs measured in 20 cells of CZTPA-1. (**d**) Histograms of PCEs measured in 20 cells of CZTPA-2.

Table 2. Photovoltaic data of PSCs based on the two HTMs, the Spiro-OMeTAD, and the Spiro-OMeTAD (dopant-free).

	V_{OC} (V)	J_{SC} (mA·cm^{-2})	FF (%)	PCE (%)	PCE Ave (%)
Spiro-OMeTAD (dopant-free)	1.02	17.26	66.14	11.74	10.02% ± 0.98
CZTPA-2	0.99	21.80	54.59	11.79	10.15% ± 0.90
CZTPA-1	0.77	20.58	38.01	6.05	5.27% ± 0.57
Spiro-OMeTAD	1.05	21.51	74.24	16.77	15.65% ± 0.71

When preparing the solution, we found the CZTPA-1 dissolution rate to be slower than the CZTPA-2 dissolution rate. What is more, the solubility of CZTPA-2 was found to be better than that of CZTPA-1 with an increasing alkyl chain length and CZTPA-2 deposits in perovskite with better crystallization. These results also demonstrate that CZTPA-2 achieves better performance.

To test the reproducibility of the devices, we fabricated 20 devices in several different batches. The devices are shown in Figure 4c,d. As shown in the PCE histogram of the corresponding device data, the average PCE of CZTPA-2 and CZTPA-1 is 10.15% and 5.27%, respectively.

Moreover, the EQE spectrum of PSCs with CZTPA-2 is also shown in Figure 4b. The integral of the current densities calculated from the EQE spectra is 21.32 mA cm^{-2} for CZTPA-2 according predominantly to the experimental data.

To calculate the hole mobility of the two HTMs, we constructed a device with a configuration of ITO/PEDOT:PSS/HTM/Au using the space-charge-limited current (SCLC) method, and the *J–V* characteristics of this device were studied in the dark. Hole mobility is calculated by the Mott–Gurney equation of $J = 9\varepsilon_r\varepsilon_0\mu Va^2/8L^3$, so we change the form of the formula to obtain $\mu = 8d^3/9\varepsilon_r\varepsilon_0(J^{1/2}/Va)^2$, where ε_r is the relative dielectric constant of the transport medium ($\varepsilon_r = 3$ for organic materials), ε_0 is the permittivity of free space (8.85×10^{-12} C V^{-1} m^{-1}), *J* is the dark current density (mA cm^{-2}), and d is the thickness of the active layer [31]. d is 48 nm for CZTPA-1 and 56 nm for CZTPA-2. Figure 5 shows that the hole mobility of Spiro-OMeTAD (doped) is 1.01×10^{-3} cm^2 V^{-1} s^{-1}. The hole mobility of CZTPA-1 is 4.68×10^{-5} cm^2 V^{-1} s^{-1}, and CZTPA-2 has the higher hole mobility of 8.06×10^{-5} cm^2 V^{-1} s^{-1}. However, they are all lower than the hole mobility of Spiro-OMeTAD (doped). Compared to CZTPA-1, CZTPA-2 has a higher hole mobility, which leads to good hole transport and enhances the charge transport in a planar PSC. CZTPA-2's higher hole mobility can be attributed to its high hole transport capability. These results suggest that both a fast charge transfer and high hole transport capability contribute to a high PCE.

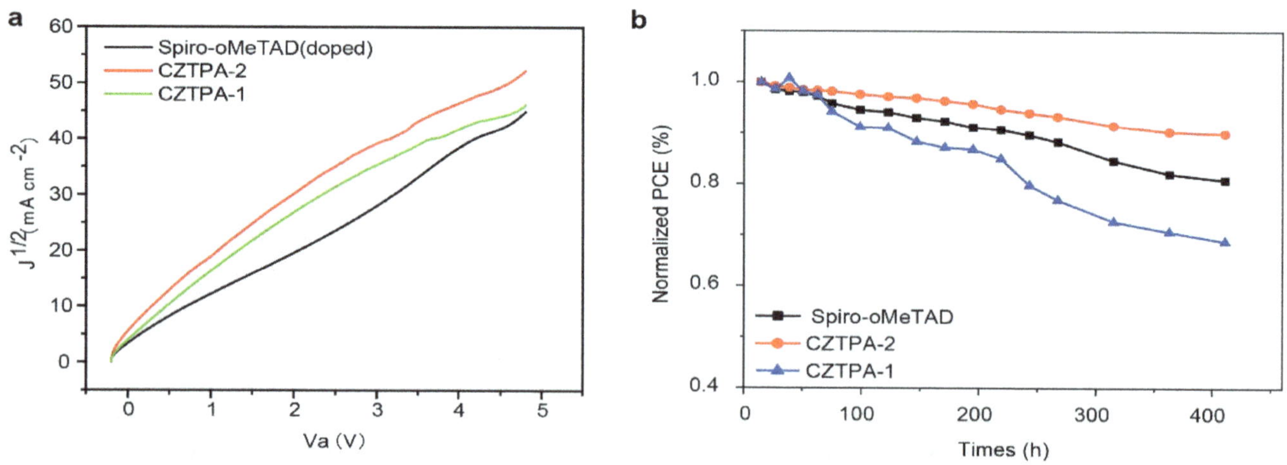

Figure 5. (**a**) Space-charge-limited current (SCLC) $J^{1/2}$–V characteristics of CZTPA-1-, CZTPA-2-, and Spiro-OMeTAD-based hole-only devices measured in the dark. (**b**) Stability of CZTPA-1, CZTPA-2, and Spiro-oMeTAD (dopant-free) without encapsulation.

We tested the stability of the three HTM devices without encapsulation by storing them in the dark under air conditions for at least 400 h. The PCE over time curve was plotted and is shown in Figure 5b. The PCE of the CZTPA-2 device was still over 10%. In comparison, the PCE of Spiro-OMeTAD (dopant-free) device dropped below 10%. CZTPA-1 exhibited low performance, with 68.2% of the original PCE. CZTPA-2 and Spiro-OMeTAD (dopant-free) maintained 89.7% and 81.6% of their initial PCE, respectively. This test verified that the device based on CZTPA-2 has the best stability of the three HTMs. The long alkyl chain, which has good morphology, may influence the stability of a PSC device.

Finally, we also tested the capacitance of the three HTMs. The capacitance versus frequency was plotted and is shown in Figure 6. The capacitance is mainly caused by charge or ion accumulation at the perovskite interface, which leads to interfacial recombination. We can see the capacitance of the device based on CZTPA-2 is obviously smaller than that of CZTPA-1 and Spiro-oMeTAD (dopant-free), which confirms that the CZTPA-2 device has less interfacial recombination and a higher PCE.

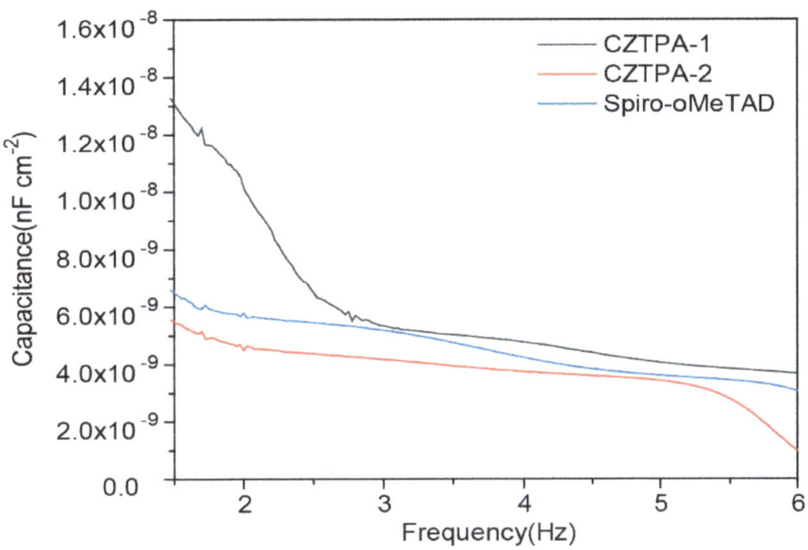

Figure 6. Capacitance versus frequency based on the devices of ITO/SnO_2/perovskite/HTMs/Au.

4. Conclusions

Two new and low-cost hole transporting materials based on a carbazole core were designed and synthesized using a simple synthesis process. CZTPA-2 (dopant-free) achieved the best performance, with a PCE of 11.79%, a J_{sc} of 21.80 mA/cm^2, a V_{oc} of 0.99 V, and a FF of 54.59%, which was slightly higher than that of Spiro-OMeTAD (dopant-free) and CZTPA-1 (dopant-free) and attributed to its higher hole transport mobility. The PL spectra, scanning electron microscopy images, and photoelectric properties indicate that CZTPA-2 with a longer alkyl chain has better optoelectrical properties. The CZTPA-2 (dopant-free) device also had the best stability, which remained at 89.7% of its original PCE after 400 h compared to CZTPA-1 and Spiro-OMeTAD (dopant-free). Besides this, its hole transport layer thickness is 35 nm. Therefore, we think that CZTPA-2 can also be used to modify an interface when compared to traditional HTMs (above 100 nm). The device based on CZTPA-2 exhibited good stability. We conclude that a longer alkyl chain may promote solubility and enhance the perovskite layer's crystallinity.

Author Contributions: K.W. conceived the idea; K.W. designed the experiment; H.C. conducted device fabrication; X.G. and T.N. helped in device measurement; K.W. conducted device fabrication and data collection and wrote the manuscript; K.W. and S.W. revised the manuscript; Writing—review & editing, H.W.; Supervision, H.W.

References

1. Kojima, A.; Teshima, K.; Shirai, Y.; Miyasaka, T. Organometal Halide Perovskites as Visible-Light Sensitizers for Photovoltaic Cells. *J. Am. Chem. Soc.* **2009**, *131*, 6050–6051. [CrossRef]
2. Michael, M.; Lee, J.L.T.T. Efficient Hybrid Solar Cells Based on Meso-Superstructured Organometal Halide Perovskites. *Science* **2012**, *338*, 643–647.
3. Kim, H.; Lee, C.; Im, J.; Lee, K.; Moehl, T.; Marchioro, A.; Moon, S.; Humphry-Baker, R.; Yum, J.; Moser, J.E.; et al. Lead Iodide Perovskite Sensitized All-Solid-State Submicron Thin Film Mesoscopic Solar Cell with Efficiency Exceeding 9%. *Sci. Rep.* **2012**, *2*, 591. [CrossRef] [PubMed]
4. Laura Calil, S.K.M.G. Hole-Transport Materials for Perovskite Solar Cells. *Angew. Chem. Int. Ed.* **2016**, *55*, 14522–14545.
5. Bi, D.; Tress, W.; Dar, M.I.; Gao, P.; Luo, J.; Renevier, C.M.; Schenk, K.; Abate, A.; Giordano, F.; Correa Baena, J.; et al. Efficient luminescent solar cells based on tailored mixed-cation perovskites. *Sci. Adv.* **2016**, *2*, 1501170–1501176. [CrossRef] [PubMed]
6. NREL Best Research Cell Efficiency Chart. Available online: https://www.nrel.gov/pv/assets/pdfs/best-research-cell-efficiencies-190416.pdf (accessed on 27 June 2019).
7. Bach, U.; Lupo, D.; Comte, P.; Moser, J.E.; Weissortel, F.; Salbeck, J.; Spretzer, H.; Gratzel, M. Solid-state dye-sensitized mesoporous TiO2 solar cells with high photon-to-electron conversion efficiencies. *Nature* **1998**, *395*, 583–585. [CrossRef]
8. Jeon, N.J.; Lee, H.G.; Kim, Y.C.; Seo, J.; Noh, J.H.; Lee, J.; Seok, S.I. o-Methoxy Substituents in Spiro-OMeTAD for Efficient Inorganic—Organic Hybrid Perovskite Solar Cells. *J. Am. Chem. Soc.* **2014**, *136*, 7837–7840. [CrossRef]
9. Gatti, T.; Menna, E.; Meneghetti, M.; Maggini, M.; Petrozza, A.; Lamberti, F. The Renaissance of fullerenes with perovskite solar cells. *Nano Energy* **2017**, *41*, 84–100. [CrossRef]
10. Wang, R.; Mujahid, M.; Duan, Y.; Wang, Z.; Xue, J.; Yang, Y. A Review of Perovskites Solar Cell Stability. *Adv. Funct. Mater.* **2019**, 1808843. [CrossRef]
11. Castro, E.; Murillo, J.; Fernandez-Delgado, O.; Echegoyen, L. Progress in fullerene-based hybrid perovskite solar cells. *J. Mater. Chem. C* **2018**, *6*, 2635–2651. [CrossRef]
12. Christians, J.A.; Fung, R.C.M.; Kamat, P.V. An Inorganic Hole Conductor for Organo-Lead Halide Perovskite Solar Cells. Improved Hole Conductivity with Copper Iodide. *J. Am. Chem. Soc.* **2014**, *136*, 758–764. [CrossRef] [PubMed]
13. Kim, J.H.; Liang, P.; Williams, S.T.; Cho, N.; Chueh, C.; Glaz, M.S.; Ginger, D.S.; Jen, A.K.Y. High-Performance and Environmentally Stable Planar Heterojunction Perovskite Solar Cells Based on a Solution-Processed Copper-Doped Nickel Oxide Hole-Transporting Layer. *Adv. Mater.* **2015**, *27*, 695–701. [CrossRef] [PubMed]
14. Qin, P.; Tanaka, S.; Ito, S.; Tetreault, N.; Manabe, K.; Nishino, H.; Nazeeruddin, M.K.; Gr tzel, M. Inorganic hole conductor-based lead halide perovskite solar cells with 12.4% conversion efficiency. *Nat. Commun.* **2014**, *5*, 3834. [CrossRef] [PubMed]
15. Heo, J.H.; Im, S.H.; Noh, J.H.; Mandal, T.N.; Lim, C.; Chang, J.A.; Lee, Y.H.; Kim, H.; Sarkar, A.; Nazeeruddin, M.K.; et al. Efficient inorganic-organic hybrid heterojunction solar cells containing perovskite compound and polymeric hole conductors. *Nat. Photonics* **2013**, *7*, 486–491. [CrossRef]
16. Huang, X.; Zhao, Z.; Cao, L.; Chen, Y.; Zhu, E.; Lin, Z.; Li, M.; Yan, A.; Zettl, A.; Wang, Y.M.; et al. High-performance transition metal-doped Pt3Ni octahedra for oxygen reduction reaction. *Science* **2015**, *348*, 1230–1234. [CrossRef]
17. Edri, E.; Kirmayer, S.; Cahen, D.; Hodes, G. High Open-Circuit Voltage Solar Cells Based on Organic?—CInorganic Lead Bromide Perovskite. *J. Phys. Chem. Lett.* **2013**, *4*, 897–902. [CrossRef] [PubMed]
18. Cabau, L.; Garcia-Benito, I.; Molina-Ontoria, A.; Montcada, N.F.; Martin, N.; Vidal-Ferran, A.; Palomares, E. Diarylamino-substituted tetraarylethene (TAE) as an efficient and robust hole transport material for 11% methyl ammonium lead iodide perovskite solar cells. *Chem. Commun.* **2015**, *51*, 13980–13982. [CrossRef]
19. Li, H.; Fu, K.; Hagfeldt, A.; Gr tzel, M.; Mhaisalkar, S.G.; Grimsdale, A.C. A Simple 3,4-Ethylenedioxythiophene Based Hole-Transporting Material for Perovskite Solar Cells. *Angew. Chem. Int. Ed.* **2014**, *53*, 4085–4088. [CrossRef]

20. Li, H.; Fu, K.; Boix, P.P.; Wong, L.H.; Hagfeldt, A.; Gr tzel, M.; Mhaisalkar, S.G.; Grimsdale, A.C. Hole-Transporting Small Molecules Based on Thiophene Cores for High Efficiency Perovskite Solar Cells. *ChemSusChem* **2014**, *7*, 3420–3425. [CrossRef]
21. Park, S.H.; Roy, A.; Beaupr, S.; Cho, S.; Coates, N.; Moon, J.S.; Moses, D.; Leclerc, M.; Lee, K.; Heeger, A.J. Bulk heterojunction solar cells with internal quantum efficiency approaching 100%. *Nat. Photonics* **2009**, *3*, 297–302. [CrossRef]
22. Qin, R.; Li, W.; Li, C.; Du, C.; Veit, C.; Schleiermacher, H.; Andersson, M.; Bo, Z.; Liu, Z.; Ingana s, O.; et al. A Planar Copolymer for High Efficiency Polymer Solar Cells. *J. Am. Chem. Soc.* **2009**, *131*, 14612–14613. [CrossRef] [PubMed]
23. Blouin, N.; Leclerc, M. Poly(2,7-carbazole)s: Structure—Property Relationships. *Acc. Chem. Res.* **2008**, *41*, 1110–1119. [CrossRef] [PubMed]
24. Blouin, N.; Michaud, A.; Leclerc, M. A Low-Bandgap Poly(2,7-Carbazole) Derivative for Use in High-Performance Solar Cells. *Adv. Mater.* **2007**, *19*, 2295–2300. [CrossRef]
25. Zhu, X.D.; Ma, X.J.; Wang, Y.K.; Li, Y.; Gao, C.H.; Wang, Z.K.; Jiang, Z.Q.; Liao, L.S. Hole-Transporting Materials Incorporating Carbazole into Spiro-Core for Highly Efficient Perovskite Solar Cells. *Adv. Funct. Mater.* **2018**, *29*, 1807094. [CrossRef]
26. Chen, Z.; Li, H.; Zheng, X.; Zhang, Q.; Li, Z.; Hao, Y.; Fang, G. Low-Cost Carbazole-Based Hole-Transport Material for Highly Efficient Perovskite Solar Cells. *ChemSusChem* **2017**, *10*, 3111–3117. [CrossRef] [PubMed]
27. Xu, B.; Sheibani, E.; Liu, P.; Zhang, J.; Tian, H.; Vlachopoulos, N.; Boschloo, G.; Kloo, L.; Hagfeldt, A.; Sun, L. Carbazole-Based Hole-Transport Materials for Efficient Solid-State Dye-Sensitized Solar Cells and Perovskite Solar Cells. *Adv. Mater.* **2014**, *26*, 6629–6634. [CrossRef] [PubMed]
28. Abate, A.; Paek, S.; Giordano, F.; Correa-Baena, J.P.; Saliba, M.; Gao, P.; Matsui, T.; Ko, J.; Zakeeruddin, S.M.; Dahmen, K.H.; et al. Silolothiophene-linked triphenylamines as stable hole transporting materials for high efficiency perovskite solar cells. *Energy Environ. Sci.* **2015**, *8*, 2946–2953. [CrossRef]
29. Yin, X.; Guan, L.; Yu, J.; Zhao, D.; Wang, C.; Shrestha, N.; Han, Y.; An, Q.; Zhou, J.; Zhou, B.; et al. One-step facile synthesis of a simple carbazole-cored hole transport material for high-performance perovskite solar cells. *Nano Energy* **2017**, *40*, 163–169. [CrossRef]
30. Zimmermann, I.; Urieta-Mora, J.; Gratia, P.; Arag, J.; Grancini, G.; Molina-Ontoria, A.N.; Ort, E.; Mart n, N.; Nazeeruddin, M.K. High-Efficiency Perovskite Solar Cells Using Molecularly Engineered, Thiophene-Rich, Hole-Transporting Materials: Influence of Alkyl Chain Length on Power Conversion Efficiency. *Adv. Funct. Mater.* **2017**, *7*, 1601674. [CrossRef]
31. Wang, Y.; Su, T.; Tsai, H.; Wei, T.; Chi, Y. Spiro-Phenylpyrazole/Fluorene as Hole-Transporting Material for Perovskite Solar Cells. *Sci. Rep.* **2017**, *7*, 7859–7867. [CrossRef]

3

Influence of the Carrier Selective Front Contact Layer and Defect State of a-Si:H/c-Si Interface on the Rear Emitter Silicon Heterojunction Solar Cells

Sunhwa Lee [1], Duy Phong Pham [1], Youngkuk Kim [1], Eun-Chel Cho [1], Jinjoo Park [2,*] and Junsin Yi [1,*]

1 School of Information and Communication Engineering, Sungkyunkwan University, Suwon 16419, Korea; ssunholic@skku.edu (S.L.); pdphong@skku.edu (D.P.P.); bri3tain@skku.edu (Y.K.); echo0211@skku.edu (E.-C.C.)

2 Division of Energy and Optical Technology Convergence, Cheongju University, Cheongju 28503, Korea

* Correspondence: jwjh3516@cju.ac.kr (J.P.); junsin@skku.edu (J.Y.)

Abstract: In this research, simulations were performed to investigate the effects of carrier selective front contact (CSFC) layer and defect state of hydrogenated amorphous silicon passivation layer/n-type crystalline silicon interface in silicon heterojunction (SHJ) solar cells employing the Automat for Simulation of hetero-structure (AFORS-HET) simulation program. The results demonstrated the effects of band offset determined by band bending at the interface of the CSFC layer/passivation layer. In addition, the nc-SiOx: H CSFC layer not only reduces parasitic absorption loss but also has a tunneling effect and field effect passivation. Furthermore, it increased the selectivity of contact. In the experimental cell, nc-SiO_x:H was used as the CSFC layer, where efficiency of the SHJ solar cell was 22.77%. Our investigation shows that if a SiO_x layer passivation layer is used, the device can achieve efficiency up to 25.26%. This improvement in the cell is mainly due to the enhancement in open circuit voltage (V_{oc}) because of lower interface defect density resulting from the SiO_x passivation layer.

Keywords: carrier selective contact; rear emitter heterojunction; passivation

1. Introduction

Crystalline silicon (c-Si) solar cells dominate the global photovoltaic market, accounting for more than 90% of production [1,2]. Persistent efforts have been undertaken to achieve their theoretical efficiency threshold of 29.4% by improving the fundamental limiting factors. Recently, an efficiency level of 26.7% for a silicon heterojunction (SHJ) cell was reported by Kaneka corporation using the integrated back contact technique [3]. This result shows the potential of SHJ cell technology to achieve extremely high efficiency. Hence, this technology is expected to not only improve cell efficiency but also reduce manufacturing cost. The performance improvement of SHJ solar cells is dependent on reduction in the following: 1) carrier collection losses using thin wafer [4]; 2) surface recombination losses by surface passivation [5,6]; and 3) parasitic absorption loss of carrier selective front contact (CSFC) layers by controlling their thickness and using wide band-gap materials [7,8]. Interface defects and the optical parasitic absorption loss of the CSFC layers remain the key restrictions on the efficiency of SHJ solar cells.

Conventional SHJ solar cells are generally configured with front p-type hydrogenated amorphous silicon (p-a-Si:H) contact layers to collect better minority carriers [9]. However, due to high activation energy and the low band-gap of such front layers, a Schottky barrier is formed between the transparent conductive oxide (TCO) and p-a-Si:H layers [10,11]. Heavily doped p-layers can be used to reduce the Schottky barrier, but this causes parasitic absorption and increased carrier recombination. Hence, a rear emitter SHJ solar cell (RE-SHJ) with a p-layer located on the rear side has been analyzed [12]. Conventional RE-SHJ solar cells use doped a-Si:H material as both the front and back carrier selective contact layers. The internal electric field is formed according to the doping concentration of these doped layers and, as a result, the carriers are separated and collected. As the doping concentration increases, doping-induced defects also increase, which limits the efficiency of the device. Additionally, the a-Si:H materials reduce transmittance owing to the inherent low band-gap. A wide-gap hydrogenated nanocrystalline silicon oxide (nc-SiO_X:H) layer, which compensates for the disadvantages of a-Si:H [13], is a suitable alternative for the efficiency improvement of RE-SHJ devices. Compared with a-Si:H, nc-SiO_X:H has a low absorption coefficient in the short wavelength region, as well as excellent conduction properties owing to the crystalline phase and quantum confinement effects [14]. In this paper, we present the numerical simulation of RE-SHJ cells using Automat for simulation hetero-structures (AFORS-HET) software. The influence of a-Si:H and silicon oxide materials functionalized as passivation and CSFC layers for the RE-SHJ cells is discussed under experimental and simulation results. The simulation is performed to identify the essential factors for enhancing the efficiency of the device. A numerical simulation is performed in conjunction with experimental data (absorption coefficient, extinction coefficient and optical band gap for each layer) to evaluate the conditions required to increase the efficiency of the solar cells.

2. Materials and Methods

2.1. Experimental Methods

In the device fabrication, a Czochralski (CZ)-grown n-type silicon wafer (148 μm thickness and 1.5 Ωcm resistivity) was used. The surface texturing was performed using sodium hydroxide (8% NaOH) solution. Radio Corporation of America (RCA) standard cleaning was used to remove the surface impurities from the texture wafers and was applied in two stages: standard cleaning 1 and 2. Subsequently, a thin a-Si:H passivation layer was prepared on both interfaces of the Si-wafer. The doped n-type layer was deposited at the front side, and the p-type layer was deposited at the rear. These layers were deposited by employing a standard plasma-enhanced chemical vapor deposition system. The front and back TCO layers were deposited using the radio frequency (RF) sputtering system, which requires Ar feed gas and approximately 200 °C substrate heating during deposition. Additionally, screen printing with Ag paste was used for the front metal grid and back contact electrodes. The RE-SHJ cell structure consisted of front electrode/TCO/n-type CSFC layer/a-Si:H passivation layer/c-Si wafer/a-Si:H passivation layer/p-layer emitter/TCO/back electrode, as shown in Figure 1.

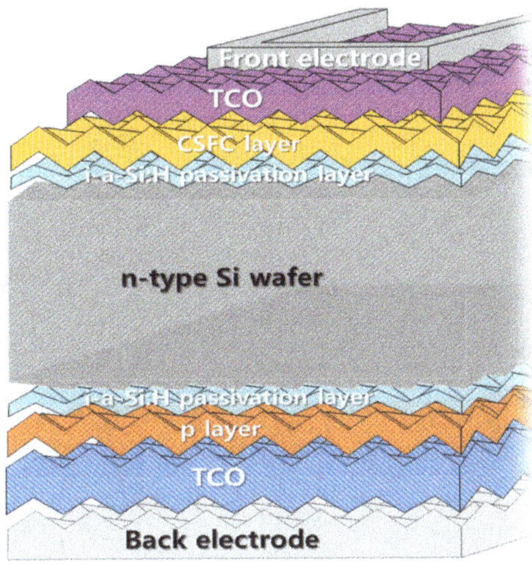

Figure 1. Schematic structure of a heterojunction solar cell.

For the wafer surface passivation with SiO_x layers, the SiO_x passivation layer was formed by varying the pressure of CO_2 plasma post-treatment from 1600 to 2200 mTorr on the a-Si/Si-wafer interface. The minority carrier lifetime (τ_{eff}) was examined using quasi-steady-state photo conductance (QSSPC, WCT-120). For measuring spectroscopic ellipsometry (SE) and X-ray photoelectron spectroscopy, a <100> n-type CZ Si wafer (1–10 Ωcm) with one-side polished was used. The thickness, refractive index, absorption coefficient and optical bandgap were measured by SE (VASE, J. A. Woollam) in the 240 to 1700 nm wavelength range. An X-ray photoelectron spectroscopy (XPS, K-Alpha, Thermo Scientific) system was utilized to estimate the chemical composition using Al Kα radiation.

2.2. Modeling of HIT Solar Cells

The recombination in the semiconductor was calculated using Shockley–Read–Hall statistics [15]. The defect state was investigated at the passivation layer/Si-wafer interface by varying the defect density in the range of 1×10^{10} to 1×10^{14} cm^{-2}eV^{-1}. The illuminated current–voltage characteristics were measured at AM 1.5G (100 mW/cm^2). The Gaussian defect distribution was utilized to describe the band tail states of a-Si:H and nc-SiO_x:H layers in both valence and conduction band. Due to the n-type c-Si substrate, oxide energy levels with an average defect density of 1×10^{11} cm^{-2}eV^{-1} were settled at 0.53 eV from the conduction band. The simulated parameters are listed in Table 1 for all layers. The RE-SHJ solar cell with n-a-Si:H CSFC layer (Figure 2a) consisted of front electrode/TCO/n-a-Si:H CSFC layer/i-a-Si:H passivation layer/n-type c-Si wafer/i-a-Si:H passivation layer/p-a-Si:H layer/TCO/back electrode. The RE-SHJ solar cell with nc-a-SiO_x:H CSFC layer (Figure 2b) consisted of front electrode/TCO/n-a-Si:H CSFC layer/i-a-Si:H passivation layer/n-type c-Si wafer/i-a-Si:H passivation layer/p-nc-SiO_x:H layer/TCO/back electrode.

Table 1. Material parameter values for SHJ solar cell simulation.

Parameters	n-a-Si:H	n-nc-SiOx:H	i-a-Si:H	p-a-Si:H	p-nc-SiOx:H	n-c-Si
Thickness (nm)	5	20	3	5	5	1.5×10^5
Dielectric constant	11.9	11.9	11.9	11.9	11.9	11.9
Electron affinity (eV)	3.9	3.95	3.9	3.9	3.9	4.05
Bandgap (eV)	1.72	2.1	1.70	1.72	2.05	1.124
Effective conduction band density (cm^{-3})	1×10^{21}	1×10^{20}	1×10^{20}	1×10^{20}	1×10^{20}	2.8×10^{19}
Effective valence band density (cm^{-3})	1×10^{21}	1×10^{20}	1×10^{20}	1×10^{20}	1×10^{20}	2.2×10^{19}
Effective electron mobility (cm^2/Vs)	5	50	5	5	20	858
Effective hole mobility (cm^2/Vs)	1	5	1	1	5	355
Doping concentration acceptors (cm^{-3})	0	0	0	1×10^{18}	1×10^{18}	0
Doping concentration donators (cm^{-3})	1×10^{18}	1.42×10^{19}	0	0	0	5.63×10^{16}
Electron thermal velocity (cm/s)	1×10^7	1×10^7	1×10^7	1×10^7	1×10^7	1×10^7
Hole thermal velocity (cm/s)	1×10^7	1×10^7	1×10^7	1×10^7	1×10^7	1×10^7
Parameters	**n-a-Si:H**	**n-nc-SiOx:H**	**i-a-Si:H**	**p a-Si:H**	**p-nc-SiOx:H**	
Defect density at conduction band (CB) edge (cm^{-3}eV^{-1})	1×10^{21}	2×10^{19}	1×10^{21}	1×10^{21}	1×10^{21}	
Defect density at valence band (VB) edge (cm^{-3}eV^{-1})	1×10^{21}	2×10^{19}	1×10^{21}	1×10^{21}	1×10^{21}	
Urbach energy for CB tail (eV)	0.025	0.025	0.025	0.035	0.035	
Urbach energy for VB tail (eV)	0.03	0.03	0.025	0.025	0.025	
$\sigma_e(\sigma_h)$ for CB tail (cm^{-2})	7×10^{-16} (7×10^{-16})	1×10^{-17} (1×10^{-15})	7×10^{-16} (7×10^{-16})	7×10^{-16} (7×10^{-16})	7×10^{-16} (7×10^{-16})	
$\sigma_e(\sigma_h)$ for VB tail (cm^{-2})	7×10^{-16} (7×10^{-16})	1×10^{-15} (1×10^{-17})	7×10^{-16} (7×10^{-16})	7×10^{-16} (7×10^{-16})	7×10^{-16} (7×10^{-16})	
Donor (Acceptor) like Gaussian density of states (cm^{-3}eV^{-1})	2.5×10^{-17} (2.5×10^{-17})	1×10^{-17} (1×10^{-17})	2.5×10^{-17} (2.5×10^{-17})	2.5×10^{-17} (2.5×10^{-17})	1.1×10^{-17} (1.1×10^{-17})	
Gaussian peak energy for donor (eV)	0.45	0.46	0.7	1.02	1.02	
Gaussian peak energy for acceptor (eV)	0.65	0.65	1	1.2	1.2	
$\sigma_e(\sigma_h)$ for acceptor like Gaussian states (cm^{-2})	3×10^{-15} (3×10^{-14})	3×10^{-15} (3×10^{-14})	3×10^{-15} (3×10^{-14})	3×10^{-15} (3×10^{-14})	3×10^{-15} (3×10^{-14})	
$\sigma_e(\sigma_h)$ for donor like Gaussian states (cm^{-2})	3×10^{-14} (3×10^{-15})	3×10^{-14} (3×10^{-15})	3×10^{-14} (3×10^{-15})	3×10^{-14} (3×10^{-15})	3×10^{-14} (3×10^{-15})	

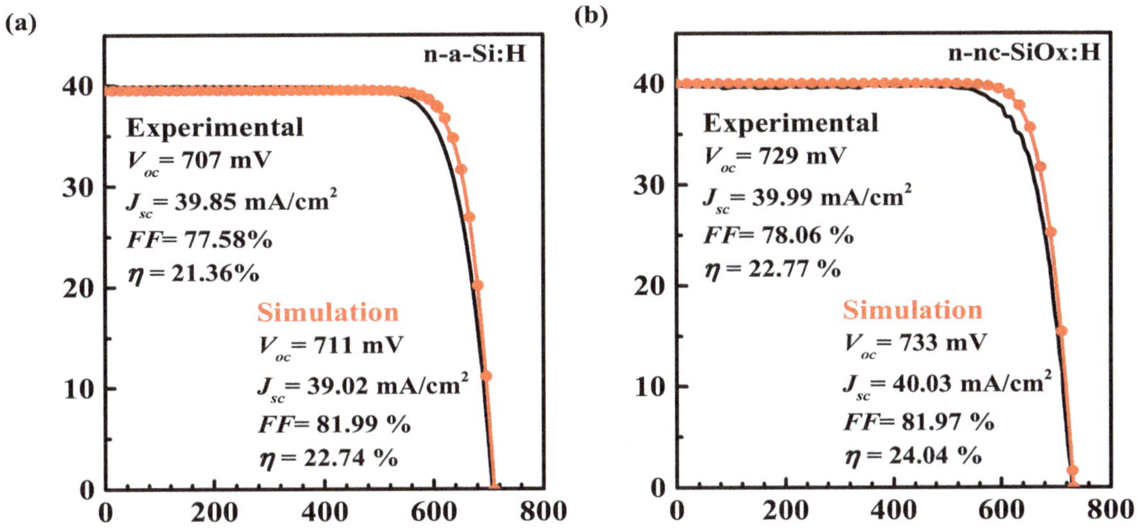

Figure 2. J-V curve characteristics of rear emitter silicon heterojunction (SHJ) solar cells with different carrier selective front contact (CSFC) layers: (**a**) a-Si:H; (**b**) a-SiO_x:H. Experimental results are indicated by the continuous line, while the simulation results are indicated by the symbol.

3. Results and Discussion

3.1. Numerical Simulation of HIT Solar Cell

The SHJ solar cells use a-Si:H materials as the passivation and carrier selective contact layers at both sides for separate electron-hole pairs. The conventional SHJ solar cell uses the p-a-Si:H layer as a CSFC layer. This layer absorbs light in the short wavelength region and has low conductivity, which leads to parasitic absorption losses. The simulation of the rear emitter with various CSFC layers and interface defect states was investigated. The simulations were performed using optical parameters of layers, including extinction coefficient (k), refractive index (n), and optical bandgap (E_g) obtained via ellipsometry, as well as electrical parameter measurements. The J-V curve characteristics of the RE-SHJ cells with different CSFC layers were compared to that of real cells, as shown in Figure 3. It can be observed that the characteristics of the simulated results are very similar to that of the real cell; for example, while fill factor (*FF*) depicted a higher deviation, the other parameters were very close to those of the real cells. The nc-SiO_X:H CSFC layer achieved a conversion efficiency of 24.04%. The decrease in parasitic absorption of the wide-gap nc-SiO_X:H CSFC layer assisted the increase in short circuit current density (J_{SC}). The increase in efficiency of the RE-SHJ cell using the nc-SiO_X:H CSFC layer is explained through the band diagram, electric field and generation/recombination rate.

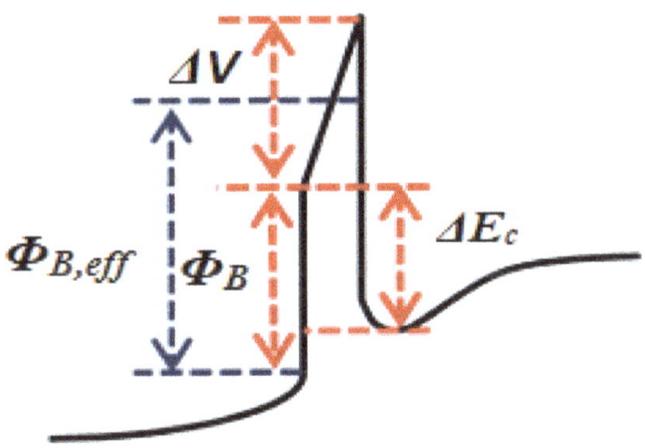

Figure 3. Construction of an effective tunnel barrier by considering the potential drop through the passivation layer.

Figure 3 can be represented by the following equation:

$$\phi_{B,eff} = \phi_B + \frac{\Delta V}{2} \tag{1}$$

where $\phi_{B,eff}$ is the effective barrier height and ΔV is the electrostatic potential drop [15]. $\phi_{B,eff}$ is 0.038 eV and 0.073 eV with a-Si:H CSFC and nc-SiOx:H CSFC layer, respectively.

Figure 4 shows the band diagram in the equilibrium condition for different CSFC layers. To observe the comparative change in the band structure, the band offsets at the valence (ΔE_v) and conduction (ΔE_c) bands of the RE-SHJ with various CSFC layers are depicted in Figure 4b,c, respectively. The values of ΔE_c are 0.082 eV and 0.046 eV for Figure 4b,c, respectively. The smaller value of ΔE_c yielded higher electron collection. It can be observed that the nc-SiO$_x$:H CSFC layer created a barrier for tunneling the electrons between the CSFC layer and n-type c-Si wafer. When using the nc-SiO$_x$:H CSFC layer, the barrier is considered triangular, as shown in Figure 4.

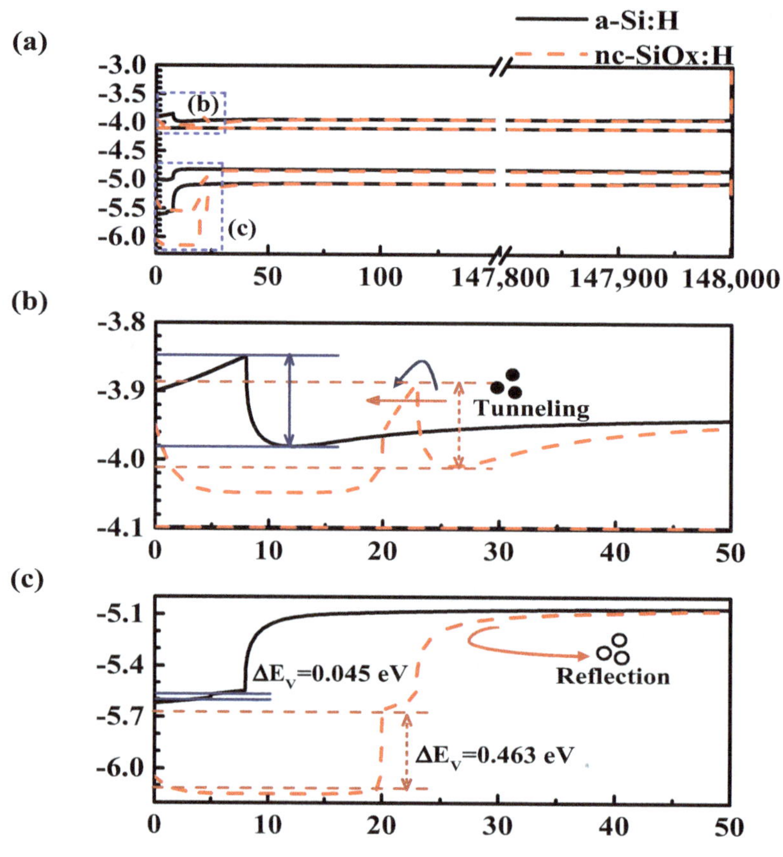

Figure 4. Band diagram of the rear emitter SHJ solar cell: (**a**) band diagram; (**b**) conduction band; (**c**) valence band with different CSFC layers.

The increase in ΔE_v acts as a hole reflector at the interface. The nc-SiO$_x$:H CSFC layer acts as an effective hole reflector. This leads to reduction in interface recombination. As a result, the nc-SiO$_x$:H CSFC layer-enhanced V_{oc} and J_{sc} are approximately 0.733 V and 40.03 mA/cm^2, respectively. To assess the effect of the nc-SiO$_x$:H CSFC layer, the electric field and generation/recombination rate were investigated using the numerical simulation as depicted in Figure 5. The electric field was 7.69×10^8 V/cm for the cell using the a-Si:H as passivation and the CSFC layer. For the cell using the nc-SiO$_x$:H CSFC layer, the electric field was increased to 2.78×10^9 V/cm between the passivation and the nc-SiO$_x$:H CSFC layers. The high electric field with nc-SiO$_x$ CSFC layer is due to high band bending, as depicted in Figure 4 [16]. The nc-SiO$_x$:H CSFC layer can increase V_{oc} as well as lead to field effect passivation. As shown in Figure 5, the interface recombination rate increased over 1×10^{21} m^{-3}S^{-1} at the n-a-Si:H CSFC /a-Si:H passivation interface but drastically reduced under 1×10^{16} $m^{-3}S^{-1}$ between nc-SiO_x:H CSFC layer/a-Si:H passivation layer.

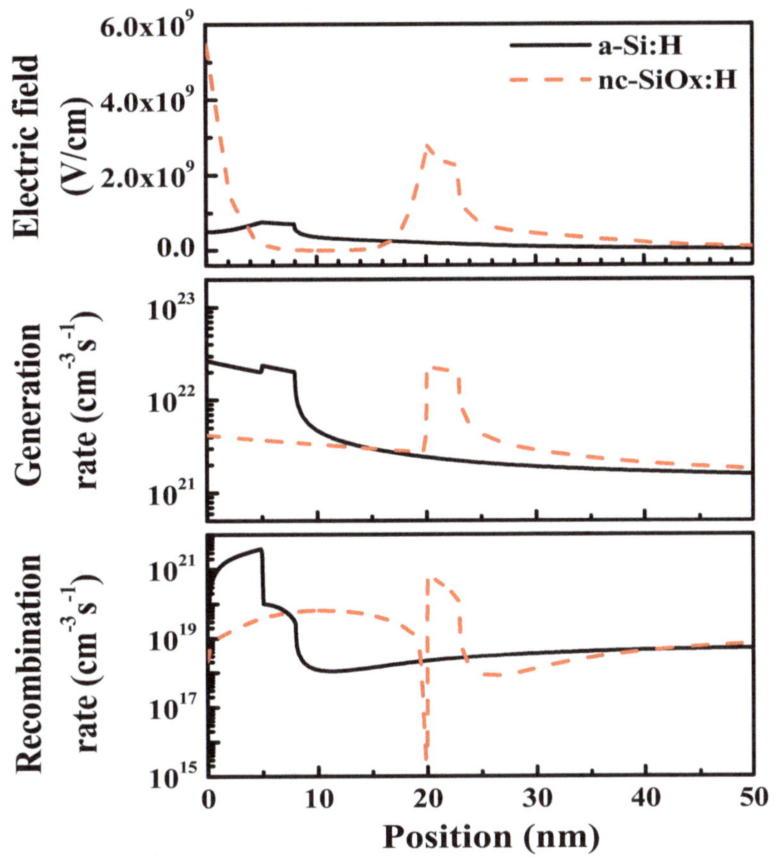

Figure 5. Influence of the CSFC layer on electric field, generation rate and recombination rate.

The field effect passivation causes decrease of recombination rate at the interface. The influence of the interface defect state of the RE-SHJ solar cell was investigated by simulation. Figure 6a shows the cell parameters as a function of the interface defect density (D_{it}) for different CSFC layers of RE-SHJ cells. When D_{it} was increased from 5×10^8 cm^{-2} to 1×10^{13} cm^{-2}, the cell efficiency and J_{sc} decreased slightly at first, and then became almost zero. These results indicate that D_{it} is required to be lower than 2×10^{10} cm^{-2} and 8×10^{10} cm^{-2} in case of using a-Si:H and nc-SiO_x:H layers, respectively, to obtain a V_{oc} over 700 mV. The V_{oc} is affected by interface recombination of the SHJ solar cell as follows [17]:

$$V_{OC} = \frac{1}{q}\left(\phi_c - AkTln\frac{qN_V S_{it}}{J_{SC}}\right) \quad (2)$$

where q is the elementary charge, ϕ_c is the effective barrier height in crystalline silicon, A is the ideality factor, k is the Boltzmann constant, T is the temperature, S_{it} is the interface recombination velocity, and N_v is the effective valence band density. The interface defect state affects surface recombination. The high D_{it} leads to increase of S_{it}, and therefore decreased V_{OC}. The interface defect state affects the quasi Fermi level and band bending. In Figure 6b, the band diagram of a SHJ solar cell using a-Si:H CSFC layer is shown with different D_{it}. When D_{it} increases, the hole quasi-Fermi energy (E_{Fp}) shifts to a high energy level, which can result in a decrease of V_{oc}. E_{Fp} can be described by the-one dimensional continuity equation as [18]:

$$E_{Fp}(x,t) = E_V(x) - kTln\frac{p(x,t)}{N_V(x)} = -q\chi(x) - q\varphi(x,t) - E_g - kTln\frac{p(x,t)}{N_V(x)} \quad (3)$$

where E_v is the valence band energy, χ is the electron affinity, $\varphi(x,t)$ is the local electric potential in the semiconductor layers and E_g is the bandgap. The value of E_{Fp} is increased due to the decrease in electric potential and an increase in N_v. The high D_{it} leads to a decrease in band bending and hence a decrease in electric potential. Additionally, it causes an improvement in the majority carrier concentration. The nc-SiO_x:H CSFC layer also follows the same trend as shown in Figure 6b. In contrast, the nc-SiO_x:H CSFC layer shows lower performance degradation even with 8×10^{10} cm^{-2} due to the tunneling effect and field effect passivation of the CSFC layer/a-Si:H passivation layer at the interface.

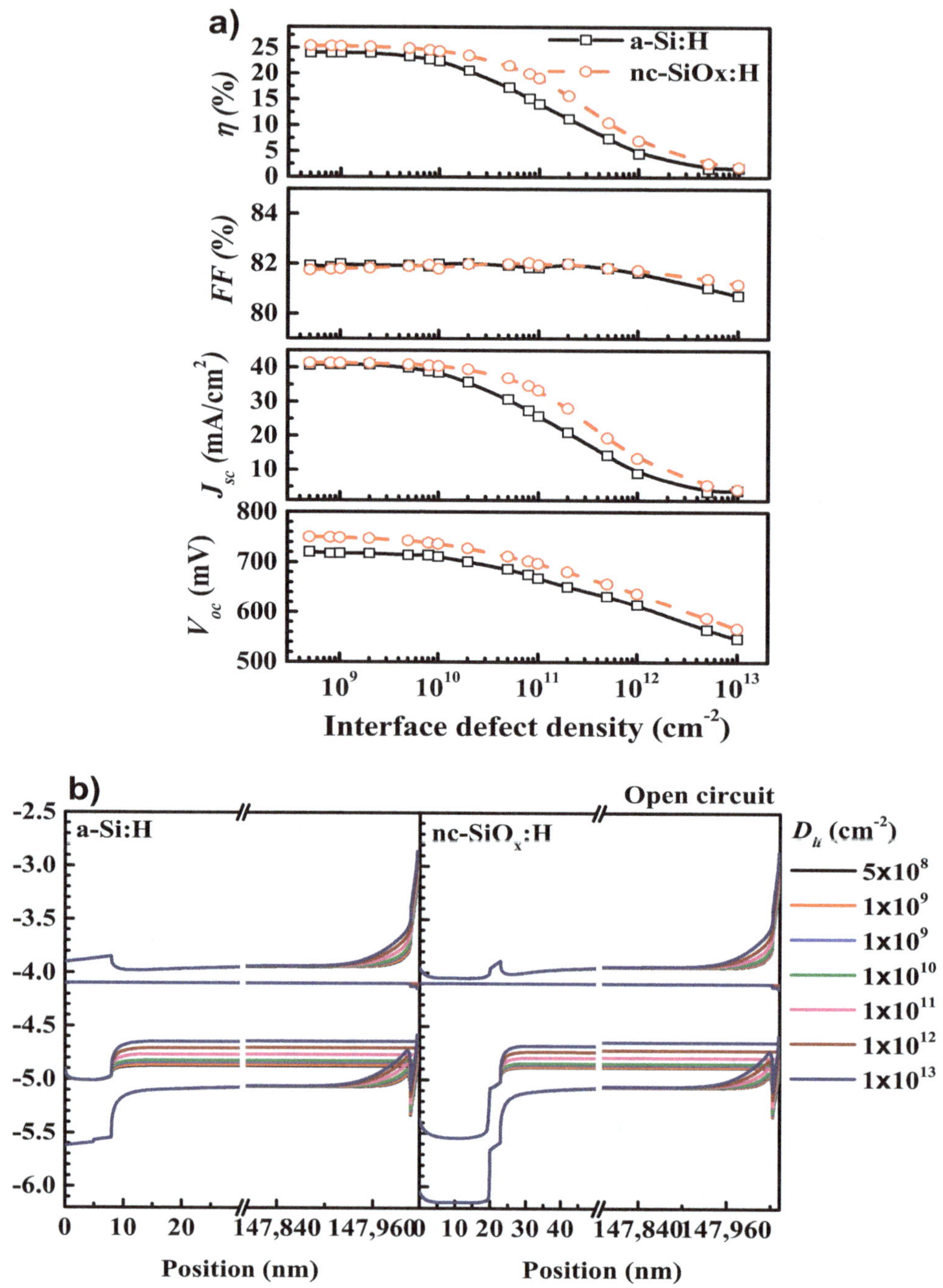

Figure 6. Schemes of rear emitter SHJ solar cells including: (**a**) The V_{oc}, J_{sc}, FF and η cell parameters as a function of interface defect density and (**b**) band diagrams with different interface defect density of a-Si:H CSFC (left) and nc-SiO_x:H CSFC (right) layer.

3.2. Hydrogenated Silicon Oxide Passivation Layer

The results in the previous section suggest that a low D_{it} can enhance V_{oc} and thus efficiency of the SHJ solar cell. A low D_{it} can lead to high band bending at the a-Si:H/c-Si interface, which increases the V_{oc} of the cell. A further low D_{it} can result from a hydrogenated silicon oxide (a-SiO_x:H) passivation layer formed through CO_2 plasma post-treatment. The Figure 7 shows the structural and optical properties of the a-SiO_x:H layers under CO_2 plasma post-treatment.

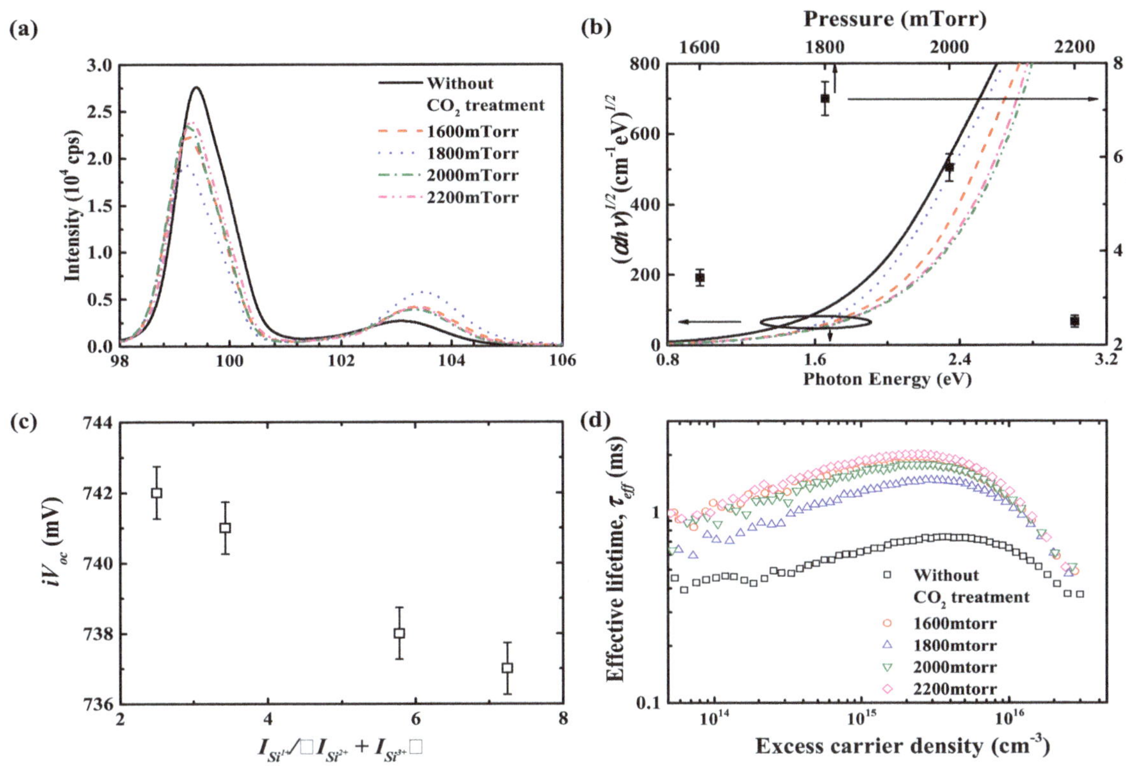

Figure 7. Structural and optical properties of a-SiO_x:H layers under different CO_2 plasma post-treatment including: (**a**) Si 2p core-level XPS spectra of a-Si:H and a-SiO_x:H different CO_2 plasma post-treatment pressure, (**b**) $I_{Si}{}^{1+}/(I_{Si}{}^{2+}+I_{Si}{}^{3+})$ ratio derived from the XPS spectra as a function of plasma post-treatment pressure and optical band gap of the a-Si:H and a-SiOx:H plotted as a function of photon energy, (**c**) iV_{oc} with respect to the $I_{Si}{}^{1+}/(I_{Si}{}^{2+}+I_{Si}{}^{3+})$ ratio, and (**d**) effective carrier lifetime measured by quasi-steady-state photo conductance (QSSPC) as a function of the treatment pressure.

Chemical composition of the a-Si:H and a-SiO_x:H layers was confirmed by employing XPS analysis, as shown in Figure 7a. After CO_2 plasma post-treatment, the pure Si peak decreased and the SiO_2 peak increased with the increase of treatment pressure. Increasing the treatment pressure led to the formation of oxygen-rich silicon oxide states. The absorption coefficient of the a-SiO_x:H layer with different CO_2 plasma treatment pressures were investigated using SE analysis, as shown in Figure 7b. The optical band gaps were determined through absorption coefficients. At higher CO_2 plasma treatment pressure, the a-SiO_x:H layer exhibited a high optical band gap, owing to the oxygen-rich components. The XPS spectrum demonstrated the increase in the formation of Si^{1+}, Si^{2+}, Si^{3+}, and Si^{4+} peaks. The XPS spectra were deconvoluted into six doublets, Si $2p_{3/2}$ (93.90 eV), Si $2P_{1/2}$ (99.90 eV), Si_2O (100.15 eV), SiO (101.05 eV), Si_2O_3 (101.80 eV) and SiO_2 (103 eV), as shown in Figure 8. In Figure 7b, the oxide state ratio ($I_{Si}{}^{1+}/(I_{Si}{}^{2+}+I_{Si}{}^{3+})$) is presented for the different working pressures during CO_2 plasma post-treatment. The oxide state ratio increased to a treatment pressure of 1800 mTorr and decreased above 1800 mTorr. The oxygen-rich components (Si^{2+} and Si^{3+}) further increased at a higher treatment pressure. To determine the optimal oxide state for high V_{oc}, as shown in Figure 7c, the effect of the oxide state ratio on iV_{oc} was investigated. The optimum value of iV_{oc} was 742 mV when the oxide state ratio was 2.5. The surface passivation properties of the a-SiO_X:H layer on textured n-type c-Si wafer was investigated. As a function of excess carrier density, the plot of τ_{eff} with a variation of plasma treatment pressure is shown in Figure 7d. After CO_2 plasma treatment, τ_{eff} increased with the increase in treatment pressure. At low injection levels of carrier density, the improvement of τ_{eff} can be explained through field effect passivation. It can be observed that a-SiO_X:H increased ΔE_v at the a-SiOx:H/c-Si interface. This phenomenon leads to enhancement of the reflection of minority carriers (hole) at the interface, which helps in low interface recombination.

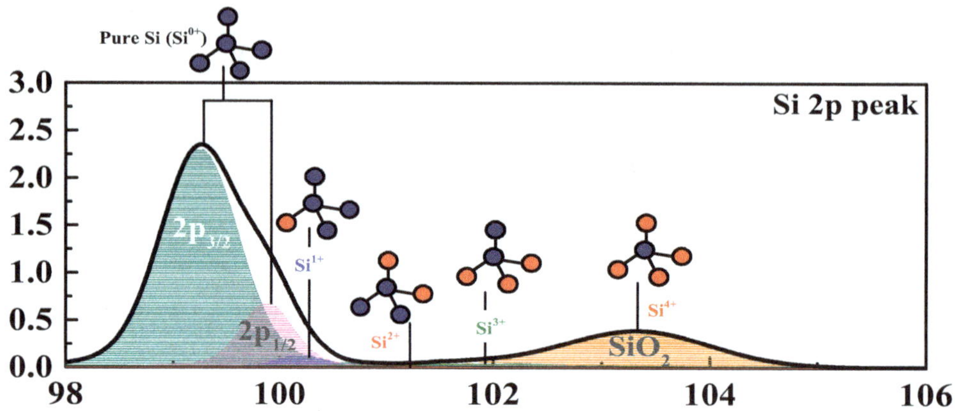

Figure 8. Peak fit for an a-SiO_X:H layer XPS spectra, showing the different oxidation states.

To confirm the cell results, we simulated the RE-SHJ solar cells using a-SiO_X:H passivation layer. In Figure 9, the J-V parameters are presented as a function of the various working pressures for the CO_2 plasma post-treatment. As depicted in Figure 9, the efficiency increases with CO_2 plasma post-treatment. The optical band gap of the a-SiO_X:H layer increased with CO_2 plasma treatment. This optical band gap created a higher ΔE_v at the a-SiO_X:H/c-Si interface, which may have assisted in hole reflection and decreased recombination. We investigated the effect of using a-SiO_X:H passivation layer. This is because the a-SiO_X:H has a wider band gap and good passivation properties. The result shows that such an increase of the V_{oc} is feasible by employing an appropriate a-SiO_X:H as the passivation layer.

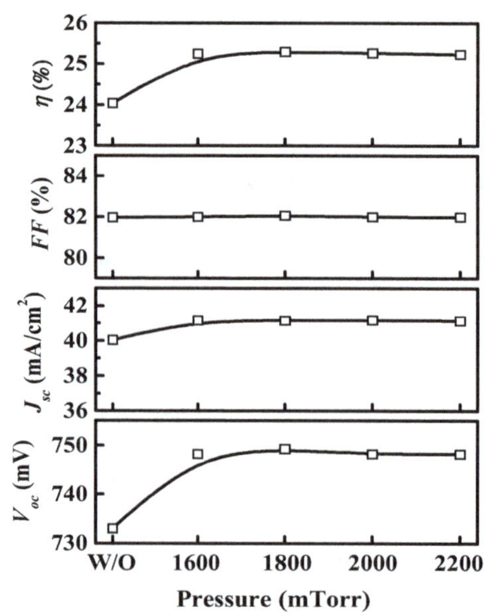

Figure 9. Estimated J-V parameters of rear emitter HIT cells as a function of CO_2 plasma post-treatment.

4. Conclusions

In this paper, the impact of various CSFC layers and interface defect densities on the efficiency of RE-SHJ solar cells was discussed through simulations and experiments. To compare the effect of the various CSFC layers, simulations were performed using a-Si:H and nc-SiO_x:H CSFC materials. Using a nc-SiO_x:H CSFC layer results in the decreased parasitic absorption and recombination loss through field effect passivation, thus enhancing RE-SHJ cell efficiency. The output parameters of RE-SHJ solar cell, such as V_{oc} = 733 mV, J_{sc} = 40.03 mA/cm^2, FF = 81.97% and η = 24.04%, were obtained using a nc-SiO_x:H CSFC layer. The increase in interface defect states resulted in a decrease in band bending and a shift in E_{Fp} to high energy and hence a reduction in V_{oc}. Finally, the nc-SiO_x:H CSFC layer has better performance in the same interface defect state than the a-Si:H CSFC layer. This improvement is attributed to the tunneling effect and field effect passivation when using the nc-SiO_x:H CSFC layer. To improve the surface passivation quality of the RE-SHJ cell, the a-SiO_x:H layer was used as the passivation layer. The V_{oc} increased due to the decrease of the recombination rate at the interface. This investigation suggests that a suitable a-SiO_x:H passivation layer can significantly improve the V_{oc} and efficiency of SHJ solar cells.

Author Contributions: S.L. is the main author. S.L. and D.P.P. carried out the investigations and undertook the simulation and data analysis. Y.K., E.-C.C. and J.Y. supervised the project as was also responsible for funding acquisition. S.L. and J.P. were responsible for the writing—original draft preparation while all authors were involved in writing—review and editing. All authors have read and agreed to the published version of the manuscript.

References

1. Zheng, C.; Kammen, D.M. An innovation-focused roadmap for a sustainable global photovoltaic industry. *Energy Policy* **2014**, *67*, 159–169. [CrossRef]
2. Battaglia, C.; Cuevas, A.; De Wolf, S. High-efficiency crystalline silicon solar cells: Status and perspectives. *Energy Environ. Sci.* **2016**, *9*, 1552–1576. [CrossRef]
3. Yoshikawa, K.; Kawasaki, H.; Yoshida, W.; Irie, T.; Konishi, K.; Nakano, K.; Uto, T.; Adachi, D.; Kanematsu, M.; Uzu, H.; et al. Silicon heterojunction solar cell with interdigitated back contacts for a photoconversion efficiency over 26%. *Nat. Energy* **2017**, *2*, 17032. [CrossRef]
4. Tohoda, S.; Fujishima, D.; Yano, A.; Ogane, A.; Matsuyama, K.; Nakamura, Y.; Tokuoka, N.; Kanno, H.; Kinoshita, T.; Sakata, H.; et al. Future directions for higher-efficiency HIT solar cells using a Thin Silicon Wafer. *J. Non-Cryst. Solids* **2012**, *358*, 2219–2222. [CrossRef]
5. Deligiannis, D.; Van Vliet, J.; Vasudevan, R.; Van Swaaij, R.A.C.M.M.; Zeman, M. Passivation mechanism in silicon heterojunction solar cells with intrinsic hydrogenated amorphous silicon oxide layers. *J. Appl. Phys.* **2017**, *121*, 085306. [CrossRef]
6. Liu, W.; Zhang, L.; Cong, S.; Chen, R.; Wu, Z.; Meng, F.; Shi, Q.; Liu, Z. Controllable a-Si:H/c-Si interface passivation by residual SiH4 molecules in H2 plasma. *Sol. Energy Mater. Sol. Cells* **2018**, *174*, 233–239. [CrossRef]
7. Holman, Z.C.; Descoeudres, A.; Barraud, L.; Fernandez, F.Z.; Seif, J.P.; De Wolf, S.; Ballif, C. Current losses at the front of silicon heterojunction solar cells. *IEEE J. Photovolt.* **2012**, *2*, 7–15. [CrossRef]
8. Mazzarella, L.; Kirner, S.; Stannowski, B.; Korte, L.; Rech, B.; Schlatmann, R. P-type microcrystalline silicon oxide emitter for silicon heterojunction solar cells allowing current densities above 40 mA/cm2. *Appl. Phys. Lett.* **2015**, *106*, 023902. [CrossRef]
9. Taguchi, M.; Terakawa, A.; Maruyama, E.; Tanaka, M. Obtaining a higher voc in HIT cells. *Prog. Photovolt. Res. Appl.* **2005**, *13*, 481–488. [CrossRef]
10. Bivour, M.; Schröer, S.; Hermle, M. Numerical analysis of electrical TCO/a-Si:H(p) contact properties for silicon heterojunction solar cells. *Energy Procedia* **2013**, *38*, 658–669. [CrossRef]
11. Varache, R.; Kleider, J.P.; Gueunier-Farret, M.E.; Korte, L. Silicon heterojunction solar cells: Optimization of emitter and contact properties from analytical calculation and numerical simulation. *Mater. Sci. Eng. B Solid-State Mater. Adv. Technol.* **2013**, *178*, 593–598. [CrossRef]

12. Bivour, M.; Schröer, S.; Hermle, M.; Glunz, S.W. Silicon heterojunction rear emitter solar cells: Less restrictions on the optoelectrical properties of front side TCOs. *Sol. Energy Mater. Sol. Cells* **2014**, *122*, 120–129. [CrossRef]
13. Mazzarella, L.; Morales-Vilches, A.B.; Korte, L.; Schlatmann, R.; Stannowski, B. Ultra-thin nanocrystalline n-type silicon oxide front contact layers for rear-emitter silicon heterojunction solar cells. *Sol. Energy Mater. Sol. Cells* **2018**, *179*, 386–391. [CrossRef]
14. Richter, A.; Smirnov, V.; Lambertz, A.; Nomoto, K.; Welter, K.; Ding, K. Versatility of doped nanocrystalline silicon oxide for applications in silicon thin-film and heterojunction solar cells. *Sol. Energy Mater. Sol. Cells* **2018**, *174*, 196–201. [CrossRef]
15. Varache, R.; Leendertz, C.; Gueunier-Farret, M.E.; Haschke, J.; Muñoz, D.; Korte, L. Investigation of selective junctions using a newly developed tunnel current model for solar cell applications. *Sol. Energy Mater. Sol. Cells* **2015**, *141*, 14–23. [CrossRef]
16. Gudovskikh, A.S.; Ibrahim, S.; Kleider, J.P.; Damon-Lacoste, J.; Roca i Cabarrocas, P.; Veschetti, Y.; Ribeyron, P.J. Determination of band offsets in a-Si:H/c-Si heterojunctions from capacitance-voltage measurements: Capabilities and limits. *Thin Solid Films* **2007**, *515*, 7481–7485. [CrossRef]
17. Dao, V.A.; Heo, J.; Choi, H.; Kim, Y.; Park, S.; Jung, S.; Lakshminarayan, N.; Yi, J. Simulation and study of the influence of the buffer intrinsic layer, back-surface field, densities of interface defects, resistivity of p-type silicon substrate and transparent conductive oxide on heterojunction with intrinsic thin-layer (HIT) solar cell. *Sol. Energy* **2010**, *84*, 777–783. [CrossRef]
18. Bashiri, H.; Karami, M.A.; Mohammadnejad, S. A theoretical investigation of quantum confinement effects in heterojunction silicon solar cells. *Indian J. Phys.* **2018**, *92*, 349–356. [CrossRef]

Nanostructured Perovskite Solar Cells

Calum McDonald [1,*], Chengsheng Ni [2], Paul Maguire [3], Paul Connor [4], John T. S. Irvine [4], Davide Mariotti [3] and Vladimir Svrcek [1]

1 Research Center for Photovoltaics, National Institute of Advanced Industrial Science and Technology (AIST), Tsukuba, Ibaraki 305-8568, Japan; vladimir.svrcek@aist.go.jp
2 College of Resources and Environment, Southwest University, Beibei, Chongqing 400715, China; nichengsheg@163.com
3 School of Engineering, Ulster University, Newtownabbey BT37 0QB, UK; pd.maguire@ulster.ac.uk (P.M.); d.mariotti@ulster.ac.uk (D.M.)
4 School of Chemistry, University of St Andrews, North Haugh, St Andrews KY16 9AJ, UK; pac5@st-andrews.ac.uk (P.C.); jtsi@st-andrews.ac.uk (J.T.S.I.)
* Correspondence: calummcdonaldpv@gmail.com

Abstract: Over the past decade, lead halide perovskites have emerged as one of the leading photovoltaic materials due to their long carrier lifetimes, high absorption coefficients, high tolerance to defects, and facile processing methods. With a bandgap of ~1.6 eV, lead halide perovskite solar cells have achieved power conversion efficiencies in excess of 25%. Despite this, poor material stability along with lead contamination remains a significant barrier to commercialization. Recently, low-dimensional perovskites, where at least one of the structural dimensions is measured on the nanoscale, have demonstrated significantly higher stabilities, and although their power conversion efficiencies are slightly lower, these materials also open up the possibility of quantum-confinement effects such as carrier multiplication. Furthermore, both bulk perovskites and low-dimensional perovskites have been demonstrated to form hybrids with silicon nanocrystals, where numerous device architectures can be exploited to improve efficiency. In this review, we provide an overview of perovskite solar cells, and report the current progress in nanoscale perovskites, such as low-dimensional perovskites, perovskite quantum dots, and perovskite-nanocrystal hybrid solar cells.

Keywords: solar cells; perovskites; perovskite nanocrystals; perovskite quantum dots; low-dimensional perovskites; nanocrystal solar cells; organic–inorganic hybrid solar cells; lead halide solar cells; hybrid solar cells

1. Introduction

In the search of high-efficiency, low-cost solar cells, a multitude of new materials and architectures are currently being explored. Over the past decade, organometal halide perovskites (OHPs) have emerged as a highly promising photovoltaic material and have been demonstrated as the active layer in perovskite solar cells (PSCs) with efficiencies over 25% for laboratory-based devices (~0.1 cm^2) [1] and around 10–15% in modules [2] and are recently being employed in high-efficiency tandem devices [3]. The performance of PSCs has seen a meteoric rise over the past decade and they are already comparable with or superior to well-established photovoltaic technologies [1]. OHPs are attractive particularly due to their ease of processing [4], large absorption coefficients [5], long carrier diffusion lengths [6], low exciton binding energies [7], and low non-radiative recombination rates [8]. These properties also make OHPs an attractive material for various other optoelectronic devices, such as light emitting diodes [9], lasers [10,11], and photodetectors [12].

OHPs have a perovskite crystal structure with the general stoichiometry ABX_3 as shown in Figure 1. The A-site is occupied by a monovalent cation e.g., methylammonium (MA, $CH_3NH_3^+$), formamidinium (FA, $CH_3(NH_2)_2^+$), Cs^+ etc. The B-site is usually occupied by a Pb^{2+} divalent metal cation and can be substituted by a similarly-sized divalent cation such as Sn^{2+}. The X-site is usually occupied by a halide anion e.g., I^-, Cl^-, Br^-. OHPs with mixed cations and/or anions are now the standard for high efficiency cells, particularly due to improved structural stability [13–15]. Their high compositional tunability, whereby the bandgap can be easily modified through ion substitution [16] and low-cost facile deposition procedures [17] makes OHPs excellent candidates for tandem solar cells, where two materials of different bandgaps are employed in conjunction to absorb different parts of the solar spectrum. OHPs can be employed either as the top cell in a tandem device (with e.g., silicon, cadmium telluride, copper indium gallium diselenide etc. bottom cell) or in a stacked perovskite–perovskite tandem device. The successful fabrication of tandem cells with OHPs has the potential to achieve efficiencies in excess of 40% [3].

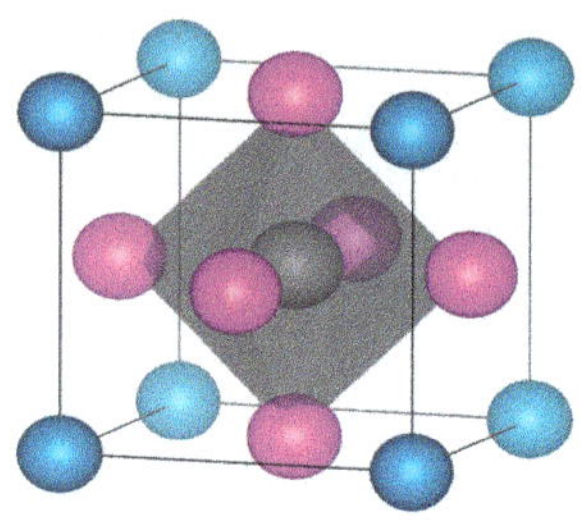

Figure 1. Cubic perovskite unit cell.

While OHPs have demonstrated remarkable efficiencies in laboratory solar cells, there remains significant challenges regarding long-term suitability and feasibility of commercialization [18]. OHPs are extremely susceptible to moisture-induced degradation, and therefore devices must be fabricated in controlled nitrogen atmospheres to avoid trapped moisture in the active layer. Furthermore, devices must be sufficiently encapsulated to prevent external moisture ingress, and the fragility of OHPs along with weak inter-layer adhesion may demand rigid glass substrates to avoid delamination or fractures in the OHP. Even so, heat and light cycling can still induce degradation in encapsulated devices due to thermal mismatch [19]. The use of encapsulants, which can be expensive, along with rigid glass supports, makes OHPs less attractive due to increased costs [3]. It is therefore highly desirable to develop perovskite materials which are stable and tolerant to moisture and other environmental stresses.

Forming nanostructured OHPs (also referred to as low-dimensional OHPs) can be a potential route towards increasing the stability. So far, various types of low-dimensional OHPs have been demonstrated in solar cells, and typically show far superior stability to bulk OHPs [20–22]. This is achieved particularly due to higher formation energies of the low-dimensional perovskite structure and the possibility of encapsulating low-dimensional OHPs in long-chain polymers, essentially providing a protective barrier to moisture [22]. However, carrier transport tends to be restricted in nanostructured perovskites due to the presence of potential barriers within the nanostructured OHP, while quantum confinement also tends to widen the bandgap towards values typically in excess of 2 eV. This therefore comes at a cost to the performance, with the best nanostructured OHPs performing between 10–18% [20–24].

Considering the recent advances in nanostructured perovskites, here we will provide an insight into the important developments and progress in photovoltaics. First, an introduction to the use of bulk OHPs in solar cells will be provided while discussing the challenges and issues facing these materials in order to provide a context for the recent direction towards nanostructured perovskites. This review will then provide a perspective into nanostructured perovskite solar cells as a possible route towards overcoming the issues pertaining to bulk OHPs. Furthermore, hybrid devices formed with OHPs and nanocrystals (NCs) will be discussed, along with high-stability metal oxide perovskite nanocrystals. We hope this will provide the reader with a basis for understanding the current status of PSCs and the potential opportunities of stable, low-dimensional perovskites.

2. Overview of Bulk Perovskite Solar Cells

PSCs were initially inspired by the dye-sensitized solar cell (DSSC), where simply replacing the dye in a DSSC with an OHP immediately yielded efficiencies of ~3% [25]. The OHPs used were either $MAPbI_3$ or $MAPbBr_3$, where MA is the small organic cation methylammonium ($CH_3NH_3^+$). Since the liquid electrolyte, which is used in DSSCs as a redox mediator, dissolved the OHP, these devices had very short lifetimes on the order of seconds. The rapid dissolution of the OHP was overcome by replacing the liquid electrolyte with a polymer which did not dissolve the OHP. Subsequently, devices were reported using the polymer spiro-MeOTAD for hole transport, quickly achieving efficiencies of ~10% with improved device lifetime [26,27]. It was demonstrated that electron and hole transport occurs in the OHP, indicating that free-carriers are generated in the OHP with long diffusion lengths and lifetimes, contrary to suspicion that photocarriers would be excitonic as for organic solar cells, and therefore the sensitized architecture was in fact not necessary [26].

The main PSC device architectures are shown in Figure 2. The OHP is sandwiched between two selective contacts, an electron transport layer (ETL) such as TiO_2, and a hole transport layer (HTL) such as spiro-OMeTAD. Metallic contacts are formed on either side of the transport layers: a window contact is formed using a transparent conducting oxide (TCO) such as indium-doped tin oxide (ITO), and a back contact is formed using either gold, silver, aluminum etc. The first architecture employed in the research timeline was the sensitized architecture using a thick mesoporous layer of TiO_2 (Figure 2a). This was quickly replaced with bi-layer devices, where the mesoporous-TiO_2 was reduced in thickness and a thicker OHP layer was deposited to allow for greater absorption of light and longer crystalline order with larger grain sizes (Figure 2b). A planar device architecture can also be used, with either n-i-p configuration (Figure 2c) or p-i-n configuration (Figure 2d). The planar device eliminates the necessity for the mesoporous TiO_2 layer, further reducing fabrication costs and complexity. Planar devices show greater potential for low-cost roll-to-roll printing of PSCs at low temperatures due to the elimination of mesoporous-TiO_2 which must typically be annealed at high temperatures during device fabrication (~500 °C) for high-efficiency PCSs, and is therefore unattractive for large-scale production while also eliminating the possibility of fabricating devices on flexible plastic substrates. Furthermore, the high-temperature annealing of TiO_2 is not suitable for the fabrication of tandem devices with silicon or perovskite bottom cells since such high-temperature annealing process will damage the silicon bottom cell [3]. Planar devices using an SnO_2 electron transport layer can be fabricated via low-temperature methods and demonstrate superior stability to mesoporous-TiO_2 devices, however the best efficiency of 21.6% is somewhat lower than mesoporous-TiO_2 devices (25.2%) [1,28]. Since PSCs employing mesoporous-TiO_2 transport layers have shown greater efficiencies than planar devices thus far [29], ideally low-temperature fabrication techniques should be developed for mesoporous-TiO_2 transport layers to enable their incorporation into tandem devices.

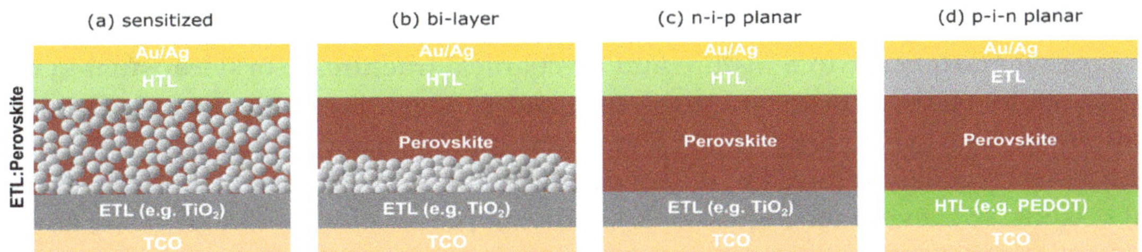

Figure 2. Various device architectures for organometal trihalide perovskite solar cells. (**a**) Mesoporous sensitized, (**b**) bi-layer, (**c**) n-i-p planar and (**d**) p-i-n planar. ETL, HTL, and TCO stand for electron transport layer, hole transport layer, and transparent conducting oxide, respectively.

2.1. Stability of Perovskite Solar Cells

While exceptional efficiencies have been demonstrated with Pb-based perovskites [13–15], significant challenges exist such as poor stability, toxicity, and rate-dependent current-voltage hysteresis. Stability is an important consideration when assessing commercialization viability of new materials given that silicon solar cells can easily operate for >25 years, even when exposed to a broad range of temperatures and intense solar irradiance. OHPs tend to degrade rapidly in open air conditions and must be fabricated in controlled atmospheres to avoid moisture contamination. The rapid degradation of $MAPbI_3$ in open-air conditions is shown in Figure 3, where the majority of the $MAPbI_3$ layer degraded to PbI_2 within 13 days [30]. Although the exact mechanism of degradation remains unclear; it is generally understood that an intermediate phase is first formed via hydration of the OHP [31,32]. Considering the decomposition of $MAPbI_3$, the hydration of $MAPbI_3$ leads to its conversion to $MA_4PbI_6{\cdot}2H_2O$ and PbI_2, followed by phase separation and the subsequent loss of MA, with the final products being CH_3NH_3I, PbI_2, and H_2O [31]. The degradation has been shown first to occur at the grain boundaries and is assisted by the presence of trapped charges which usually exist at defect sites, surfaces, and grain boundaries [33]. Ions can easily migrate within OHPs, causing charge accumulation, phase segregation, lattice distortions, and strain in the perovskite structure [34–38]. The degradation of OHPs is enhanced under illumination, and degradation can be accelerated even under moderate temperatures of ~60 °C [39,40]. Furthermore, I_2, which is generated within the OHP due to exposure to moisture, can easily migrate and leads to the self-sustaining and irreversible degradation of the OHP [41]. The degradation of OHPs leads to the release of the gaseous products CH_3NH_2, HX, CH_3X, and NH_3 (where X is a halide), and the release of these gases can be observed at temperatures below 70 °C [42].

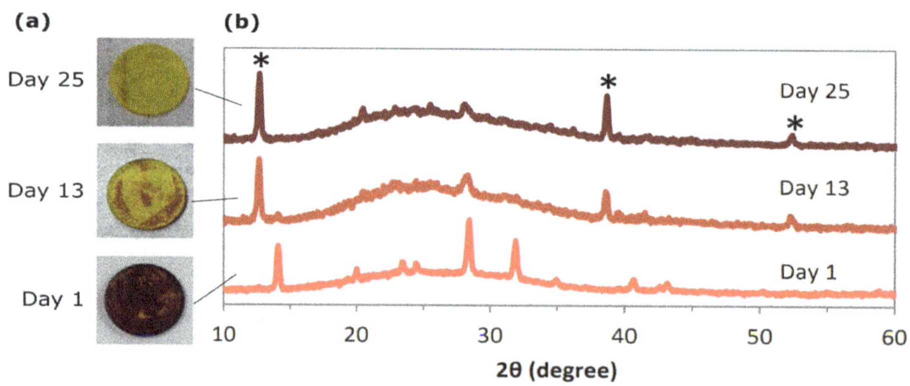

Figure 3. Degradation of $MAPbI_3$. (**a**) Photographs of $MAPbI_3$ degradation and (**b**) corresponding X-ray diffraction (XRD) spectra of the same samples after 1, 13, and 26 days stored in ambient conditions. The starred peaks in the XRD spectra correspond to PbI_2. Reproduced from ref. [30], with permission from John Wiley and Sons, 2016.

Due to the high susceptibility of OHPs to degrade when exposed to moisture, it is therefore necessary to carefully control the atmosphere during fabrication. Entire device encapsulation is necessary to prevent exposure to moisture and mechanical fractures. For encapsulated devices, the formation of bubbles has been observed in the encapsulant layer due to the release of gaseous species. Encapsulation prevents gaseous products from escaping, creating a thermodynamically enclosed system which is expected to reduce the rate of degradation [42]. Encapsulation is therefore essential for several reasons: to prevent the ingress of moisture; to prevent the release of gases; and to prevent the release of toxic materials to the environment. However, due to the thermal expansion coefficient mismatch between the various layers, including the encapsulant, temperature cycling of the PSC (i.e., day and night temperature variations) can lead to significant delamination and device failure. Careful selection of the encapsulant and various device layers is therefore necessary to minimize delamination caused by temperature cycling. This eliminates the possibility of flexible, low-weight modules, and the low stability and Pb-contamination necessitates careful recycling of PSCs. In spite of these measures, the question of whether the lifetime of OHPs can match silicon PV remains dubious.

2.2. Toxicity of Perovskite Solar Cells

Pb-containing OHPs' decomposition results in the formation of Pb-halide compounds, metallic Pb, and various carbonated molecules [43]. Although PSCs contain small amounts of Pb (~0.4 g/m^2 for a 400 μm-thick OHP layer) [44], the harmful Pb-halides generated via degradation are highly water-soluble and therefore pose a significant risk to the environment [45]. The contamination of Pb can be addressed either by replacing Pb with other non-toxic elements or by stabilizing the structure of the perovskite so as to avoid the formation of PbI_2. Unfortunately, computational studies have suggested that there is no viable alternative to Pb in PSCs to achieve the similarly high efficiencies which are in excess of 20% [46]. The high efficiencies of OHPs is attributed to the favorable Pb^{2+} orbital hybridization with I^- and Br^- halide ions which results in high absorption coefficients and long carrier diffusion lengths [47]. Sn is a potential alternative to Pb, and whilst still toxic to animals and humans, it is less harmful than Pb. [43] Sn-OHPs have been produced by the direct replacement of Pb with Sn, but the best efficiency achieved to date is 7.14% [23]. In addition, the stability of Sn-based devices is usually worse than Pb-OHPs due to the tendency of tin to easily oxidize from Sn^{2+} to Sn^{4+}. This can be mitigated to some extent by the addition of SnF_2 and ethylenediammonium during fabrication to inhibit the formation of Sn^{4+} [23,48]. While pure Sn-OHPs are unstable, the oxidation of Sn^{2+} becomes less energetically favorable when less than 50% of the B-site in the perovskite structure is occupied by Sn^{2+} (i.e., $MAPb_{\geq 0.5}Sn_{\leq 0.5}I_3$) and the stability is significantly improved [49]. Notably, Zn, which is a 2+ ion with a slightly smaller ionic radius than Pb, has also been investigated for the partial replacement of Pb and has demonstrated an improvement in the power conversion efficiency (PCE) for small amounts of Zn (~1% to 5%). The introduction of Zn into $MAPbI_3$ leads to the formation of larger grains which are more homogeneous, and layers which are more compact and with fewer pinholes. This is achieved through a lattice contraction induced by the smaller Zn ion, along with stronger coordination with the organic cation, leading to a reduction in the amount of point defects [50–53]. However, this work only serves to reduce Pb contamination without eliminating it entirely, and the contamination of toxic Pb and Sn remains and degradation is still observed [49].

2.3. Hysteresis in PSCs

A common issue exhibited by nearly all PSCs is a hysteresis present during solar cell characterization. Hysteresis, defined as the dependence of the state of a system on its history, is frequently observed during current density-voltage (J-V) measurements, where a change in the voltage scan direction between forward and backward results in a differing J-V response, as shown in Figure 4a. A device without J-V hysteresis is shown in Figure 4b. The observed hysteresis is largely attributed to ion mobility within the OHP [54–56], whilst other mechanisms have also been proposed, see reference [57]. Hysteresis is problematic as it primarily introduces difficulties in accurately measuring device performance, but can also be indicative of stability issues [41,58]. Recent work [13,15] has shown that high-efficiency mesoscopic devices possess low hysteresis in the forward and backward J-V scans with the same scan rates from 10 mV/s to 50 mV/s; however, hysteresis is still well observed particularly for fast scans [56,59,60]. Selecting appropriate contacts and forming high-quality OHP layers appears to negate most of the hysteresis observed during standard performance measurements with slow scan speeds; however, the J-V character for fast scans is often unreported and ionic motion and charge accumulation are still likely to be present in the perovskite layer. Furthermore, hysteresis is often intensified as devices are scaled to active areas over 1 cm^2, particularly due to issues with controlling morphology when depositing OHPs over larger areas [61]. The hysteresis observed in OHPs depends on various measurement conditions during the J-V characterization, in particular: the voltage scan rate and scan range [56,62]; the delay time between applying the bias voltage and measuring the current [63]; and the poling voltage prior to measurement [57]. Hysteresis has also been shown to vary with the grain size of the perovskite [57,64], the A-site cation [65], and device architecture [62,63].

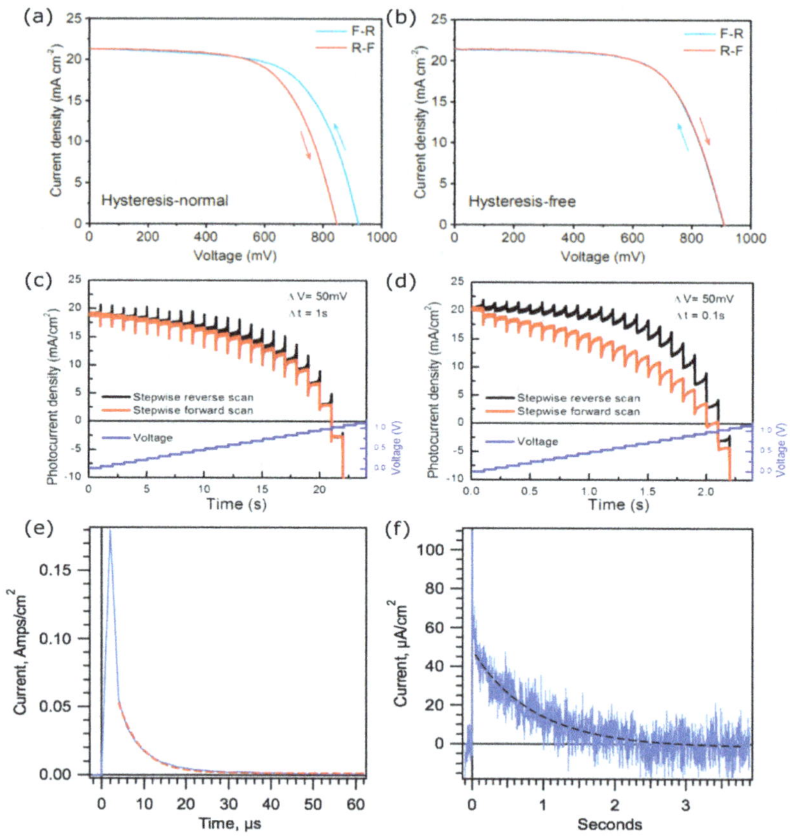

Figure 4. (**a**,**b**) Current density-voltage curves with forward (R-F) and reverse (F-R) voltage scan direction for a device with hysteresis (**a**) and without (**b**). Reproduced from ref. [66], with permission from The Royal Society of Chemistry, 2017. (**c**,**d**) Time-dependent photocurrent response under reverse and forward stepped scans with (b) 1 s step time and (c) 0.1 s step time. Reproduced from ref. [67], with permission from American Chemical Society, 2015. (**e**,**f**) Current decay after removing device from illumination showing two discharging events occurring over different timescales. Reproduced from ref. [68], with permission from American Chemical Society, 2015.

The hysteresis is well-described by Figure 4c,d whereby the voltage is scanned forward and backward in a stepwise fashion with different delay times between the steps: 1 s in Figure 4c and 0.1 s in Figure 4d [67]. It is clear that at least two processes are involved: one is an ultrafast process which leads to an almost instantaneous (microsecond) change in photocurrent, followed by a slower response on the timescale of milliseconds to seconds. There is a large difference in the forward and reverse J-V scans observed for a 0.1 s voltage step time: this arises because when the step speed is too fast, the photocurrent is not able to stabilize and there is a remnant charge stored in the device. This was further investigated and it was shown that there are at least two ways in which charge is stored in OHPs (Figure 4e,f) [68]. After removing an OHP device from illumination, the photogenerated current decayed from 180 mA/cm^2 to less than 50 μA/cm^2 within 50 μs (Figure 4e). This was followed by a second, longer decay event which occurred over the next ~3 s (Figure 4f). Although the peak current in the second decay event (~50 μA/cm^2) accounted for less than 1% of the initial photocurrent (~180 mA/cm^2), the lifetime of the second current was far longer and therefore the total charge associated with this slower decay was calculated to be ~50 times larger than the charge associated with the initial microsecond-discharge event. Therefore, at least two types of capacitive electronic charges were confirmed in OHPs: the first one is small (~0.2 μC cm^{-2}) and likely due to charge trapping; and the second one is much larger (~40 μC cm^{-2}), which could be the result of mobile ions or dipole realignment [68]. Furthermore, it is also known that large differences in the carrier mobility of the electron and hole transport layers can lead to charge accumulation resulting in hysteresis [68].

Understanding the origin and mechanism of hysteresis could lead to the improvement of the performance and stability of PSCs. The main mechanisms which have been proposed to contribute to the effect are: ion migration [56,67,69], charge trapping and accumulation [70,71], and polarization of dipoles [57,62,72]. These mechanisms are represented in Figure 5 and are described briefly in order of the legend:

1. Charge traps: Charges can become trapped at defects on surfaces or at grain boundaries and induce recombination, reducing the photocurrent.
2. Ferroelectric dipoles: Some reports have indicated that OHPs such as $MAPbI_3$ are ferroelectric, and the polarization of domains would modify carrier transport through the perovskite, resulting in the observed hysteresis [62,73–75].
3. Electrons and holes: Similar to charge trapping, electrons and holes can accumulate in transport layers due to defects or imbalances in the carrier mobilities of the electron and hole transport layers.
4. Ion migration: Iodide ions and methylammonium ions can migrate to interfaces under applied bias and alter the internal field reducing the efficiency of carrier separation.
5. Interfacial electrode polarization: A capacitive polarization may arise due to the accumulation of charges or ions at interfaces and cause an energy barrier to carrier extraction.

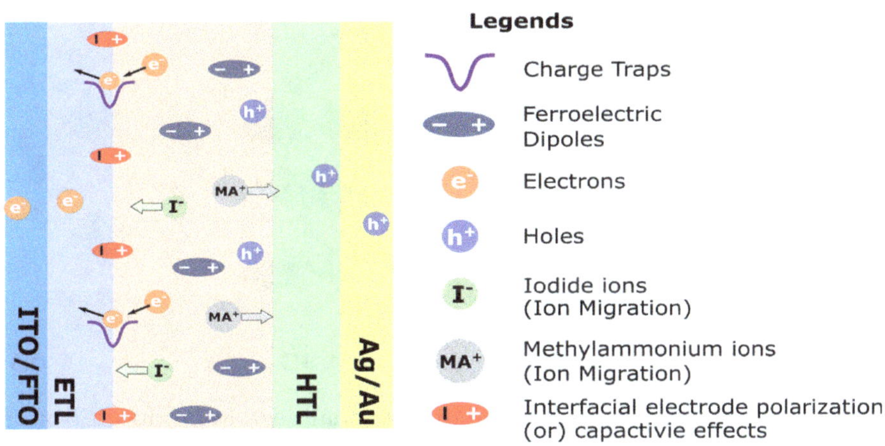

Figure 5. Schematic of the proposed contributions to hysteresis. ITO, FTO, ETL and HTL stand for indium-doped tin oxide, fluorine-doped tin oxide, electron transport layer, and hole transport layer, respectively. Reproduced from ref. [76], with permission from Elsevier, 2016.

These processes may occur simultaneously, and each process will have a different impact on the hysteresis depending on various parameters such as the device structure, interfacial quality, and the properties of the perovskite layer (grain size, defect density, composition etc.), amongst others.

3. Nanostructured Perovskite Absorbers

3.1. Introduction

Non-toxic and/or stable materials with similar properties to bulk Pb-OHPs are a high priority and are currently being explored, such as the replacement of Pb with Sn or Bi [23,77], lead-free halide double perovskites [78], and low-dimensional materials [22]. The efficiencies of these solar cells are often far lower than bulk Pb-OHPs and a large amount of development is still required. Nanostructured perovskites include perovskite quantum dots, nanoparticles, nanosheets, nanorods, and perovskites with nanoscale internal ordering. These materials are often termed low-dimensional perovskites (LDPs) and can generally be envisioned by reducing the bulk perovskite structure to the nanoscale in at least one structural dimension.

Figure 6 shows schematically how a bulk perovskite with ABX_3 structure transforms from a three-dimensional perovskite (3DP) to an LDP. In 3DPs, i.e., the typical bulk perovskites used in record-efficiency devices, each $BX_6{}^{4-}$ octahedra is connected along all three axes and is anisotropic. It is rather important that this octahedral structure is mostly preserved since the orbital hybridization of B and X sites is responsible for many of the favorable optoelectronic properties of OHPs. For two-dimensional perovskites (2DPs), e.g., nanoplatelets and nanosheets, the $BX_6{}^{4-}$ octahedra is connected along two axes and consists of 2D slabs of octahedra with the organic cation occupying the A-site in the voids between slabs. Surrounding the nanosheets are organic 'barrier' molecules which prevent the sheets from crystallizing into a larger 3D structure whilst also providing encapsulation and protection against degradation. For one-dimensional perovskites (1DPs), e.g., nanowires and nanorods, the $BX_6{}^{4-}$ octahedral network extends along only one axis and is encapsulated with organic barrier molecules. For 1DPs and 2DPs, various organic barriers can be selected, and a wide range of choices exist. Hydrophobic organic barriers can be selected which protect the structure against moisture. For zero-dimensional perovskites (0DPs), the $BX_6{}^{4-}$ octahedra is disconnected in all directions and consists of isolated octahedral clusters stabilized by a cationic sublattice. A distinction is often made between 0DPs and quantum dots (QDs), where for a perovskite QD (PQD), the $BX_6{}^{4-}$ octahedra remains connected in all three axes and the radius of the particle is below the Bohr exciton radius, whereas for a 0DP each octahedra is completely disconnected from adjacent octahedra, as shown in Figure 6.

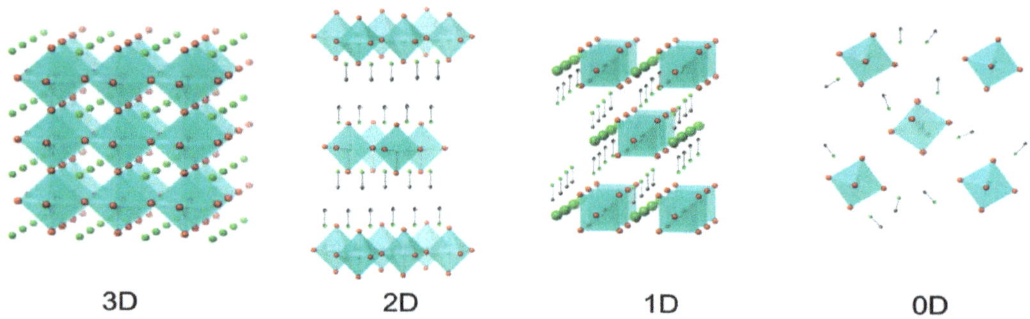

Figure 6. Overview of the different perovskite dimensionalities. Reproduced from ref. [79], with permission from John Wiley and Sons, 2015.

Low-dimensional materials can also be produced which are not strictly perovskites yet follow a similar set of design rules; being based on a large heavy metal ion bonded ionically with halide ions, and stabilized by a sublattice of 1+ cations: For example, B-site 3+ cations such as Bi^{3+} form $B_2X_9{}^{3-}$ bioctahedra instead of a $BX_6{}^{4-}$ octahedra for 2+ cations, forming the 0DP material $(CH_3NH_3)_3Bi_2I_9$. These materials, which can be produced very similarly to standard perovskites (i.e., from solution) whilst also possessing similar properties, are discussed later. The perovskite term is used loosely to describe these materials, as in some cases the perovskite structure is disturbed.

LDPs exhibit quantum confinement effects which are particularly noticeable through a widening of the bandgap [22]. Although 3DPs already have a bandgap close to the optimum value of ~1.4 eV for a single junction solar cell, a wider bandgap is advantageous for forming tandem devices or for indoor photovoltaics [80]. Furthermore, quantum confinement effects introduce the possibility to reduce losses via carrier multiplication which has already been demonstrated in $CsPbI_3$ quantum dots [81] and in the 0DP material $(CH_3NH_3)_3Bi_2I_9$ [82]. The effective use of carrier multiplication in a single-junction solar cell can potentially increase efficiency to ~44% [83], far beyond the Shockley-Queisser (SQ) efficiency limit for a single junction cell of ~33% [84]. In addition, both 3DPs and LDPs are capable of incorporating a low concentration of inorganic nanocrystals into their lattice to form internal energy band alignments which can be used to increase carrier collection and absorption. These hybrid devices can potentially harvest a wide range of the solar spectrum through quantum confinement effects without significantly altering the device architecture, and will be discussed later [85,86].

LDPs often exhibit excitonic behavior as carriers become localized. Since LDPs are often stabilized with organic barriers or a cationic sub-lattice which behaves as an insulating spacer layer, this results in a potential barrier surrounding the individual sheets, rods, or clusters. Carriers therefore become localized on the sheets, rods, or clusters, which often inhibits carrier extraction. The strength of the exciton binding energy is strongly dependent on the dimensionality, with 0DPs usually exhibiting the highest exciton binding energies [87,88].

3.2. One- and Two-Dimensional Perovskites

Along with PQDs, perovskite nanosheets and nanorods are the most successful types of LDPs demonstrating the highest efficiencies in photovoltaic devices. The main advantage of reduced dimensionality is that the OHP can be encapsulated with a more stable long chain organic molecule which reduces the rate of degradation. In reference [22] it was shown via simulations that the stability of $MAPbI_3$ perovskites can be improved by producing a 2D perovskite encapsulated by larger cations. Further to the benefit of the protective ligands, the 2D perovskite structure has a higher formation energy, which therefore yields a more stable perovskite material. A single 2D slab of the perovskite structure, i.e., a monolayer, encapsulated with organic barrier, is termed $n = 1$, as shown in Figure 7. The bandgap is strongly dependent on the number of perovskite slabs (n); as n increases, the bandgap narrows and the strength of quantum confinement reduces, and the dimensionality tends towards a quasi-2D structure ($n > \sim 10$), while for very large values of n the perovskite tends towards a 3D structure.

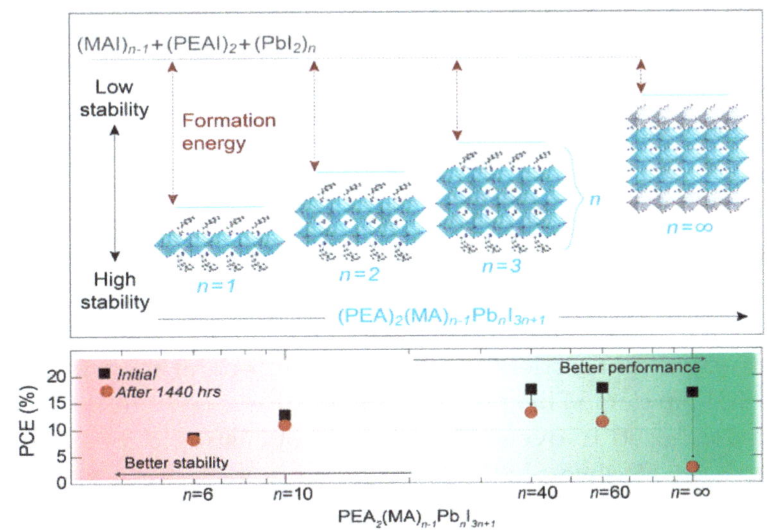

Figure 7. Reducing the dimensionality of organometal halide perovskites leads to higher stability, but lower device performance. Reproduced from ref. [22], with permission from American Chemical Society, 2016

There are a very large number of organic molecules which can potentially be used as the barrier layer, however thus far only a limited number of molecules have been investigated, e.g.: phenylethyl ammonium ($C_8H_9NH_3$, PEA) [22], benzyl ammonium ($C_6H_5CH_2NH_3$, BA) [89], 2-iodoethylammonium ($IC_2H_4NH_3$) [90], polyethylenimine ($(C_2H_5N)_n$, PEI) [91], 2-thiophenemethylammonium (C_5H_7NS, ThMA) [92], and 3-bromobenzylammonium iodide ($BrC_6H_4CH_2NH_2.HI$, 3BBA) [24]. The absorption spectra of 2DPs is weakly associated with the selection of the barrier molecule; optical properties are far more dependent on the n value [93]. As n tends towards lower values, the stability of the 2DP increases [22], yet the device performance tends to decrease dramatically due to the widening of the band gap and the higher proportion of insulating barrier molecules which have a detrimental effect on carrier transport. Whilst the high in-plane mobility of bulk OHPs is retained along the nanosheets and nanorods, the transport between nanosheets/rods is restricted due to the potential barrier created by the insulating organic barriers which reduces the overall carrier mobility [88]. However, this can be mitigated somewhat by using shorter barrier molecules [94].

In reference [22], the $MAPbI_3$ perovskite was reduced to a 2DP and a quasi-2DP structure using PEA barriers with varying n values. A quasi-2DP with $n = 40$ was capable of achieving ~15% efficiency, however the stability of quasi-2DPs is still rather poor. Reducing the n value to 6 provided high stability, yet the efficiency fell towards ~5%. It is likely that the low efficiency was due to the disordered nature of the sheets which are not aligned perpendicular to the contacts, inhibiting charge transfer. This is shown schematically in Figure 8a. When nanosheets are oriented horizontally, i.e., parallel to the contacts, the charge carrier transfer is restricted in the vertical direction, and charge carrier extraction in inhibited because the long organic barriers separating the LDP sheets inhibit transfer between the layers.

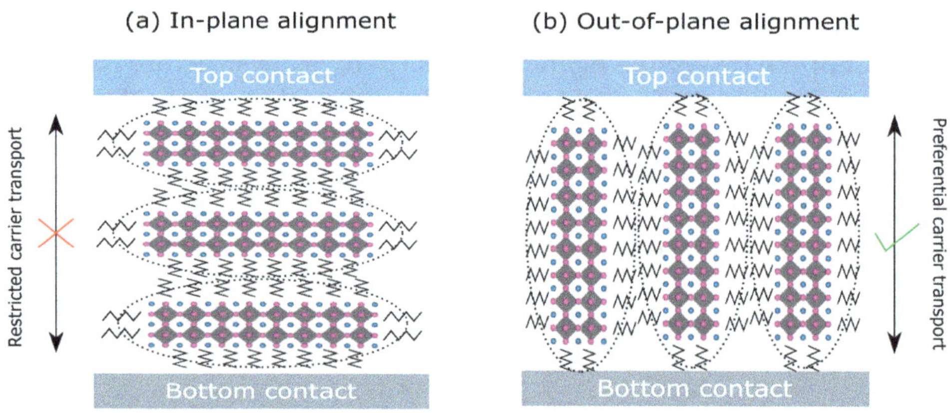

Figure 8. Solar cells based on a perovskite absorber with a two-dimensional network. (**a**) Sheets align parallel with the contacts resulting in low carrier mobility between the contacts and (**b**) sheets align perpendicular to the contacts resulting in favorable out-of-plane mobility between contacts.

Higher efficiencies can be achieved by vertically orientating the inorganic sheets, as shown schematically in Figure 8b, whereby charge transport is less restricted. If the nanosheets/rods are orientated vertically, i.e., perpendicular to the contacts (out-of-plane), charge transport is predominantly along the perovskite structure and carrier extraction is therefore far more efficient since carriers must overcome fewer potential barriers. This was initially demonstrated in BA-capped 2DPs with $n = 3$ and the efficiency was increased to over 12% using a hot casting deposition technique to achieve out-of-plane alignment of the 2D sheets [23]. However, these devices still showed rather poor stability when exposed to 65% relative humidity without encapsulation, while fully encapsulated devices demonstrated impressive stability. This has also been demonstrated in perovskite nanorods, with an increase in efficiency from 1.74% to over 15% following out-of-plane alignment [92]. This was achieved by using a methylammonium chloride (MACl) assisted film formation technique which resulted in vertically aligned perovskite nanorods, demonstrating far improved stability over 3D perovskite. Disordered (unaligned) 2DPs usually show significant hysteresis [95], which is likely due to a bias-voltage induced charging effect caused by the insulating organic molecules and poor charge transport when the 2DP sheets are not vertically aligned. However, the hysteresis is mostly eliminated when the nanosheets are aligned out-of-plane with respect to the contacts since charge transport is less restricted [23].

A problem which must be overcome in 2DPs is a stacking misalignment of the 2DP grains which reduces carrier mobility. It was shown that even when 2DPs are aligned with favorable out-of-plane alignment, stacking misalignments between grains restricts charge transfer between vertically aligned sheets [83]. In order to improve device performance, it is important to minimize stacking misalignment between grains. In addition, it was recently demonstrated that it is essential to use LDPs with at least $n > 2$, as it has been shown that exciton dissociation occurs within the nanosheets of 2DPs due to the presence of lower energy states at the edges of the nanosheets which exit only for nanosheets with $n > 2$ [96]. While these edge states are present for 2DPs with $n > 2$, for $n \leq 2$, edge-state exciton dissociation was not observed, and the device performance was significantly lower. These lower energy states exist at the edge of 2DPs and provide a favorable energy pathway for excitons to dissociate into free-carriers with longer lifetimes, which was demonstrated to significantly improve device performance. This work demonstrated that it is imperative to synthesize 2DPs with at least $n = 3$ in order to benefit from the favorable exciton dissociation mechanism, even though thinner nanosheets ($n \leq 2$) can provide higher stability.

Recent work demonstrated that mixed n value 2D perovskites can achieve both favorable carrier transport and band alignment introduced via a unique nanostructuring of the 2D perovskite film, achieving a PCE of 18.2% [24]. The introduction of the barrier molecule 3-bromobenzylammonium iodide (3BBA) leads to the oriented growth of small n value 2D perovskites perpendicular to the substrate ($n \approx 1$–4), followed by the crystallization of large n value quasi-2D perovskites in the bulk of the film, shown schematically in Figure 9a and the overall device structure in Figure 9b. This structure also introduces a favorable band alignment as shown in Figure 9c whereby the larger bandgap of the mixed low n value 2DPs provides a potential energy gradient driving carriers to the desirable extraction contacts. This demonstrates the remarkable tunability that can be achieved through nanostructuring perovskites to achieve favorable energy band alignment. The devices also showed impressive stability: Unencapsulated devices stored in a dark oven between measurements under ≈40% relative humidity retained 80% of the original PCE after 2400 h. The device could also be submerged underwater for 60 s without any immediate negative effect on the efficiency. It was stipulated that the hydrophobicity due to the presence of iodine in 3BBA results in the enhanced moisture durability of these 2DPs.In general, 2DPs have not been optimized yet via cation engineering to the same extent as 3DPs, which has led to the high performance and improved stability of 3DPs today [13]. Recently, 5% Cs^+ doping in a 2DP demonstrated an efficiency increase from 12.3% to 13.7%, which was attributed to improved crystal quality and low trap defects, increased grain size, and improved carrier transport [97]. Since most 2DPs with low n values show wide bandgaps, it is important to engineer 2DPs which absorb in the visible spectrum. Material engineering and optimization as such demonstrates that there is still great potential for work on improving the 2DPs' material properties.

Finally, 2DPs may also find use in improving the stability of 3DPs by acting as a protective capping layer. A 2DP was demonstrated as the capping layer in a 3DP solar cell and displayed over 19% efficiency, along with improved stability over the 3DP alone [98]. Further work in this area showed that the deposition of a hydrophobic 2D perovskite on top of a 3D perovskite not only protects against moisture, but also improves carrier extraction. The formation of the 2D perovskite on the surface of the 3D perovskite consumes detrimental and undesirable non-perovskite phases present at the surface of the 3D perovskite and resulted in faster injection of holes into the HTL [99]. More recently, an ammonium salt post-treatment of a 3D OHP film increased the PCE from 20.5% to 22.3% via the formation of a 1DP passivation layer [100]. Devices retained 95% of the initial PCE after continuous illumination for 550 h. This area of work presents a route towards avoiding the necessity for encapsulants in PSCs, therefore reducing costs and avoiding issues pertaining to thermal expansion mismatch.

3.3. Zero-Dimensional Perovskites

Pb-based 0DPs have been previously studied but so far seem unsuitable for photovoltaics [101,102]. For example, when the typical perovskite $MAPbI_3$ is transformed into a 0DP with the chemical formula $(CH_3NH_3)_4PbI_6$, the structure is extremely unstable [101]. Alternatively, more stable inorganic Pb-based 0DPs can be produced such as Cs_4PbBr_6, however, the bandgap is very large: Pb- based 0DPs tend to have very large bandgaps which are unsuitable for photovoltaics, typically in the UV-range, irrespective of the halide anion selected [102].

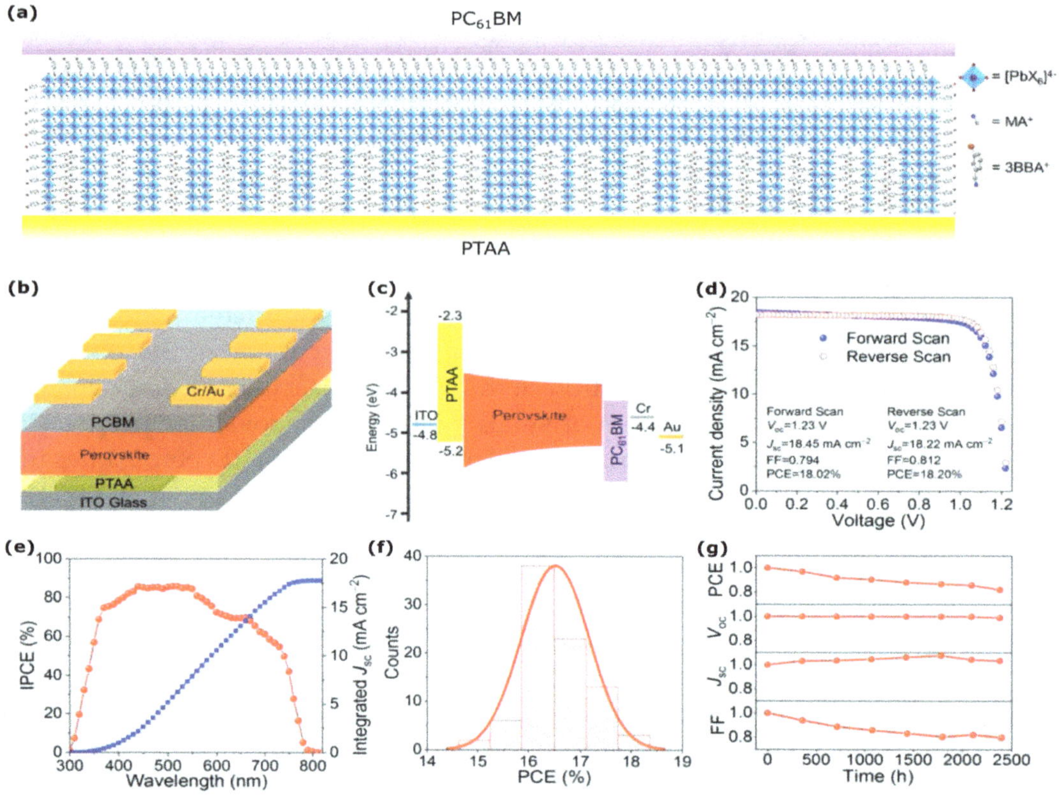

Figure 9. 2D perovskite solar cells using 3-bromobenzylammonium iodide barrier molecule. (**a**) Schematic of the device nanostructuring, (**b**) schematic of the device architecture, (**c**) energy band alignment relative to the vacuum level in eV, (**d**) current density-voltage measurement, (**e**) incident photon conversion efficiency (ICPE), (**f**) histogram showing reproducibility of the power conversion efficiency (PCE) and (**g**) solar cell stability for devices stored in the dark between measurements under ≈40 relative humidity. Reproduced with modifications for clarity from ref. [24], with permission from John Wiley and Sons, 2018.

Alternatively, Bi-based 0DPs have bandgaps closer to 2 eV and have been demonstrated as the absorber in photovoltaic cells [86,103–105]. Bi, which is adjacent to Pb in the periodic table, has a similar atomic radius to Pb yet with one additional valence electron yielding 3+ instead of 2+, resulting in a $B_2X_9^{3-}$ bioctahedral structure rather than the BX_6^{4-} octahedral structure. These perovskite structures have the formula $A_3B_2X_9$, but can also be expressed as $AB_{2/3}X_3$, i.e., a metal-deficient perovskite. Figure 10 shows the structure of a 0DP with the chemical formula $(CH_3NH_3)_3Bi_2I_9$, where $Bi_2I_9^{3-}$ clusters are separated by a $CH_3NH_3^+$ cationic lattice. Here, $CH_3NH_3^+$ can be replaced with a range of organic and inorganic cations. Whilst these materials are often referred to as 'perovskites', their crystallographic structure is slightly different to the perovskite structure, whereby the BX_6^{4-} octahedra is instead replaced with a $B_2X_9^{3-}$ bioctahedra.

Bi-0DPs have been studied and the best devices have achieved efficiencies of 1.64% [105]. These materials, with bandgaps of ~2 eV, generally exhibit high exciton binding energies (~300 meV) and high effective masses for carriers [88]. Because of the excitonic nature of these materials with quantum confinement effects, 0DPs have been shown to exhibit carrier multiplication [82]. However, due to the high exciton binding energy, the rates of electron-hole recombination is high which limits device performance. 0DPs also exhibit anisotropic carrier mobilities if the cluster is non-symmetrical and/or if the spacing between clusters varies between planes [88]. It is therefore necessary to try to overcome the high exciton binding energy and carrier transport issues by a range of possible methods, such as modifying the cationic sub-lattice, using semiconducting polymers which enhance carrier mobility between the clusters, or by forming hybrids with inorganic nanocrystals which assist in exciton dissociation.

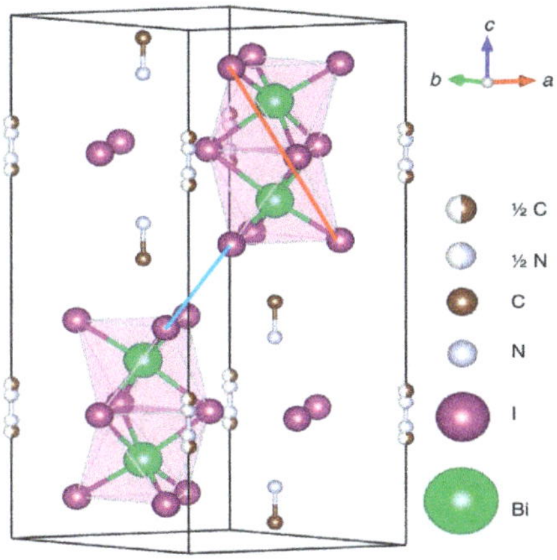

Figure 10. Schematic of the structure of $(CH_3NH_3)_3Bi_2I_9$ which forms a zero-dimensional network. $Bi_2I_9^{3-}$ clusters are stabilized within a $(CH_3NH_3)^+$ ionic lattice.

$(CH_3NH_3)_3Bi_2I_9$ solar cells can be processed and stored entirely in ambient conditions and have demonstrated far superior stability to 3DPs, likely due to the formation of a native surface layer of Bi_2O_3/BiOI which provides self-encapsulation of the perovskite [30]. This layer does not inhibit carrier extraction, and is also likely responsible for the negligible hysteresis observed in these devices [86]. If 0DPs were employed as the wide-bandgap top cell in a tandem solar cell, their high stability can provide encapsulation for the less-stable OHP bottom cell to prevent moisture ingress. Furthermore, the absorption can be modified by incorporating optically active organic molecules or forming hybrids with nanocrystals with suitable band alignment [86], and the large bandgap of 2 eV can be reduced to values as low as 1.45 eV through doping and/or changing the A-site cation [106–108].

Sb-based 0DPs have also been demonstrated with the formula $(CH_3NH_3)_3Sb_2I_9$ and have so far achieved higher efficiencies than Bi-0DPs, with the best devices so far achieving 2.77% efficiency [109]. The higher efficiencies of these devices is likely due to the intrinsically lower exciton binding energy of Sb-0DPs [110]. Since the bandgap of Sb-0DPs is still quite large (~1.9 eV), researchers have attempted to lower the bandgap through Sn-doping, and successfully reduced the bandgap to 1.53 eV with 40% replacement of Sb with Sn to form $(CH_3NH_3)_3Sb_{0.6}Sn_{0.4}I_9$. Doping with Sn increased the efficiency of the devices from 0.57% (without Sn, bandgap = 2.0 eV) to 2.7% (40% Sn, bandgap = 1.53 eV). Since the starting efficiency of the undoped Sb-0DP reference device was quite low (0.57%) compared to the highest reported in the literature (~2.77%), it is likely that through device optimization of the Sn-doped Sb-0DP will quickly lead to higher efficiencies in the near future, likely exceeding 5%. These Sn-doped Sb-0DPs demonstrated impressive stability with no change in the XRD spectra after 15 days of exposure to ambient conditions. Although inorganic 0DPs have also been produced with the formula $Cs_3Sb_2I_9$ and $Cs_3Bi_2I_9$, these devices tend to show very low efficiencies below 0.1% [111,112], likely due to their large bandgaps and high exciton binding energy, and have therefore not been pursued to the same extent.

3.4. Perovskite Quantum Dot Solar Cells

High exciton binding energies and inefficient charge transfer are significant issues associated with LDPs which limit carrier extraction, therefore inhibiting device performance. This can potentially be overcome in PQDs through close-packing with electronic coupling between QDs. Colloidal PQDs can be readily synthesized from solution using organic capping molecules, such as oleic acid, oleylamine, octadecene, etc. which prevent the perovskite from forming into a larger crystal [113]. These long chain molecules must be removed during device fabrication for efficient solar cell performance. However, PQDs with organic A-site cations are often highly unstable, and it is therefore not possible to remove these barrier molecules as they are essential for preventing rapid degradation. As discussed previously, the issue of long chain organic barrier molecules in perovskite nanorods and nanosheets can be overcome by aligning the sheets and rods perpendicular to the contacts, minimizing the number of potential barriers that must be overcome by charge carriers. However, due to the spherical shape of QDs, this type of favorable alignment is not possible, and researchers must therefore look towards inorganic PQDs which do not require encapsulation in protective organic barriers or use a different architecture [85,114].

All-inorganic perovskites can be formed by replacing the A-site with an inorganic cation, such as Cs^+, e.g., $CsPbI_3$. Inorganic PQDs such as $CsPbI_3$ are the most favorable perovskite material since the bandgap of bulk $CsPbI_3$ is the smallest of the inorganic perovskites (1.73 eV for the cubic phase) [115]. However, accessing the desired cubic phase of $CsPbI_3$ is challenging: For bulk $CsPbI_3$, the orthorhombic phase is thermodynamically preferred at room temperature, but the large bandgap of 2.82 eV renders orthorhombic $CsPbI_3$ unsuitable for photovoltaics [115]. The cubic phase exhibits a more favorable bandgap of 1.73 eV; however, this phase is unstable at room temperature. Forming $CsPbI_3$ quantum dots enabled researchers to achieve the cubic phase at room temperature, as the contribution of the surface energy for $CsPbI_3$ quantum dots was shown to retain the favorable cubic perovskite phase [114].

$CsPbI_3$ PQD solar cells were fabricated with 10.77% efficiency [114]. These devices could be fabricated at ambient conditions and showed impressive stability when stored in a desiccator, with no decrease in performance after 60 days. However, when stored in relative humidity of 40–60% there was a significant decrease in the device performance after just 2 days, although QD devices demonstrated improved stability over bulk $CsPbI_3$. Furthermore, $CsPbI_3$ QD devices showed significant hysteresis, likely due to difficulties associated with charge transfer between quantum dots, ion migration, and charge trapping at QD surfaces.

These devices were later improved by a post treatment of the $CsPbI_3$ QDs, and increased the efficiency to 13.43%, as shown in Figure 11 [21]. This was achieved through efficient QD coupling via a post-treatment of the film, allowing improved change transfer between the QDs in the film. The post-treatment involved soaking the $CsPbI_3$ QD thin film in a formamidinium iodide in ethyl acetate solution for 10 s. The post-treatment creates a coating on the $CsPbI_3$ QDs and does not alter their nanocrystalline character. It was confirmed that the post-treatment improved the carrier mobility from 0.23 to 0.50 $cm^2\ V^{-1}\ s^{-1}$. However, the poor stability of $CsPbI_3$ at ambient conditions has not yet been addressed, and it is likely that these materials will require encapsulation. Alternatively, a Cs- salt post-treatment was reported achieving PCE of 14.1% [116]. The Cs-salt treatment is performed after the removal of ligands from the $CsPbI_3$ QDs. When the ligands are removed, Cs vacancies are left behind on the $CsPbI_3$ QDs. These vacancies are filled by Cs via a Cs-salt post treatment, resulting in improved free carrier mobility, lifetime, and diffusion length, as well as greater stability over untreated $CsPbI_3$ QDs.

One of the advantages of $CsPbI_3$ QDs is the possibility of carrier multiplication, which has already been demonstrated in $CsPbI_3$ QDs with a high carrier multiplication quantum yield of 98% [81]. While the bandgap for quantum confined materials scales as $E_g \sim \frac{1}{r}$ where r is the radius, the rate of Auger recombination scales as $\frac{1}{r^6}$ and therefore forming smaller QDs is more favorable for carrier multiplication. The average radius of the QDs in this work was 5.75 nm and the exciton Bohr radius for $CsPbI_3$ QDs is 6 nm. The QDs are therefore in the weak quantum confinement regime, yet still exhibited highly efficient carrier multiplication indicating that strong quantum confinement is not necessary in these materials for carrier multiplication [81].

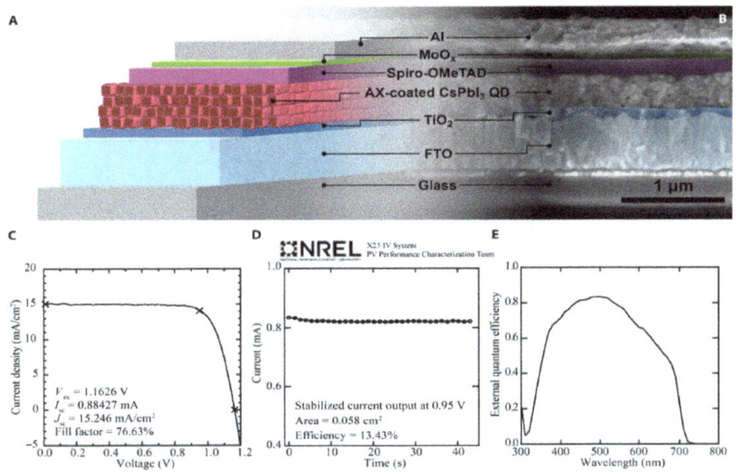

Figure 11. $CsPbI_3$ quantum dot solar cells. (**A**) Schematic of the device structure, (**B**) cross-sectional scanning electron microscopy image, (**C**) current density-voltage scans under solar simulated light, (**D**) stabilized current at a constant voltage of 0.95 V, and (**E**) external quantum efficiency. Reproduced from ref. [21], with permission from AAAS, 2017.

The band energy structure of the active layer can be tuned to achieve improved carrier extraction by using PQDs with varying condition band, valence band, and Fermi level positions. The sequential deposition of PQDs with varying band energy positions has been shown to improve carrier extraction [117] and is reproduced in Figure 12. A schematic of the sequential deposition of PQDs is shown in Figure 12a and the band energy positions of the PQDs studied in this work are shown in Figure 12b. PQDs were synthesized in the series $Cs_xFA_{1-x}PbI_3$ and PQD heterojunction devices were fabricated with the structure ITO/TiO_2/PQDs I/PQDs II/spiro-MeOTAD/MoO_x/Al. The best device performance was obtained using either $Cs_{0.5}FA_{0.5}PbI_3$ or $Cs_{0.25}FA_{0.75}PbI_3$ as the bottom layer and $CsPbI_3$ on the top. Devices based on a $Cs_{0.25}FA_{0.75}PbI_3$:$CsPbI_3$ heterojunction were investigated further for optimization. Figure 12c shows the SEM cross section of the device and Figure 12d shows the effect of varying the thickness ratio of $Cs_{0.25}FA_{0.75}PbI_3$:$CsPbI_3$ on the EQE spectra. A ratio of 1:3 ($Cs_{0.25}FA_{0.75}PbI_3$:$CsPbI_3$) retained most of the short wavelength EQE contribution from $CsPbI_3$ whilst also red-shifting the EQE onset slightly. Higher proportions of $Cs_{0.25}FA_{0.75}PbI_3$ lead to a fall in EQE at shorter wavelengths, despite red-shifting the EQE onset more. Figure 12e shows that varying the bottom layer composition, i.e., by fabricating devices with the structure ITO/TiO_2/$Cs_xFA_{1-x}PbI_3$/$CsPbI_3$/spiro-MeOTAD/MoO_x/Al for x = 0.25, 0.5 and 0.75 leads to a similar red-shift in the EQE as the bandgap of the $Cs_xFA_{1-x}PbI_3$ PQDs is decreased. The J-V characteristics are shown in Figure 12f, and ratios of 1:3 and 2:2 achieve the highest PCEs, however due to the large hysteresis present in these devices, the SPO was also presented and revealed that devices with a 1:3 ratio of $Cs_{0.25}FA_{0.75}PbI_3$:$CsPbI_3$ achieved the highest SPO at 15.52%. Finally, bulk heterojunction architecture devices were also fabricated by mixing the PQDs. These devices did not exhibit the same enhanced performance confirming that a bi-layer heterojunction of PQDs is essential for achieving improved carrier collection.

A summary has been provided in Table 1 comparing a selection of the most notable results since 2018 for 0D, 1D, 2D and QD perovskites, as well as also including some of the notable heterojunctions formed between 3D perovskites and LDPs. This table also provides a summary of the stability of the solar cell devices, noting the storage conditions and the solar cell J-V measurement type (i.e. continuous or intermittent, where continuous measurements typically involve the device remaining under constant solar simulated light, whilst for intermittent measurements the device is removed from illumination and stored in specified storage conditions between measurements).

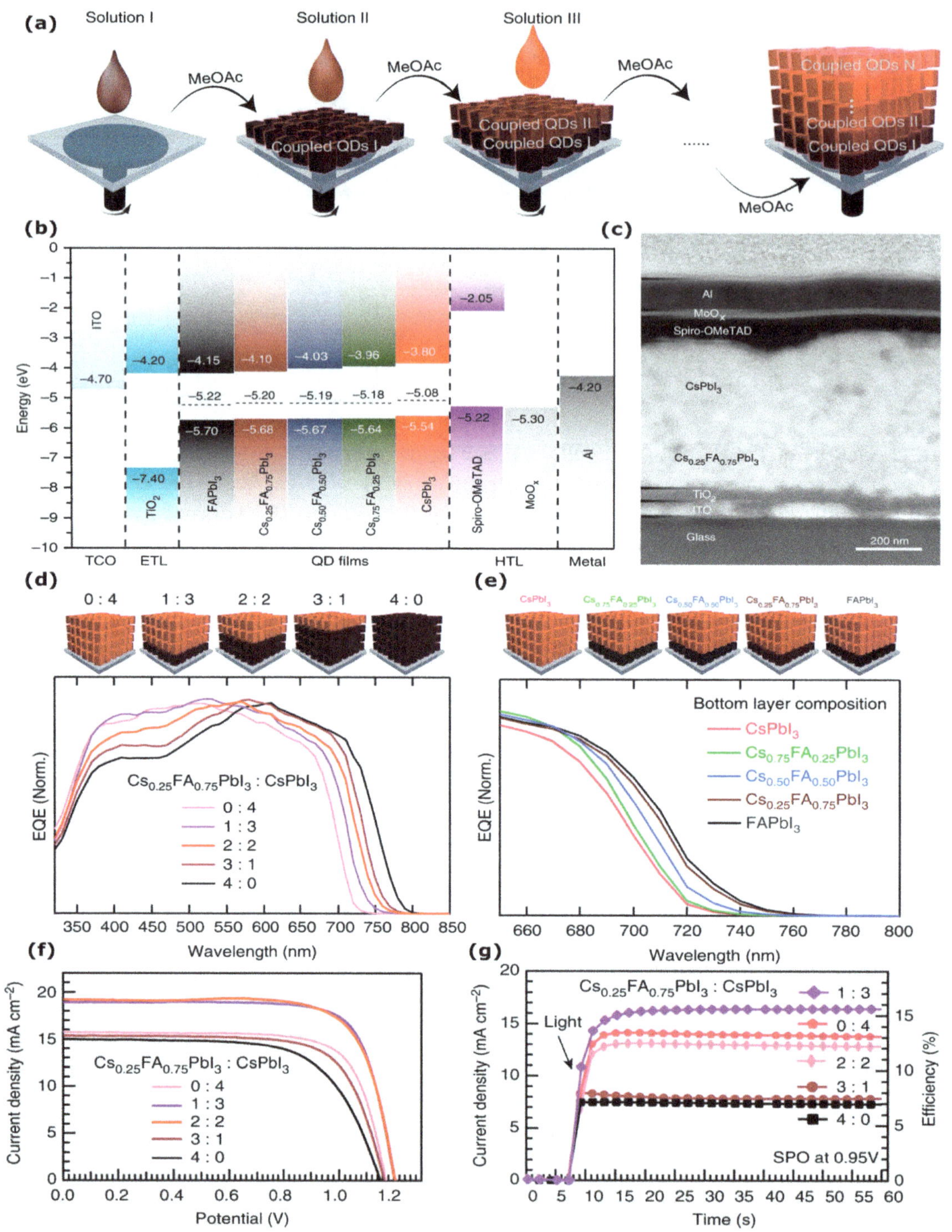

Figure 12. Perovskite quantum dot (PQD) solar cells with charge separating heterostructure. (**a**) Schematic of the device fabrication via spin coating, (**b**) energy band structure of the various PQDs used in the study, (**c**) cross-sectional scanning electron microscope of a typical device, (**d**) the external quantum efficiency (EQE) of solar cells made with various ratios of $Cs_{0.25}Fa_{0.75}PbI_3$ to $CsPbI_3$ quantum dots, (**e**) EQE at the absorption edge of various quantum dots in the series $Cs_xFA_{1-x}PbI_3$ as the bottom layer. (**f**) current density-voltage (JV) curves for the devices shown in (**d**) and (**g**) stabilized power output (SPO) of the varying compositions shown in (**f**).

Table 1. A selection of notable reports on low-dimensional perovskite solar cells. QDs, PCE, RT, and RH stand for quantum dots, power conversion efficiency, room temperature, relative humidity, respectively.

Dimensionality	Material (*n* Value) [1]	PCE (%)	Stability	Reference (Year)
0D	$(CH_3NH_3)_3Sb_2I_9$	2.77	Retained 80% of initial PCE after 3 h under constant illumination, ambient conditions, encapsulated.	[109] (2018)
0D	$Cs_3Sb_2I_9$	1.21	Retained 95% of initial PCE after 60 days, intermittent measurements, stored at RT, 50% RH, unencapsulated.	[118] (2019)
0D	$Cs_3Sb_2I_9$	1.49	Retained >80% of initial PCE after 30 days, intermittent measurements, storage conditions unspecified.	[119] (2018)
1D	$(ThMA)_2(MA)_{n-1}Pb_nI_{3n+1}$ (n = 3)	15.42	Retained 90% of initial PCE after 100 h, intermittent, stored in N_2 in the dark, unencapsulated.	[92] (2018)
1D, mixed with 3D $MAPbI_3$	1,4-benzene diammonium (BDA)-PbI_4 (n = 1)	14.1	Retained 95% of original PCE after >1000 h, intermittent measurement, stored in dark at RT, 85% RH, encapsulated.	[120] (2019)
1D/3D heterostructure	ethylammonium iodide (EAI)-treated $FA_{0.93}Cs_{0.07}PbI_3$	22.3	Retained 95% of initial PCE after 550 h, continuous measurement under constant illumination in N_2 atmosphere at RT, unencapsulated.	[100] (2019)
2D	$(FPEA)_2MA_4Pb_5I_{16}$ (n = 5)	13.64	Retained 65% of initial PCE after 576 h, ambient air at 70 °C unencapsulated.	[121] (2019)
2D	$(BzDA)A_9Pb_{10}(I_{0.93}Br_{0.07})_{31}$ (n = 10)	15.6	Retained 80% of initial PCE after 84 h, intermittent measurements, kept in dark at RT in ambient air, RH = 20–50% unencapsulated.	[122] (2019)
Quasi-2D	3-bromobenzylammonium iodide (BBAI)- (n = 2)	18.2	Retained 82% of initial PCE after 2400 h, intermittent measurements, stored in dark at RT, ~40% RH, unencapsulated.	[24] (2018)
Quasi-2D	$(BE)_2(FA)_8Pb_9I_{28}$ (n value not reported)	17.4	Retained 80% of initial PCE after 50 h, stored in the dark at RT, RH = 80%, unencapsulated.	[123] (2018)
QDs	$CsPbI_3$	14.1	Retained 70% of initial PCE after 50 h, intermittent measurements, stored in the dark at RT and 40% RH, unencapsulated.	[116] (2019)
QDs	$CsPbI_3$:$Cs_{0.25}FA_{0.75}PbI_3$	17.4	Retained 80% of initial PCE after 10 h, intermittent measurements under constant illumination at 40 °C and 25% RH, encapsulated.	[117] (2019)

[1] *n* values are only applicable for 1D and 2D materials.

3.5. Perovskite-Nanocrystal Hybrid Devices

The formation of hybrid layers and devices through incorporating nanocrystals into the OHP layer have also been explored, both in bulk 3DPs [85] and in 0DPs [86]. The introduction of quantum-confined NCs to bulk 3D OHPs enables the possibility of carrier multiplication. Thus far, SiNCs have been primarily studied in this context, since most nanocrystals studied for organic-inorganic hybrid photovoltaics are toxic Pb- or Cd-based [20], and it would be counterintuitive to add a toxic material to a lead-free perovskite. SiNCs are an environmentally-friendly material which are non-toxic and can be synthesized through a wide variety of methods [124,125]. The properties of SiNCs can also be easily modified by surface engineering and the absorption and emission properties can be influenced by the surface terminations [125–127]. Surface engineering can also improve carrier transport in SiNCs by passivating surface defects [128]. While SiNCs do present their own challenges, they represent an important model NC material.

It was previously demonstrated that the incorporation of silicon nanocrystals (SiNCs) into the 0DP with the formula $(CH_3NH_3)_3Bi_2I_9$ led to an enhancement in the device performance [86]. It was proposed that the SiNCs may act as a dissociation pathway for tightly-bound excitons on the nanoclusters of $Bi_2I_9^{3-}$ bioctahedra. An electronic junction formed between the perovskite material and the inorganic nanocrystal can provide an energetically favorable pathway for excitons to overcome the potential barrier created by the cationic sublattice, providing exciton dissociation before the carrier recombines. Once the exciton is dissociated it becomes a free-carrier which can be extracted. This is commonly employed in organic–inorganic hybrid solar cells using SiNCs to enhance exciton dissociation [129]. These types of hybrids may present a route towards significantly improving the efficiency of LDPs.

Hybrid $MAPbI_3$-SiNC devices also exhibit improved device performance and stability [85]. X-ray photoelectron spectroscopy (XPS) indicated that $MAPbI_3$ bonds with SiNCs via intermediate oxide bonds with nitrogen in methylammonium (N–O–Si). The oxidation of SiNCs was also observed in XPS and is likely responsible for the improved stability, whereby SiNCs may act as a 'sponge' absorbing oxidizing species in the $MAPbI_3$ layer resulting in slight oxidation of the SiNCs. Furthermore, hybrid devices with SiNCs exhibited improved device performance after light soaking for 8 min (Figure 13), whilst the performance of $MAPbI_3$-only devices decreased. This is commonly observed in $MAPbI_3$ devices and is attributed to light-activated trap states with inhibited photocarrier extraction [130]. The observation of the inverse behavior in hybrid devices suggests that SiNCs may inhibit defect migration possibly via bonding with the perovskite structure.

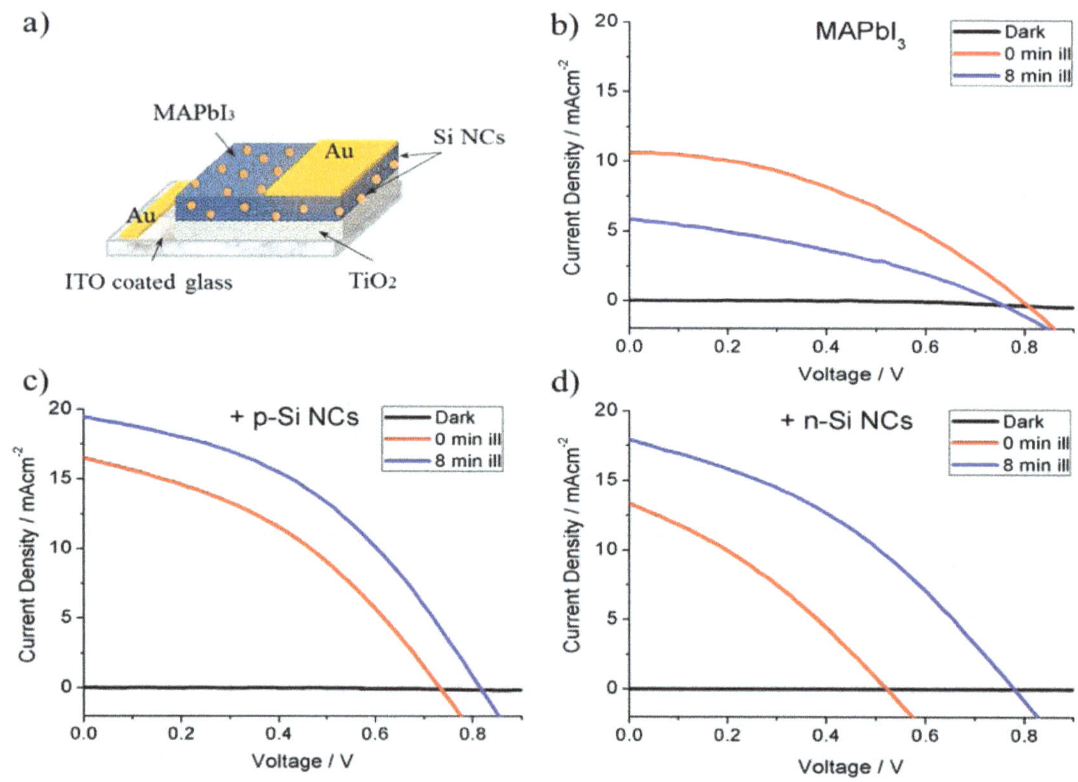

Figure 13. Perovskite-silicon nanocrystal (SiNC) hybrid solar cells show improved device performance especially after light-soaking. (**a**) Schematic of device structure, and current-density voltage (JV) curves for (**b**) $MAPbI_3$ alone, (**c**) $MAPbI_3$ with p-type SiNCs, and (**d**) n-type SiNCs. Reported from ref. [85], with permission from Elsevier, 2018.

In addition, incorporating nanocrystals into OHPs presents the opportunity to create various types of favorable band alignment between the OHP and the nanocrystal. Coupling the properties of nanocrystals with perovskites can lead to improvements in device performance and opens up an avenue of possibilities to exceed the SQ-limit. Forming an inverted type-I junction can potentially improve carrier collection either through optical coupling or electronic coupling. $MAPbI_3$-SiNC hybrid devices form an inverted type-I band alignment (Figure 14), where wider-bandgap SiNCs were incorporated into the perovskite layer with electronic and/or optical coupling with the OHP, depending on whether or not the SiNCs are oxidized [85]. In an electronically coupled inverted type- I junction, the absorption in the wider-bandgap nanocrystal generates carriers which can be transferred into the adjacent conduction and valence bands of the smaller bandgap perovskite. In an optically coupled system, the nanocrystal behaves as a 'interpenetrated' down converter for high energy photons, where radiative carrier recombination via photoluminescence (PL) results in excitations in the narrow bandgap perovskite. It is therefore important that the peak PL emission is tailored to the bandgap of the perovskite to maximize the conversion efficiency. In $MAPbI_3$:SiNC hybrid devices, it is expected that the structure initially forms an electronically coupled junction whereby carriers generated in the SiNCs can transfer into the OHP. After oxidation, carriers generated in SiNCs are trapped by the oxide potential barrier and recombine via photoluminescence, thus generating an optically coupled junction. It was found using Kelvin probe and XPS that the type-I band alignment is preserved even after the SiNCs became oxidized [85]. These new architectures represent new opportunities for exploring different combinations of materials with perovskite structures.

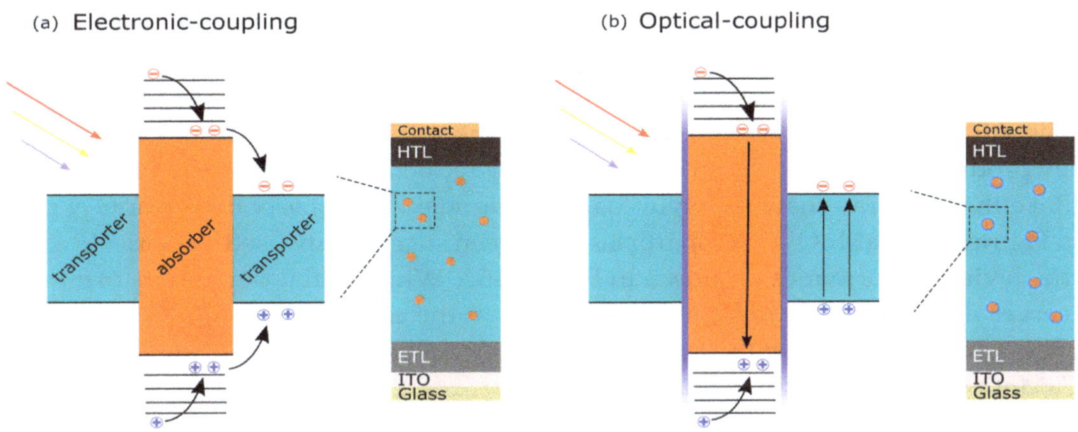

Figure 14. Inverted type-I band alignment: (**a**) electronically coupled and (**b**) optically coupled. Reproduced from ref. [85], with permission from Elsevier, 2018..

3.6. Perovskite Oxide Nanoparticles

Perovskite oxides (ABO_3) are attractive materials for photovoltaics because of the possibility of low-cost, non-toxic photovoltaics with high stability [131]. However, most semiconducting perovskite oxides have large bandgaps (~3–5 eV) due to oxygen-metal transitions with large differences in their electronegativities [132,133], and are therefore generally unsuitable for absorbing light within the solar spectral range. Attempts to reduce the bandgap of perovskite oxides include doping [134], intrinsic defects [135], forming oxynitrides [136], solid solutions [137], and cationic ordering [132].

Perovskite oxides and their derivatives (layered perovskite oxides) represent a large family of materials which exhibit a multitude of properties, and have been investigated for applications including photovoltaics [138]. Perovskite oxides possess a high-degree of flexibility given that 90% of the metallic natural elements in the periodic table can adopt a stable perovskite-type oxide structure [139]. There remains a significant opportunity for exploring the use of metal oxides in photovoltaics to achieve affordable solar cell devices with high efficiency and tunability, whilst easily meeting the often elusive requirement of high stability. The use of metal oxides with highly-tunable absorption properties via the introduction of vacancies [135] and doping [140] would allow for the facile fabrication of multi-junction devices with high stability.

Ferroelectric perovskite oxides have been demonstrated in photovoltaics [133], however they tend to possess large bandgaps (~3–5 eV) and low conductivities, and therefore efficiencies are low (~1%). Plasmonic perovskite oxides have not been explored to the same extent for photovoltaics. Perovskite oxides can be heavily doped to be plasmonic or can be achieved through structural vacancies to strongly modify electronic properties [140]. However, one of the issues associated with plasmonic materials is carrier extraction, and therefore forming extremely thin absorber layers using nano-sized plasmonic oxides is necessary to rapidly extract carriers before recombining.

The perovskite oxide $CaMnO_3$ is an orthorhombic perovskite, and upon reduction in flowing Ar gas the structure can be transformed to an oxygen-deficient perovskite with the structure $CaMnO_{2.5}$. The structure of $CaMnO_{2.5}$ is essentially an orthorhombic perovskite with an internal 1D nanostructure ordering as shown in Figure 15a. The introduction of oxygen vacancies removes one oxygen atom from each MnO_6 octahedra and results in a square pyramid of MnO_5. This structural transformation reveals many interesting properties, such as plasmonic behavior and significantly improved electrocatayltic and photocatalytic activity [141]. $CaMnO_{2.5}$ can be described as an orthorhombic perovskite because the Ca and Mn perovskite sub-lattice is preserved. Since the resulting powder is phase pure, the oxygen vacancies are expected to be ordered resulting in 1D chains of MnO_5 square pyramids [141]. The MnO_5 square pyramids are connected along a one- dimensional network extending through the crystal connected by oxygen atoms, which may be favorable for carrier extraction. This oxygen deficiency creates an internal molecular level porosity. The one-dimensional network of MnO_5 pyramids may also enhance charge transport and enable efficient carrier separation whereby photoexcited carriers are transported along segregated Mn–O carrier transport channels. $CaMnO_{2.5}$ displays broad absorption of light from the infra-red through to the visible region of the solar spectrum. Nanoparticles of $CaMnO_{2.5}$ can be easily produced via a sol-gel process followed by reductive annealing, and then deposited as an ultrathin film either by spray coating or spin coating. The shape and size of the $CaMnO_{2.5}$ nanoparticles is shown in Figure 15b. $CaMnO_{2.5}$ nanoparticles have been successfully used to fabricate a photovoltaic cell, and the device performance is shown in Figure 15c. While initial device performance was low, this work serves as a proof of concept and it is likely that the efficiency can be significantly improved, primarily through optimization of the layer thickness and interfacial engineering to improve coupling between $CaMnO_{2.5}$ nanoparticles and transport layers.

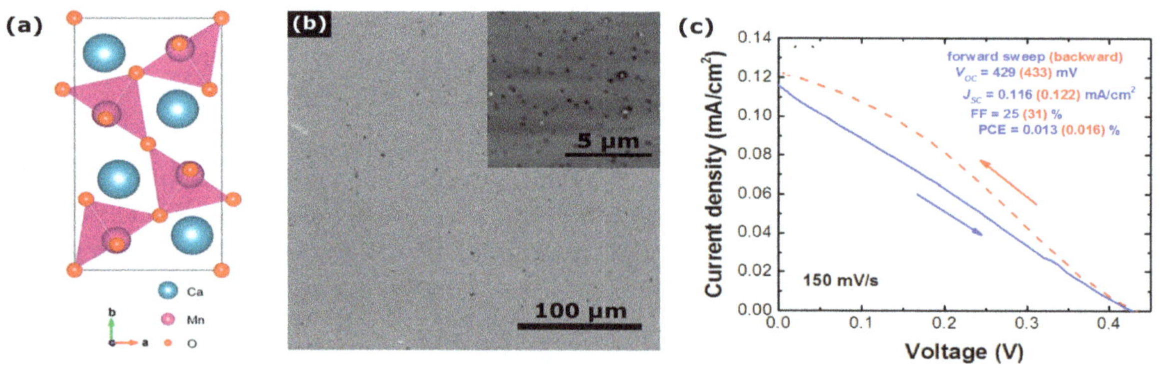

Figure 15. (**a**) Structure of $CaMnO_{2.5}$ reproduced from ref. [141], with permission from American Chemical Society, 2014, (**b**) optical microscope images of $CaMnO_{2.5}$ after laser fragmentation, the inset shows a high-magnification optical microscope image, and (**c**) current density-voltage characteristic of a $CaMnO_{2.5}$ solar cell under solar simulated light.

4. Conclusions and Outlook

This review article has provided a summary of 3D bulk OHPs and an overview of the recent direction and progress towards LDPs. To date, 1DPs and 2DPs have shown the highest efficiencies, yet it is unclear whether these materials will suffer the same long-term stability issues as bulk OHPs. Nanosheets with $n \leq 2$ tend to show impressive stabilities but suffer from low performance issues, particularly due to their very large bandgaps. It is still unclear whether 2DPs and 1DPs with $n > 2$ can demonstrate the long-term stability required for commercialization. Furthermore, the issue of stacking faults between grains, which inhibits charge transfer through the layer, must be overcome to increase the efficiency towards 20%. Exploring conductive organic barriers could be a possible route towards overcoming carrier transport issues.

0DPs tend to be highly stable, however their efficiencies are often very low due to issues associated with carrier extraction, where excitons tend to be strongly localized on BX_6^{4-} or $B_2X_9^{3-}$ clusters. Methods to enhance exciton dissociation and carrier transport need to be further explored if these materials are to demonstrate noteworthy efficiencies in photovoltaics, particularly through forming hybrids with nanocrystals to promote exciton dissociation, and by exploring various ion substitutions at the A-site to lower the exciton binding energy. Provided that these challenges can be overcome, 0DPs with large bandgaps can be incorporated as a top cell in a tandem solar cell. For use in single junction cells, it is important to explore doping along with varying the A-site ion with the aim of discovering 0DPs with smaller bandgaps, which are currently often >2 eV.

PQDs have shown impressive performance so far, yet the choice of materials is rather limited due to the poor stability of organometal PQDs, which are unstable unless capped with long chain organic barriers which inhibit carrier transport. Inorganic $CsPbI_3$ QDs do not require capping molecules and have demonstrated improved short-term stability along with impressive solar cell efficiencies over 13%. Despite this, $CsPbI_3$ QDs are highly unstable in ambient conditions and encapsulation of the entire solar cell device is essential. As research in this field is still in its infancy, there are limited studies on the stability of $CsPbI_3$ QDs and the extent to which the stability can be improved remains unclear. It is therefore currently difficult to predict the potential for $CsPbI_3$ QDs in photovoltaics.

Hybrid devices can be formed by adding NCs to bulk 3DPs or 0DPs. These hybrid device architectures have been explored using SiNCs, demonstrating an improvement in the device performance due to the possibility of a type-I band alignment which can be optically and/or electronically coupled to improve carrier collection. Furthermore, adding SiNCs indicates a route towards extending the device lifetime, whereby SiNCs are oxidized by the residual moisture in the layer rather than degrading the OHP. This was shown to preserve the favorable type-I band alignment without affecting the device performance.

Due to the significant stability issues suffered by OHPs, occurring both in bulk and low-dimensional forms, we have also briefly introduced the field of perovskite oxide nanomaterials, studying the oxygen-deficient perovskite $CaMnO_{2.5}$. This material, which absorbs a broad range of light in the solar spectrum from infrared to ultra-violet, has a one-dimensional internal structure which may promote carrier transport. Although the efficiency of the solar cell device is low, there remains significant opportunities for tuning the properties, optimizing devices, and exploring doping to improve device performance.

Finally, while the efficiencies of LDPs are still often far lower than bulk-OHPs, it is encouraging that higher device efficiencies are continually being reported. Provided that these devices can be fabricated with efficiencies of >20%, it is likely that they will be attractive to the market assuming they can be produced at very low cost and with far superior stability to 3D OHPs.

References

1. National Renewable Energy Laboratory Best Research-Cell Efficiency Chart. Available online: https://www.nrel.gov/pv/assets/pdfs/best-research-cell-efficiencies.20190923.pdf (accessed on 25 September 2019).

2. Green, M.A.; Hishikawa, Y.; Dunlop, E.D.; Levi, D.H.; Hohl-Ebinger, J.; Ho-Baillie, A.W.Y. Solar cell efficiency tables (version 52). *Prog. Photovolt. Res. Appl.* **2018**, *26*, 427–436. [CrossRef]
3. Leijtens, T.; Bush, K.A.; Prasanna, R.; McGehee, M.D. Opportunities and challenges for tandem solar cells using metal halide perovskite semiconductors. *Nat. Energy* **2018**, 1–11. [CrossRef]
4. Jung, H.S.; Park, N.-G. Perovskite Solar Cells: From Materials to Devices. *Small* **2015**, *11*, 10–25. [CrossRef] [PubMed]
5. Xing, G.; Mathews, N.; Sun, S.; Lim, S.S.; Lam, Y.M.; Gratzel, M.; Mhaisalkar, S.; Sum, T.C. Long-Range Balanced Electron- and Hole-Transport Lengths in Organic-Inorganic $CH_3NH_3PbI_3$. *Science* **2013**, *342*, 344–347. [CrossRef] [PubMed]
6. Stranks, S.D.; Eperon, G.E.; Grancini, G.; Menelaou, C.; Alcocer, M.J.P.; Leijtens, T.; Herz, L.M.; Petrozza, A.; Snaith, H.J. Electron-hole diffusion lengths exceeding 1 micrometer in an organometal trihalide perovskite absorber. *Science* **2013**, *342*, 341–344. [CrossRef] [PubMed]
7. Lin, Q.; Armin, A.; Nagiri, R.C.R.; Burn, P.L.; Meredith, P. Electro-optics of perovskite solar cells. *Nat. Photonics* **2014**, *9*, 106–112. [CrossRef]
8. Xing, G.; Mathews, N.; Lim, S.S.; Yantara, N.; Liu, X.; Sabba, D.; Grätzel, M.; Mhaisalkar, S.; Sum, T.C. Low-temperature solution-processed wavelength-tunable perovskites for lasing. *Nat. Mater.* **2014**, *13*, 476–480. [CrossRef]
9. Xing, J.; Zhao, Y.; Askerka, M.; Quan, L.N.; Gong, X.; Zhao, W.; Zhao, J.; Tan, H.; Long, G.; Gao, L.; et al. Color-stable highly luminescent sky-blue perovskite light-emitting diodes. *Nat. Commun.* **2018**, *9*. [CrossRef]
10. Li, Z.; Moon, J.; Gharajeh, A.; Haroldson, R.; Hawkins, R.; Hu, W.; Zakhidov, A.; Gu, Q. Room-temperature continuous-wave operation of organometal halide perovskite lasers. *ACS Nano* **2018**, *12*, 10968–10976. [CrossRef]
11. Wang, K.; Wang, S.; Xiao, S.; Song, Q. Recent Advances in Perovskite Micro- and Nanolasers. *Adv. Opt. Mater.* **2018**, *6*, 1800278. [CrossRef]
12. Feng, J.; Gong, C.; Gao, H.; Wen, W.; Gong, Y.; Jiang, X.; Zhang, B.; Wu, Y.; Wu, Y.; Fu, H.; et al. Single-crystalline layered metal-halide perovskite nanowires for ultrasensitive photodetectors. *Nat. Electron.* **2018**, *1*, 404–410. [CrossRef]
13. Saliba, M.; Matsui, T.; Seo, J.-Y.; Domanski, K.; Correa-Baena, J.-P.; Mohammad, K.N.; Zakeeruddin, S.M.; Tress, W.; Abate, A.; Hagfeldt, A.; et al. Cesium-containing Triple Cation Perovskite Solar Cells: Improved Stability, Reproducibility and High Efficiency. *Energy Environ. Sci.* **2016**, *9*, 1989–1997. [CrossRef] [PubMed]
14. Jeon, N.J.; Noh, J.H.; Yang, W.S.; Kim, Y.C.; Ryu, S.; Seo, J.; Seok, S. Il Compositional engineering of perovskite materials for high-performance solar cells. *Nature* **2015**, *517*, 476–480. [CrossRef] [PubMed]
15. Li, X.; Bi, D.; Yi, C.; Decoppet, J.-D.; Luo, J.; Zakeeruddin, S.M.; Hagfeldt, A.; Gratzel, M. A vacuum flash-assisted solution process for high-efficiency large-area perovskite solar cells. *Science* **2016**, *353*, 58–62. [CrossRef] [PubMed]
16. Eperon, G.E.; Stranks, S.D.; Menelaou, C.; Johnston, M.B.; Herz, L.M.; Snaith, H.J. Formamidinium lead trihalide: A broadly tunable perovskite for efficient planar heterojunction solar cells. *Energy Environ. Sci.* **2014**, *7*, 982–988. [CrossRef]
17. Seok, S.I.; Grätzel, M.; Park, N.G. Methodologies toward Highly Efficient Perovskite Solar Cells. *Small* **2018**, *14*. [CrossRef]
18. Huang, F.; Li, M.; Siffalovic, P.; Cao, G.; Tian, J. From scalable solution fabrication of perovskite films towards commercialization of solar cells. *Energy Environ. Sci.* **2019**, *12*, 518–549. [CrossRef]
19. Cheacharoen, R.; Rolston, N.; Harwood, D.; Bush, K.A.; Dauskardt, R.H.; McGehee, M.D. Design and understanding of encapsulated perovskite solar cells to withstand temperature cycling. *Energy Environ. Sci.* **2018**, *11*, 144–150. [CrossRef]
20. Fan, X.; Zhang, M.; Wang, X.; Yang, F.; Meng, X.; Chen, C.W.M.; Tang, C.Y.; Chen, C.W.M.; Cao, J.; Wang, Z.G.; et al. Recent progress in organic–inorganic hybrid solar cells. 2013, 1, 8694–8709. *J. Mater. Chem. A* **2013**, *1*, 8694–8709. [CrossRef]
21. Sanehira, E.M.; Marshall, A.R.; Christians, J.A.; Harvey, S.P.; Ciesielski, P.N.; Wheeler, L.M.; Schulz, P.; Lin, L.Y.; Beard, M.C.; Luther, J.M. Enhanced mobility $CsPbI_3$ quantum dot arrays for record-efficiency, high-voltage photovoltaic cells. *Sci. Adv.* **2017**, *3*, eaao4204. [CrossRef]

22. Quan, L.N.; Yuan, M.; Comin, R.; Voznyy, O.; Beauregard, E.M.; Hoogland, S.; Buin, A.; Kirmani, A.R.; Zhao, K.; Amassian, A.; et al. Ligand-Stabilized Reduced-Dimensionality Perovskites. *J. Am. Chem. Soc.* **2016**, *138*, 2649–2655. [CrossRef] [PubMed]
23. Tsai, H.; Nie, W.; Blancon, J.-C.; Stoumpos, C.C.; Asadpour, R.; Harutyunyan, B.; Neukirch, A.J.; Verduzco, R.; Crochet, J.J.; Tretiak, S.; et al. High-efficiency two-dimensional Ruddlesden–Popper perovskite solar cells. *Nature* **2016**, *536*, 312–316. [CrossRef] [PubMed]
24. Yang, R.; Li, R.; Cao, Y.; Wei, Y.; Miao, Y.; Tan, W.L.; Jiao, X.; Chen, H.; Zhang, L.; Chen, Q.; et al. Oriented Quasi-2D Perovskites for High Performance Optoelectronic Devices. *Adv. Mater.* **2018**, *30*, 1804771. [CrossRef] [PubMed]
25. Kim, M.R.; Ma, D. Quantum-Dot-Based Solar Cells: Recent Advances, Strategies, and Challenges. *J. Phys. Chem. Lett.* **2015**, *6*, 85–99. [CrossRef]
26. Lee, M.M.; Teuscher, J.; Miyasaka, T.; Murakami, T.N.; Snaith, H.J. Efficient hybrid solar cells based on meso-superstructured organometal halide perovskites. *Science* **2012**, *338*, 643–647. [CrossRef]
27. Kim, H.-S.; Lee, C.-R.; Im, J.-H.; Lee, K.-B.; Moehl, T.; Marchioro, A.; Moon, S.-J.; Humphry-Baker, R.; Yum, J.-H.; Moser, J.E.; et al. Lead Iodide Perovskite Sensitized All-Solid-State Submicron Thin Film Mesoscopic Solar Cell with Efficiency Exceeding 9%. *Sci. Rep.* **2012**, *2*, 591. [CrossRef]
28. Xiong, L.; Guo, Y.; Wen, J.; Liu, H.; Yang, G.; Qin, P.; Fang, G. Review on the Application of SnO2 in Perovskite Solar Cells. *Adv. Funct. Mater.* **2018**, *28*. [CrossRef]
29. Yang, W.S.; Park, B.-W.; Jung, E.H.; Jeon, N.J.; Kim, Y.C.; Lee, D.U.; Shin, S.S.; Seo, J.; Kim, E.K.; Noh, J.H.; et al. Iodide management in formamidinium-lead-halide-based perovskite layers for efficient solar cells. *Science* **2017**, *356*, 1376–1379. [CrossRef]
30. Hoye, R.L.Z.; Brandt, R.E.; Osherov, A.; Stevanovic, V.; Stranks, S.D.; Wilson, M.W.B.; Kim, H.; Akey, A.J.; Perkins, J.D.; Kurchin, R.C.; et al. Methylammonium Bismuth Iodide as a Lead-Free, Stable Hybrid Organic-Inorganic Solar Absorber. *Chem. A Eur. J.* **2016**, *22*, 2605–2610. [CrossRef]
31. Zhang, L.; Ju, M.-G.; Liang, W. The effect of moisture on the structures and properties of lead halide perovskites: A first-principles theoretical investigation. *Phys. Chem. Chem. Phys.* **2016**, *18*, 23174–23183. [CrossRef]
32. Rocks, C.; Svrcek, V.; Maguire, P.; Mariotti, D. Understanding Surface Chemistry During $MAPbI_3$ Spray Deposition and its Effect on Photovoltaic Performance. *J. Mater. Chem. C* **2017**, *5*, 902–916. [CrossRef]
33. Ahn, N.; Kwak, K.; Jang, M.S.; Yoon, H.; Lee, B.Y.; Lee, J.-K.; Pikhitsa, P.V.; Byun, J.; Choi, M. Trapped charge-driven degradation of perovskite solar cells. *Nat. Commun.* **2016**, *7*, 13422. [CrossRef] [PubMed]
34. Hoke, E.T.; Slotcavage, D.J.; Dohner, E.R.; Bowring, A.R.; Karunadasa, H.I.; McGehee, M.D. Reversible photo-induced trap formation in mixed-halide hybrid perovskites for photovoltaics. *Chem. Sci.* **2015**, *6*, 613–617. [CrossRef] [PubMed]
35. Wu, B.; Fu, K.; Yantara, N.; Xing, G.; Sun, S.; Sum, T.C.; Mathews, N. Charge Accumulation and Hysteresis in Perovskite-Based Solar Cells: An Electro-Optical Analysis. *Adv. Energy Mater.* **2015**, *5*, 1500829. [CrossRef]
36. deQuilettes, D.W.; Zhang, W.; Burlakov, V.M.; Graham, D.J.; Leijtens, T.; Osherov, A.; Bulović, V.; Snaith, H.J.; Ginger, D.S.; Stranks, S.D. Photo-induced halide redistribution in organic–inorganic perovskite films. *Nat. Commun.* **2016**, *7*, 11683. [CrossRef]
37. Choi, J.J.; Yang, X.; Norman, Z.M.; Billinge, S.J.L.; Owen, J.S. Structure of Methylammonium Lead Iodide Within Mesoporous Titanium Dioxide: Active Material in High-Performance Perovskite Solar Cells. *Nano Lett.* **2014**, *14*, 127–133. [CrossRef]
38. Jeon, N.J.; Noh, J.H.; Kim, Y.C.; Yang, W.S.; Ryu, S.; Seok, S. Il Solvent engineering for high-performance inorganic–organic hybrid perovskite solar cells. *Nat. Mater.* **2014**, *13*, 897–903. [CrossRef]
39. Misra, R.K.; Aharon, S.; Li, B.; Mogilyansky, D.; Visoly-Fisher, I.; Etgar, L.; Katz, E.A. Temperature- and Component-Dependent Degradation of Perovskite Photovoltaic Materials under Concentrated Sunlight. *J. Phys. Chem. Lett.* **2015**, *6*, 326–330. [CrossRef]
40. Bush, K.A.; Palmstrom, A.F.; Yu, Z.J.; Boccard, M.; Cheacharoen, R.; Mailoa, J.P.; McMeekin, D.P.; Hoye, R.L.Z.; Bailie, C.D.; Leijtens, T.; et al. 23.6%-efficient monolithic perovskite/silicon tandem solar cells with improved stability. *Nat. Energy* **2017**, *2*, 17009. [CrossRef]
41. Wang, S.; Jiang, Y.; Juarez-Perez, E.J.; Ono, L.K.; Qi, Y. Accelerated degradation of methylammonium lead iodide perovskites induced by exposure to iodine vapour. *Nat. Energy* **2016**, *2*, 395–398. [CrossRef]

42. Juarez-Perez, E.J.; Ono, L.K.; Maeda, M.; Jiang, Y.; Hawash, Z.; Qi, Y. Photodecomposition and thermal decomposition in methylammonium halide lead perovskites and inferred design principles to increase photovoltaic device stability. *J. Mater. Chem. A* **2018**, *6*, 9604–9612. [CrossRef]
43. Babayigit, A.; Ethirajan, A.; Muller, M.; Conings, B. Toxicity of organometal halide perovskite solar cells. *Nat. Mater.* **2016**, *15*, 247–251. [CrossRef] [PubMed]
44. Park, N.-G.; Grätzel, M.; Miyasaka, T.; Zhu, K.; Emery, K. Towards stable and commercially available perovskite solar cells. *Nat. Energy* **2016**, *1*, 16152. [CrossRef]
45. Clever, H.L.; Johnston, F.J. The solubility of some sparingly soluble lead salts: An evaluation of the solubility in water and aqueous electrolyte solution. *J. Phys. Chem. Ref. Data* **1982**, *9*, 751–784. [CrossRef]
46. Filip, M.R.; Giustino, F. Computational Screening of Homovalent Lead Substitution in Organic-Inorganic Halide Perovskites. *J. Phys. Chem. C* **2016**, *120*, 166–173. [CrossRef]
47. Yin, W.J.; Shi, T.; Yan, Y. Unique properties of halide perovskites as possible origins of the superior solar cell performance. *Adv. Mater.* **2014**, *26*, 4653–4658. [CrossRef] [PubMed]
48. Lee, S.J.; Shin, S.S.; Kim, Y.C.; Kim, D.; Ahn, T.K.; Noh, J.H.; Seo, J.; Seok, S. Il Fabrication of Efficient Formamidinium Tin Iodide Perovskite Solar Cells through SnF_2–Pyrazine Complex. *J. Am. Chem. Soc.* **2016**, *138*, 3974–3977. [CrossRef]
49. Leijtens, T.; Prasanna, R.; Gold-Parker, A.; Toney, M.F.; McGehee, M.D. Mechanism of Tin Oxidation and Stabilization by Lead Substitution in Tin Halide Perovskites. *ACS Energy Lett.* **2017**, *2*, 2159–2165. [CrossRef]
50. Muscarella, L.A.; Petrova, D.; Jorge Cervasio, R.; Farawar, A.; Lugier, O.; McLure, C.; Slaman, M.J.; Wang, J.; Ehrler, B.; Von Hauff, E.; et al. Air-Stable and Oriented Mixed Lead Halide Perovskite (FA/MA) by the One-Step Deposition Method Using Zinc Iodide and an Alkylammonium Additive. *ACS Appl. Mater. Interfaces* **2019**, *11*, 17555–17562. [CrossRef]
51. Chen, R.; Hou, D.; Lu, C.; Zhang, J.; Liu, P.; Tian, H.; Zeng, Z.; Xiong, Q.; Hu, Z.; Zhu, Y.; et al. Zinc ion as effective film morphology controller in perovskite solar cells. *Sustain. Energy Fuels* **2018**, *2*, 1093–1100. [CrossRef]
52. Kooijman; Muscarella; Williams Perovskite Thin Film Materials Stabilized and Enhanced by Zinc(II) Doping. *Appl. Sci.* **2019**, *9*, 1678. [CrossRef]
53. Shai, X.; Wang, J.; Sun, P.; Huang, W.; Liao, P.; Cheng, F.; Zhu, B.; Chang, S.Y.; Yao, E.P.; Shen, Y.; et al. Achieving ordered and stable binary metal perovskite via strain engineering. *Nano Energy* **2018**, *48*, 117–127. [CrossRef]
54. Eames, C.; Frost, J.M.; Barnes, P.R.F.; O'Regan, B.C.; Walsh, A.; Islam, M.S. Ionic transport in hybrid lead iodide perovskite solar cells. *Nat. Commun.* **2015**, *6*, 7497. [CrossRef] [PubMed]
55. Meloni, S.; Moehl, T.; Tress, W.; Franckevičius, M.; Saliba, M.; Lee, Y.H.; Gao, P.; Nazeeruddin, M.K.; Zakeeruddin, S.M.; Rothlisberger, U.; et al. Ionic polarization-induced current–voltage hysteresis in $CH_3NH_3PbX_3$ perovskite solar cells. *Nat. Commun.* **2016**, *7*, 10334. [CrossRef]
56. Tress, W.; Marinova, N.; Moehl, T.; Zakeeruddin, S.M.; Nazeeruddin, M.K.; Grätzel, M. Understanding the rate-dependent J–V hysteresis, slow time component, and aging in $CH_3NH_3PbI_3$ perovskite solar cells: The role of a compensated electric field. *Energy Environ. Sci.* **2015**, *8*, 995–1004. [CrossRef]
57. Cao, X.; Li, Y.; Li, C.; Fang, F.; Yao, Y.; Cui, X.; Wei, J. Modulating Hysteresis of Perovskite Solar Cells by a Poling Voltage. *J. Phys. Chem. C* **2016**, *120*, 22784–22792. [CrossRef]
58. Yuan, Y.; Huang, J. Ion Migration in Organometal Trihalide Perovskite and Its Impact on Photovoltaic Efficiency and Stability. *Acc. Chem. Res.* **2016**, *49*, 286–293. [CrossRef]
59. Levine, I.; Nayak, P.K.; Wang, J.T.-W.; Sakai, N.; Van Reenen, S.; Brenner, T.M.; Mukhopadhyay, S.; Snaith, H.J.; Hodes, G.; Cahen, D. Temperature-dependent Hysteresis in $MAPbI_3$ Solar Cells. *J. Phys. Chem. C* **2016**. [CrossRef]
60. Mei, A.; Li, X.; Liu, L.; Ku, Z.; Liu, T.; Rong, Y.; Xu, M.; Hu, M.; Chen, J.; Yang, Y.; et al. A hole-conductor-free, fully printable mesoscopic perovskite solar cell with high stability. *Science* **2014**, *345*, 295–298. [CrossRef]
61. Ko, Y.; Kim, Y.; Lee, C.; Kim, Y.; Jun, Y. Investigation of Hole-Transporting Poly(triarylamine) on Aggregation and Charge Transport for Hysteresisless Scalable Planar Perovskite Solar Cells. *ACS Appl. Mater. Interfaces* **2018**, *10*, 11633–11641. [CrossRef]
62. Wei, J.; Zhao, Y.; Li, H.; Li, G.; Pan, J.; Xu, D.; Zhao, Q.; Yu, D. Hysteresis Analysis Based on the Ferroelectric Effect in Hybrid Perovskite Solar Cells. *J. Phys. Chem. Lett.* **2014**, *5*, 3937–3945. [CrossRef] [PubMed]

63. Unger, E.L.; Hoke, E.T.; Bailie, C.D.; Nguyen, W.H.; Bowring, A.R.; Heumüller, T.; Christoforo, M.G.; McGehee, M.D.; Pullerits, T.; Stepanov, A.; et al. Hysteresis and transient behavior in current–voltage measurements of hybrid-perovskite absorber solar cells. *Energy Environ. Sci.* **2014**, *7*, 3690–3698. [CrossRef]
64. Kim, H.S.; Park, N.G. Parameters affecting I-V hysteresis of $CH_3NH_3PbI_3$ perovskite solar cells: Effects of perovskite crystal size and mesoporous TiO_2 layer. *J. Phys. Chem. Lett.* **2014**, *5*, 2927–2934. [CrossRef] [PubMed]
65. Sanchez, R.S.; Gonzalez-Pedro, V.; Lee, J.-W.; Park, N.-G.; Kang, Y.S.; Mora-Sero, I.; Bisquert, J. Slow Dynamic Processes in Lead Halide Perovskite Solar Cells. Characteristic Times and Hysteresis. *J. Phys. Chem. Lett.* **2014**, *5*, 2357–2363. [CrossRef]
66. Rong, Y.; Hu, Y.; Ravishankar, S.; Liu, H.; Hou, X.; Sheng, Y.; Mei, A.; Wang, Q.; Li, D.; Xu, M.; et al. Tunable hysteresis effect for perovskite solar cells. *Energy Environ. Sci.* **2017**, *10*, 2383–2391. [CrossRef]
67. Chen, B.; Yang, M.; Zheng, X.; Wu, C.; Li, W.; Yan, Y.; Bisquert, J.; Garcia-Belmonte, G.; Zhu, K.; Priya, S. Impact of Capacitive Effect and Ion Migration on the Hysteretic Behavior of Perovskite Solar Cells. *J. Phys. Chem. Lett.* **2015**, *6*, 4693–4700. [CrossRef]
68. O'Regan, B.C.; Barnes, P.R.F.; Li, X.; Law, C.; Palomares, E.; Marin-Beloqui, J.M. Optoelectronic studies of methylammonium lead iodide perovskite solar cells with mesoporous TiO_2: Separation of electronic and chemical charge storage, understanding two recombination lifetimes, and the evolution of band offsets during J–V hysteresis. *J. Am. Chem. Soc.* **2015**, *137*, 5087–5099. [CrossRef]
69. Richardson, G.; O'Kane, S.E.J.; Niemann, R.G.; Peltola, T.A.; Foster, J.M.; Cameron, P.J.; Walker, A.B.; Pace, G.; Manna, L.; Caironi, M.; et al. Can slow-moving ions explain hysteresis in the current–voltage curves of perovskite solar cells? *Energy Environ. Sci.* **2016**, *9*, 1476–1485. [CrossRef]
70. Shao, Y.; Xiao, Z.; Bi, C.; Yuan, Y.; Huang, J. Origin and elimination of photocurrent hysteresis by fullerene passivation in $CH_3NH_3PbI_3$ planar heterojunction solar cells. *Nat. Commun.* **2014**, *5*, 5784. [CrossRef]
71. Zheng, F.; Xin, Y.; Huang, W.; Zhang, J.; Wang, X.; Shen, M.; Dong, W.; Fang, L.; Bai, Y.; Shen, X.; et al. Above 1% Efficiency of a Ferroelectric Solar Cell Based on the Pb({Zr,Ti)O_3} Film. *J. Mater. Chem. A* **2013**, *2*, 1363–1368. [CrossRef]
72. Frost, J.M.; Butler, K.T.; Brivio, F.; Hendon, C.H.; van Schilfgaarde, M.; Walsh, A. Atomistic Origins of High-Performance in Hybrid Halide Perovskite Solar Cells. *Nano Lett.* **2014**, *14*, 2584–2590. [CrossRef] [PubMed]
73. Chen, B.; Zheng, X.; Yang, M.; Zhou, Y.; Kundu, S.; Shi, J.; Zhu, K.; Priya, S. Interface band structure engineering by ferroelectric polarization in perovskite solar cells. *Nano Energy* **2015**, *13*, 582–591. [CrossRef]
74. Kutes, Y.; Ye, L.; Zhou, Y.; Pang, S.; Huey, B.D.; Padture, N.P. Direct observation of ferroelectric domains in solution-processed $CH_3NH_3PbI_3$ perovskite thin films. *J. Phys. Chem. Lett.* **2014**, *5*, 3335–3339. [CrossRef] [PubMed]
75. Kim, H.; Kim, S.K.; Kim, B.J.; Shin, K.; Gupta, M.K.; Jung, H.S.; Kim, S.; Park, N. Ferroelectric Polarization in $CH_3NH_3PbI_3$ Perovskite. *J. Phys. Chem. Lett.* **2015**, *6*, 1729–1735. [CrossRef] [PubMed]
76. Elumalai, N.K.; Uddin, A. Hysteresis in organic-inorganic hybrid perovskite solar cells. *Sol. Energy Mater. Sol. Cells* **2016**, *157*, 476–509. [CrossRef]
77. Zhang, X.; Wu, G.; Gu, Z.; Guo, B.; Liu, W.; Yang, S.; Ye, T.; Chen, C.; Tu, W.; Chen, H. Active-layer evolution and efficiency improvement of $(CH_3NH_3)_3Bi_2I_9$-based solar cell on TiO_2-deposited ITO substrate. *Nano Res.* **2016**, *9*, 2921–2930. [CrossRef]
78. Giustino, F.; Snaith, H.J. Towards Lead-free Perovskite Solar Cells. *ACS Energy Lett.* **2016**, *1*, 1233–1240. [CrossRef]
79. González-Carrero, S.; Galian, R.E.; Pérez-Prieto, J. Organometal Halide Perovskites: Bulk Low-Dimension Materials and Nanoparticles. *Part. Part. Syst. Charact.* **2015**, *32*, 709–720. [CrossRef]
80. Müller, M.F.; Freunek, M.; Reindl, L.M. Maximum efficiencies of indoor photovoltaic devices. *IEEE J. Photovoltaics* **2013**, *3*, 59–64.
81. de Weerd, C.; Gomez, L.; Capretti, A.; Lebrun, D.M.; Matsubara, E.; Lin, J.; Ashida, M.; Spoor, F.C.M.; Siebbeles, L.D.A.; Houtepen, A.J.; et al. Efficient carrier multiplication in $CsPbI_3$ perovskite nanocrystals. *Nat. Commun.* **2018**, *9*. [CrossRef]
82. Ni, C.; Hedley, G.; Payne, J.; Svrcek, V.; McDonald, C.; Jagadamma, L.K.; Edwards, P.; Martin, R.; Jain, G.; Carolan, D.; et al. Charge carrier localised in zero-dimensional $(CH_3NH_3)_3Bi_2I_9$ clusters. *Nat. Commun.* **2017**, *8*. [CrossRef]

83. Nozik, A.J. Nanoscience and nanostructures for photovoltaics and solar fuels. *Nano Lett.* **2010**, *10*, 2735–2741. [CrossRef] [PubMed]
84. Shockley, W.; Queisser, H.J. Detailed Balance Limit of Efficiency of p-n Junction Solar Cells. *J. Appl. Phys.* **1961**, *32*, 510. [CrossRef]
85. Rocks, C.; Svrcek, V.; Velusamy, T.; Macias-Montero, M.; Maguire, P.; Mariotti, D. Type-I alignment in $MAPbI_3$ based solar devices with doped-silicon nanocrystals. *Nano Energy* **2018**, *50*, 245–255. [CrossRef]
86. McDonald, C.; Ni, C.; Švrček, V.; Lozac'H, M.; Connor, P.A.; Maguire, P.; Irvine, J.T.S.; Mariotti, D. Zero-dimensional methylammonium iodo bismuthate solar cells and synergistic interactions with silicon nanocrystals. *Nanoscale* **2017**, *9*. [CrossRef] [PubMed]
87. Kawai, T.; Shimanuki, S. Optical Studies of $(CH_3NH_3)_3Bi_2I_9$ Single Crystals. *Phys. Status Solidi* **1993**, *177*, 43–45. [CrossRef]
88. Pazoki, M.; Johansson, M.B.; Zhu, H.; Broqvist, P.; Edvinsson, T.; Boschloo, G.; Johansson, E.M.J. Bismuth Iodide Perovskite Materials for Solar Cell Applications: Electronic Structure, Optical Transitions, and Directional Charge Transport. *J. Phys. Chem. C* **2016**, *120*, 29039–29046. [CrossRef]
89. Cao, D.H.; Stoumpos, C.C.; Farha, O.K.; Hupp, J.T.; Kanatzidis, M.G. 2D Homologous Perovskites as Light-Absorbing Materials for Solar Cell Applications. *J. Am. Chem. Soc.* **2015**, *137*, 7843–7850. [CrossRef]
90. Koh, T.M.; Shanmugam, V.; Schlipf, J.; Oesinghaus, L.; Müller-Buschbaum, P.; Ramakrishnan, N.; Swamy, V.; Mathews, N.; Boix, P.P.; Mhaisalkar, S.G. Nanostructuring Mixed-Dimensional Perovskites: A Route Toward Tunable, Efficient Photovoltaics. *Adv. Mater.* **2016**, *28*, 3653–3661. [CrossRef]
91. Yao, K.; Wang, X.; Xu, Y.X.; Li, F.; Zhou, L. Multilayered Perovskite Materials Based on Polymeric-Ammonium Cations for Stable Large-Area Solar Cell. *Chem. Mater.* **2016**, *28*, 3131–3138. [CrossRef]
92. Lai, H.; Kan, B.; Liu, T.; Zheng, N.; Xie, Z.; Zhou, T.; Wan, X.; Zhang, X.; Liu, Y.; Chen, Y. Two-Dimensional Ruddlesden–Popper Perovskite with Nanorod-like Morphology for Solar Cells with Efficiency Exceeding 15%. *J. Am. Chem. Soc.* **2018**, *140*, 11639–11646. [CrossRef] [PubMed]
93. Cohen, B.E.; Wierzbowska, M.; Etgar, L. High efficiency quasi 2D lead bromide perovskite solar cells using various barrier molecules. *Sustain. Energy Fuels* **2017**, *1*, 1935–1943. [CrossRef]
94. Ma, C.; Shen, D.; Ng, T.W.; Lo, M.F.; Lee, C.S. 2D Perovskites with Short Interlayer Distance for High-Performance Solar Cell Application. *Adv. Mater.* **2018**, *30*, 1800710. [CrossRef] [PubMed]
95. Smith, I.C.; Hoke, E.T.; Solis-Ibarra, D.; McGehee, M.D.; Karunadasa, H.I. A Layered Hybrid Perovskite Solar-Cell Absorber with Enhanced Moisture Stability. *Angew. Chem. Int. Ed.* **2014**, *53*, 11232–11235. [CrossRef]
96. Blancon, J.C.; Tsai, H.; Nie, W.; Stoumpos, C.C.; Pedesseau, L.; Katan, C.; Kepenekian, M.; Soe, C.M.M.; Appavoo, K.; Sfeir, M.Y.; et al. Extremely efficient internal exciton dissociation through edge states in layered 2D perovskites. *Science* **2017**, *355*, 1288–1292. [CrossRef]
97. Zhang, X.; Ren, X.; Liu, B.; Munir, R.; Zhu, X.; Yang, D.; Li, J.; Liu, Y.; Smilgies, D.M.; Li, R.; et al. Stable high efficiency two-dimensional perovskite solar cells via cesium doping. *Energy Environ. Sci.* **2017**, *10*, 2095–2102. [CrossRef]
98. Liu, G.; Zheng, H.; Xu, X.; Zhu, L.; Alsaedi, A.; Hayat, T.; Pan, X.; Dai, S. Efficient solar cells with enhanced humidity and heat stability based on benzylammonium–caesium–formamidinium mixed-dimensional perovskites. *J. Mater. Chem. A* **2018**, *6*, 18067–18074. [CrossRef]
99. Liu, Y.; Akin, S.; Pan, L.; Uchida, R.; Arora, N.; Milić, J.V.; Hinderhofer, A.; Schreiber, F.; Uhl, A.R.; Zakeeruddin, S.M.; et al. Ultrahydrophobic 3D/2D fluoroarene bilayer-based water-resistant perovskite solar cells with efficiencies exceeding 22%. *Sci. Adv.* **2019**, *5*, eaaw2543. [CrossRef]
100. Alharbi, E.A.; Alyamani, A.Y.; Kubicki, D.J.; Uhl, A.R.; Walder, B.J.; Alanazi, A.Q.; Luo, J.; Burgos-Caminal, A.; Albadri, A.; Albrithen, H.; et al. Atomic-level passivation mechanism of ammonium salts enabling highly efficient perovskite solar cells. *Nat. Commun.* **2019**. [CrossRef]
101. Takeoka, Y.; Asai, K.; Rikukawa, M.; Sanui, K. Hydrothermal Synthesis and Structure of Zero-dimensional Organic–inorganic Perovskites. *Chem. Lett.* **2005**, *34*, 602–603. [CrossRef]
102. Akkerman, Q.A.; Abdelhady, A.L.; Manna, L. Zero-Dimensional Cesium Lead Halides: History, Properties, and Challenges. *J. Phys. Chem. Lett.* **2018**, *9*, 2326–2337. [CrossRef] [PubMed]
103. Park, B.W.; Philippe, B.; Zhang, X.; Rensmo, H.; Boschloo, G.; Johansson, E.M.J. Bismuth Based Hybrid Perovskites $A_3Bi_2I_9$ (A: Methylammonium or Cesium) for Solar Cell Application. *Adv. Mater.* **2015**, *27*, 6806–6813. [CrossRef] [PubMed]

104. Qiu, X.; Jiang, Y.; Zhang, H.; Qiu, Z.; Yuan, S.; Wang, P.; Cao, B. Lead-free mesoscopic Cs_2SnI_6 perovskite solar cells using different nanostructured ZnO nanorods as electron transport layers. *Phys. Status Solidi - Rapid Res. Lett.* **2016**, *10*, 587–591. [CrossRef]
105. Zhang, Z.; Li, X.; Xia, X.; Wang, Z.; Huang, Z.; Lei, B.; Gao, Y. High-Quality $(CH_3NH_3)_3Bi_2I_9$Film-Based Solar Cells: Pushing Efficiency up to 1.64%. *J. Phys. Chem. Lett.* **2017**, *8*, 4300–4307. [CrossRef] [PubMed]
106. Zhu, H.; Pan, M.; Johansson, M.B.; Johansson, E.M.J. High Photon-to-Current Conversion in Solar Cells Based on Light-Absorbing Silver Bismuth Iodide. *ChemSusChem* **2017**, *10*, 2592–2596. [CrossRef]
107. Vigneshwaran, M.; Ohta, T.; Iikubo, S.; Kapil, G.; Ripolles, T.S.; Ogomi, Y.; Ma, T.; Pandey, S.S.; Shen, Q.; Toyoda, T.; et al. Facile synthesis and characterization of sulfur doped low bandgap bismuth based perovskites by soluble precursor route. *Chem. Mater.* **2016**, *28*, 6436–6440. [CrossRef]
108. Kim, Y.; Yang, Z.; Jain, A.; Voznyy, O.; Kim, G.-H.; Liu, M.; Quan, L.N.; García de Arquer, F.P.; Comin, R.; Fan, J.Z.; et al. Pure Cubic-Phase Hybrid Iodobismuthates $AgBi_2I_7$ for Thin-Film Photovoltaics. *Angew. Chem.* **2016**, *128*, 9738–9742. [CrossRef]
109. Karuppuswamy, P.; Boopathi, K.M.; Mohapatra, A.; Chen, H.C.; Wong, K.T.; Wang, P.C.; Chu, C.W. Role of a hydrophobic scaffold in controlling the crystallization of methylammonium antimony iodide for efficient lead-free perovskite solar cells. *Nano Energy* **2018**, *45*, 330–336. [CrossRef]
110. Chatterjee, S.; Pal, A.J. Tin(IV) Substitution in $(CH_3NH_3)_3Sb_2I_9$: Toward Low-Band-Gap Defect-Ordered Hybrid Perovskite Solar Cells. *ACS Appl. Mater. Interfaces* **2018**, *10*, acsami.8b12018. [CrossRef]
111. Correa-Baena, J.P.; Nienhaus, L.; Kurchin, R.C.; Shin, S.S.; Wieghold, S.; Putri Hartono, N.T.; Layurova, M.; Klein, N.D.; Poindexter, J.R.; Polizzotti, A.; et al. A-Site Cation in Inorganic $A_3Sb_2I_9$ Perovskite Influences Structural Dimensionality, Exciton Binding Energy, and Solar Cell Performance. *Chem. Mater.* **2018**, *30*, 3734–3742. [CrossRef]
112. Ghosh, B.; Wu, B.; Mulmudi, H.K.; Guet, C.; Weber, K.; Sum, T.C.; Mhaisalkar, S.; Mathews, N. Limitations of $Cs_3Bi_2I_9$ as Lead-Free Photovoltaic Absorber Materials. *ACS Appl. Mater. Interfaces* **2018**, *10*, 35000–35007. [CrossRef] [PubMed]
113. Ha, S.T.; Su, R.; Xing, J.; Zhang, Q.; Xiong, Q. Metal halide perovskite nanomaterials: Synthesis and applications. *Chem. Sci.* **2017**, *8*, 2522–2536. [CrossRef] [PubMed]
114. Swarnkar, A.; Marshall, A.R.; Sanehira, E.M.; Chernomordik, B.D.; Moore, D.T.; Christians, J.A.; Chakrabarti, T.; Luther, J.M. Quantum dot–induced phase stabilization of α-$CsPbI_3$ perovskite for high-efficiency photovoltaics. *Science* **2016**, *354*, 92–95. [CrossRef] [PubMed]
115. Eperon, G.E.; Paternò, G.M.; Sutton, R.J.; Zampetti, A.; Haghighirad, A.A.; Cacialli, F.; Snaith, H.J. Inorganic caesium lead iodide perovskite solar cells. *J. Mater. Chem. A* **2015**, *3*, 19688–19695. [CrossRef]
116. Ling, X.; Zhou, S.; Yuan, J.; Shi, J.; Qian, Y.; Larson, B.W.; Zhao, Q.; Qin, C.; Li, F.; Shi, G.; et al. 14.1% CsPbI3 Perovskite Quantum Dot Solar Cells via Cesium Cation Passivation. *Adv. Energy Mater.* **2019**, *9*, 1900721. [CrossRef]
117. Zhao, Q.; Hazarika, A.; Chen, X.; Harvey, S.P.; Larson, B.W.; Teeter, G.R.; Liu, J.; Song, T.; Xiao, C.; Shaw, L.; et al. High efficiency perovskite quantum dot solar cells with charge separating heterostructure. *Nat. Commun.* **2019**, *10*, 2842. [CrossRef]
118. Umar, F.; Zhang, J.; Jin, Z.; Muhammad, I.; Yang, X.; Deng, H.; Jahangeer, K.; Hu, Q.; Song, H.; Tang, J. Dimensionality Controlling of $Cs_3Sb_2I_9$ for Efficient All-Inorganic Planar Thin Film Solar Cells by HCl-Assisted Solution Method. *Adv. Opt. Mater.* **2019**, *7*, 1801368. [CrossRef]
119. Singh, A.; Boopathi, K.M.; Mohapatra, A.; Chen, Y.F.; Li, G.; Chu, C.W. Photovoltaic Performance of Vapor-Assisted Solution-Processed Layer Polymorph of $Cs_3Sb_2I_9$. *ACS Appl. Mater. Interfaces* **2018**, *10*, 2566–2573. [CrossRef]
120. Ma, C.; Shen, D.; Huang, B.; Li, X.; Chen, W.C.; Lo, M.F.; Wang, P.; Hon-Wah Lam, M.; Lu, Y.; Ma, B.; et al. High performance low-dimensional perovskite solar cells based on a one dimensional lead iodide perovskite. *J. Mater. Chem. A* **2019**, *7*, 8811–8817. [CrossRef]
121. Zhang, F.; Kim, D.H.; Lu, H.; Park, J.S.; Larson, B.W.; Hu, J.; Gao, L.; Xiao, C.; Reid, O.G.; Chen, X.; et al. Enhanced Charge Transport in 2D Perovskites via Fluorination of Organic Cation. *J. Am. Chem. Soc.* **2019**, *141*, 5972–5979. [CrossRef]
122. Cohen, B.E.; Li, Y.; Meng, Q.; Etgar, L. Dion-Jacobson Two-Dimensional Perovskite Solar Cells Based on Benzene Dimethanammonium Cation. *Nano Lett.* **2019**, *19*, 2588–2597. [CrossRef] [PubMed]

123. Zheng, H.; Liu, G.; Zhu, L.; Ye, J.; Zhang, X.; Alsaedi, A.; Hayat, T.; Pan, X.; Dai, S. The Effect of Hydrophobicity of Ammonium Salts on Stability of Quasi-2D Perovskite Materials in Moist Condition. *Adv. Energy Mater.* **2018**, *8*. [CrossRef]
124. Liu, J.J.; Erogbogbo, F.; Yong, K.-T.; Ye, L.; Liu, J.J.; Hu, R.; Chen, H.; Hu, Y.; Yang, Y.; Yang, J.; et al. Assessing clinical prospects of silicon quantum dots: Studies in mice and monkeys. *ACS Nano* **2013**, *7*, 7303–7310. [CrossRef] [PubMed]
125. Mariotti, D.; Mitra, S.; Švrček, V.; Svrček, V. Surface-engineered silicon nanocrystals. *Nanoscale* **2013**, *5*, 1385. [CrossRef]
126. Dasog, M.; Bader, K.; Veinot, J.G.C. Influence of Halides on the Optical Properties of Silicon Quantum Dots. *Chem. Mater.* **2015**, *27*, 1153–1156. [CrossRef]
127. Ramos, E.; Monroy, B.M.; Alonso, J.C.; Sansores, L.E.; Salcedo, R.; Mart, A. Theoretical Study of the Electronic Properties of Silicon Nanocrystals Partially Passivated with Cl and F. *J. Phys. Chem. C* **2012**, *36*, 3988–3994. [CrossRef]
128. Ding, Y.; Sugaya, M.; Liu, Q.; Zhou, S.; Nozaki, T. Oxygen passivation of silicon nanocrystals: Influences on trap states, electron mobility, and hybrid solar cell performance. *Nano Energy* **2014**, *10*, 322–328. [CrossRef]
129. Liu, C.-Y.; Holman, Z.C.; Kortshagen, U.R. Hybrid solar cells from P3HT and silicon nanocrystals. *Nano Lett.* **2009**, *9*, 449–452. [CrossRef]
130. Nie, W.; Blancon, J.-C.; Neukirch, A.J.; Appavoo, K.; Tsai, H.; Chhowalla, M.; Alam, M.A.; Sfeir, M.Y.; Katan, C.; Even, J.; et al. Light-activated photocurrent degradation and self-healing in perovskite solar cells. *Nat. Commun.* **2016**, *7*, 11574. [CrossRef]
131. Rühle, S.; Anderson, A.Y.; Barad, H.N.; Kupfer, B.; Bouhadana, Y.; Rosh-Hodesh, E.; Zaban, A. All-oxide photovoltaics. *J. Phys. Chem. Lett.* **2012**, *3*, 3755–3764. [CrossRef]
132. Nechache, R.; Harnagea, C.; Li, S.; Cardenas, L.; Huang, W.; Chakrabartty, J.; Rosei, F. Bandgap tuning of multiferroic oxide solar cells. *Nat. Photonics* **2014**, *9*, 61–67. [CrossRef]
133. Fan, Z.; Sun, K.; Wang, J. Perovskites for photovoltaics: A combined review of organic–inorganic halide perovskites and ferroelectric oxide perovskites. *J. Mater. Chem. A* **2015**, *3*, 18809–18828. [CrossRef]
134. Lv, M.; Xie, Y.; Wang, Y.; Sun, X.; Wu, F.; Chen, H.; Wang, S.; Shen, C.; Chen, Z.; Ni, S.; et al. Bismuth and chromium co-doped strontium titanates and their photocatalytic properties under visible light irradiation. *Phys. Chem. Chem. Phys.* **2015**, *17*, 26320–26329. [CrossRef] [PubMed]
135. Xu, X.; Randorn, C.; Efstathiou, P.; Irvine, J.T.S. A red metallic oxide photocatalyst. *Nat. Mater.* **2012**, *11*, 595–598. [CrossRef] [PubMed]
136. Yashima, M.; Fumi, U.; Nakano, H.; Omoto, K.; Hester, J.R. Crystal Structure, Optical Properties, and Electronic Structure of Calcium Strontium Tungsten Oxynitrides $Ca_xSr_{1-x}WO_2N$. *J. Phys. Chem. C* **2013**, *117*, 18529–18539. [CrossRef]
137. Shi, J.; Ye, J.; Zhou, Z.; Li, M.; Guo, L. Hydrothermal Synthesis of $Na_{0.5}La_{0.5}TiO_3$-$LaCrO_3$ Solid-Solution Single-Crystal Nanocubes for Visible-Light-Driven Photocatalytic H_2 Evolution. *Chem. A Eur. J.* **2011**, *17*, 7858–7867. [CrossRef]
138. Grinberg, I.; West, D.V.; Torres, M.; Gou, G.; Stein, D.M.; Wu, L.; Chen, G.; Gallo, E.M.; Akbashev, A.R.; Davies, P.K.; et al. Perovskite Oxides for Visible-Light-Absorbing Ferroelectric and Photovoltaic Materials. *Nature* **2013**, *503*, 509–512. [CrossRef]
139. Peña, M.A.; Fierro, J.L.G. Chemical Structures and Performance of Perovskite Oxides. *Chem. Rev.* **2001**, *101*, 1981–2018.
140. Sun, C.; Searles, D.J. Electronics, Vacancies, Optical Properties, and Band Engineering of Red Photocatalyst $SrNbO_3$: A Computational Investigation. *J. Phys. Chem. C* **2014**, *118*, 11267–11270. [CrossRef]
141. Kim, J.; Yin, X.; Tsao, K.C.; Fang, S.; Yang, H. Ca2Mn2O5 as Oxygen-Deficient Perovskite Electrocatalyst for Oxygen Evolution Reaction. *J. Am. Chem. Soc.* **2014**, *136*, 14646–14649. [CrossRef]

5

Carboxylic Acid Functionalization at the *Meso*-Position of the Bodipy Core and its Influence on Photovoltaic Performance

Filip Ambroz, Joanna L. Donnelly, Jonathan D. Wilden, Thomas J. Macdonald * and Ivan P. Parkin *

Department of Chemistry, University College London 20 Gordon St., London WC1H 0AJ, UK; filip.ambroz.16@ucl.ac.uk (F.A.); joanna.donnelly.15@ucl.ac.uk (J.L.D.); j.wilden@ucl.ac.uk (J.D.W.)
* Correspondence: tom.macdonald@ucl.ac.uk (T.J.M.); i.p.parkin@ucl.ac.uk (I.P.P.)

Abstract: Two bodipy dyes with different carboxylic acids on the *meso*-position of the bodipy core were prepared and used to sensitize TiO_2 photoelectrodes. On the basis of spectroscopic characterization, the photoelectrodes were used to fabricate photoelectrochemical cells (PECs) for solar light harvesting. Photovoltaic measurements showed that both bodipy dyes successfully sensitized PECs with short-circuit current densities (J_{SC}) two-fold higher compared to the control. The increase in generated current was attributed to the gain in spectral absorbance due to the presence of bodipy. Finally, the influence of co-sensitization of bodipy and N719 dye was also investigated and photovoltaic device performance discussed.

Keywords: Dye-sensitized solar cells; bodipy dye; co-sensitization; N719 dye; photoelectrochemical cells

---

1. Introduction

Over the last few decades numerous types of photoelectrochemical cells (PECs) that convert sunlight to electricity have been extensively explored as an alternative photovoltaic technology to silicon [1,2]. The most promising include, dye-sensitized solar cells (DSSCs) [3,4], metal chalcogenide solar cells [5,6], organic/polymer solar cells [7–9], and perovskite solar cells [10,11]. While significant research has been conducted for their progress, they all have drawbacks in certain aspects that hinder large scale market employment [12]. For DSSCs in particular, the main efforts have been related to improving cell efficiencies where investigations have focused on two main components, such as the dye sensitizer and semiconducting electrode [13,14]. The foremost step in achieving high efficiencies is the light absorption by the dye molecules that should ideally cover as much of the visible spectrum as possible. To complement this, ruthenium-based dyes such as Di-tetrabutylammonium cis-bis(isothiocyanato)bis(2,2′-bipyridyl-4,4′-dicarboxylato)ruthenium(II) (best known as N719) and other ruthenium-based derivatives have been the most successful candidates thus far [15–17]. While N719 dye has been extensively used in DSSCs research since it shows promise of high efficiency, the limited supply, fluctuation in price, and environmental impact of ruthenium have all prompted investigations into other alternatives [18] Possible choices to replace ruthenium-based dyes include porphyrin [19,20], squaraine [21,22], and boron dipyrromethene (bodipy) dye complexes [23,24]. As a result of many unique features that bodipy dyes possess, they have emerged as an attractive class of functional dyes that can be used for a variety of applications [25,26]. They are known to exhibit characteristics such as ease of structural modification, minimum triplet state formation, photostability, low rates of intersystem crossing, and more [27], which renders them promising for use in the future [28]. In the past decade, a variety of bodipy dyes have been synthesized for use in solar cell applications, however most attempts were not prosperous since the efficiencies were shown to be poor [29]. This was mostly a result of insufficient light harvesting since efforts to design a metal free organic dye that would achieve panchromatic absorption proved to be rather ambitious [30]. Therefore, bodipy dyes were separately engineered to cover a particular absorption spectral range and co-sensitized together in a PEC with the aim to obtain high efficiency [31]. In addition, structural modifications and functionalization of the bodipy core was also shown to have a notable influence on the photovoltaic performance of PECs. For instance, by replacing fluorine on the bodipy core with long alkoxy groups, losses due to charge recombination can be minimized resulting in efficiencies in the range of 5.75% as reported by Ziessel et al [32].

Herein, we report the details of the synthesis and characterization of two bodipy sensitizers with different carboxylic acids at the *meso*-position [29] of the bodipy core. In addition, spectroscopy results of bodipy-sensitized TiO_2 photoelectrodes are discussed followed by PEC fabrication and characterization using simulated sunlight. Finally, the influence of bodipy and N719 dye co-sensitization on the performance of PECs is also investigated.

2. Experimental Section

2.1. Materials

All chemicals used in this study were obtained from Sigma Aldrich (St. Louis, MO, USA), Acros Organics (Geel, Belgium), Alfa Aesar (Haverhill, MA, USA) or Thermo Fischer Scientific (Waltham, MA, USA). They were used as received unless otherwise stated.

2.2. Synthesis of Bodipy Dye 1 and 2

2,4-dimethylpyrrole (0.120 mL, 1.20 mmol, 2 eq.) and glutaric or succinic anhydride (0.600 mmol, 1 eq.) were dissolved in dry dichloromethane (DCM, 10.0 mL) under argon. $BF_3{\cdot}Et_2O$ (0.500 mL, 4.00 mmol) and TEA (0.420 mL, 3.00 mmol) were added and the reaction was heated under reflux for five hours. The organic material was washed with water, dried over $MgSO_4$, filtered and concentrated in vacuo before purification by flash column chromatography on silica gel; eluent: Hexane/EtOAc/AcOH:50/50/0 to 0/97/3.

2.2.1. Bodipy Dye 1

4-(5,5-difluoro-1,3,7,9-tetramethyl-5H-4l4,5l4-dipyrrolo[1,,2-c:2′,1′-f][1,3,2]diazaborinin-10-yl) butanoic acid (22.0 mg, 0.066 mmol, 11%).^{1}H NMR (600.130 MHz, $CDCl_3$) δ_H 6.06 (s, 2H, ArH), 2.82–2.78 (m, 2H, CH_2), 2.57–2.54 (m, 2H, CH_2), 2.52 (s, 6H, $2CH_3$), 2.43 (s, 6H, $2CH_3$), 2.04–1.99 (m, 2H, CH_2) ppm; R_f (EtOAc) = 0.1 (stained with bromocresol green).

2.2.2. Bodipy Dye 2

3-(5,5-difluoro-1,3,7,9-tetramethyl-5H-4l4,5l4-dipyrrolo[1,2-c:2′,1′-f][1,3,2]diazaborinin-10-yl) propanoic acid (27.0 mg, 0.084 mmol, 14%).^{1}H NMR (600.130 MHz, $CDCl_3$) δ_H 6.06 (s, 2H, ArH), 3.36–3.33 (m, 2H, CH_2), 2.71–2.69 (m, 2H, CH_2), 2.52 (s, 6H, $2CH_3$), 2.45 (s, 6H, $2CH_3$) ppm; R_f (EtOAc) = 0.1 (stained with bromocresol green).

2.3. Preparation of Photoelectrochemical Cells (PECs)

PEC fabrication followed the procedures already reported in our previous work [14]. Briefly, TiO_2 photoelectrodes were prepared by doctor blading commercially available TiO_2 paste (GreatCell Solar, DSL-18NRT, Queanbeyan, Australia) onto pre-cleaned fluorine-doped tin oxide (FTO) glass followed by sintering at 500 °C for 30 min which resulted in a uniform layer of anatase TiO_2 nanoparticles (NPs) labelled as a transparent layer. On this layer, an additional layer of bigger-sized TiO_2 NPs (WER2-0 paste, GreatCell Solar, Queanbeyan, Australia) was deposited (doctor blading technique) followed by another sintering at 500 °C for 30 min. The second layer was labelled as light-scattering layer. The photoelectrodes were then subject to treatment with an aqueous solution of titanium tetrachloride ($TiCl_4$, 40 mM) for 30 min at ≈ 70 °C followed by another sintering at 500 °C for 30 min. It should be noted that the active solar cell area of the photoelectrodes was 0.1256 cm^2. Control (unsensitized) PECs (see procedures below for dye sensitization), were directly eclipsed with counter electrodes (Pt-coated FTO glass, GreatCell Solar, Queanbeyan, Australia) with a thermoplastic sealant (GreatCell Solar, MS004610, Queanbeyan, Australia) following by heating at ≈ 100 °C for 10 min to seal the sealant. An electrolyte consisting of iodine (I_2, 0.05 M), 1,2-dimethyl-3-propylimidazolium iodide (DMPII, 0.6 M), guanidium thiocyanate (0.10 M) and 4-tert-Butylpyridine (TBP, 0.5 M) in a mixture of acetonitrile and valeronitrile (volume ratio, 85:15) was injected between the sealed electrode via vacuum filling.

2.4. Dye Sensitization

Bodipy dye 1 and 2 sensitized photoelectrodes underwent dye sensitization with a solution of a relevant dye (0.5 mM) in acetonitrile for 20 h in the dark.

Control (sensitized only with N719) photoelectrodes underwent dye sensitization with a solution of ruthenizer 535-bisTBA (Solaronix 0.5 mM, Aubonne, Switzerland) in absolute ethanol for 20 h in the dark.

Bodipy dye 1 and 2 co-sensitized photoelectrodes underwent dye sensitization with a solution that was a mixture of bodipy dye 1 or 2 solution in absolute ethanol (0.5 mM)—5 vol% and a solution of ruthenizer 535-bisTBA (0.5 mM, Solaronix, Aubonne, Switzerland) in absolute ethanol for 20 h in the dark.

3. Characterisation

Monitoring of all reactions was achieved using 60 F254 silica coated aluminium TLC plates by Merck. Visualisation of these was carried out using UV light of wavelength of 254 and/or 365 nm. Purification was achieved by flash column chromatography using silica gel (43–60 µm) from Merck.

All ^1H, ^{13}C, ^{11}B and ^{19}F nuclear magnetic resonance spectra were recorded at 600 MHz using the Bruker Avance III 600 Cryo or at 700 MHz with the Bruker Avance Neo 700. Instruments clearly indicated for each spectra. Chemical shifts are reported in ppm relative to the internal standard TMS as follows: chemical shift, multiplicity (s = singlet, d = doublet, t = triplet, q = quartet, sep = septet, m = multiplet), coupling constant(s), integration, and assignment using either $CDCl_3$ or DMSO-d^6 as a solvent and TMS as an internal standard.

X-Ray photoelectron spectroscopy (XPS) analysis was carried out using a Thermo Scientific K-alpha photoelectron spectrometer with monochromatic Al-K_α radiation. Peak positions were calibrated to carbon (284.8 eV) and plotted using the CasaXPS software. The measurements were performed on control and sensitized photoelectrodes. X-Ray diffraction (XRD) analysis was performed with a Bruker D8 discovery X-Ray diffractometer using monochromatic Cu $K_{\alpha 1}$ and Cu $K_{\alpha 2}$ radiation of wavelengths 1.54056 and 1.54439 Å, respectively.

The UV/Vis absorption spectra were taken on a Perkin Elmer Lambda 950 instrument with a measurement interval of 1 nm.

The photoluminescence (PL) spectra were recorded using Horiba FluoroMax-4 spectrofluorometer equipped with a PMT detector.

Fourier-transform infrared (FTIR) spectra were recorded with a Bruker FTIR Spectrometer Alpha II with an attenuated total reflection (ATR) attachment. Measurements were performed in the mid-IR region (4000–400 cm^{-1}).

J–V measurements were performed under one-sun (AM 1.5G) illumination using a LOT calibrated solar simulator with a Xenon lamp. Devices were connected to a Keithley 2400 source meter to output the data. Photocurrent measurements were obtained with a halogen lamp chopped to a frequency of 188 Hz through a Newport monochromator; a 4-point probe in connection with a lock-in amplifier is used to collect data. The monochromatic beam is calibrated using a Silicon photo-diode.

4. Results and Discussion

4.1. Synthesis

The synthesis of bodipy dye 1 and 2 is depicted in Scheme 1 and is based on previously reported synthetic strategies [33,34]. (See Supplementary Materials for more details) Molecular structures of the obtained compounds were characterized by spectroscopic methods, such as NMR which is shown in Supplementary Materials Figure S1 and S2. Singlet peaks integrating to two protons at 6.06 ppm for both compounds confirm the successful construction of the BODIPY scaffold.

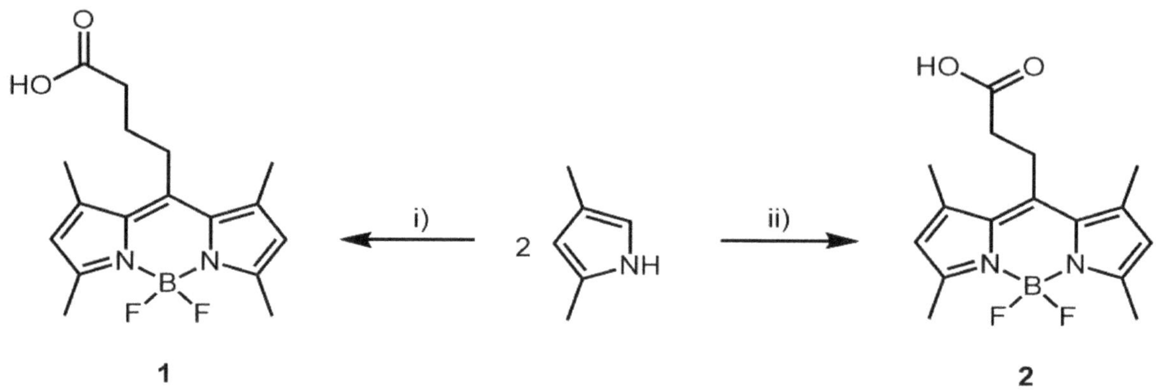

Scheme 1. Synthetic routes towards bodipy dyes 1 and 2. (**1**) Succinic anhydride, dichloromethane (DCM), $BF_3 \cdot Et_2O$, reflux 5h, $BF_3 \cdot Et_2O$, TEA, 16 h. (**2**) Glutaric anhydride, DCM, $BF_3 \cdot Et_2O$, reflux 5h, $BF_3 \cdot Et_2O$, TEA, 16 h.

4.2. Optical Properties

The absorption and PL emission spectra of solutions of bodipy dyes 1 and 2 are illustrated in Figure 1a,b. Both dyes exhibit three absorption features (more pronounced for dye 1 due to higher molar attenuation coefficients (ε) at those particular wavelengths—Table 1) with two located in the UV region and one in the visible (green) portion of the electromagnetic spectrum. From the absorption onset of the dyes, their band gap values were determined to be $\approx$ 2.4 eV (Table 1). Due to Stokes shift, where the position of the band maxima of the absorption was at $\approx$ 495 nm, the emission of the dyes resulted to be 514 nm for dye 1 and 520 nm for dye 2. The bodipy dyes were then used for PECs, by sensitizing TiO_2 photoelectrodes in a dye solution and their corresponding UV/Vis spectra are shown in Figure 1c,d. To evaluate if the TiO_2 photoelectrodes successfully absorbed the dyes, comparison was made to the untreated (unsensitized) TiO_2 photoelectrodes. As can be seen in Figure 1c,d, both sensitized TiO_2 photoelectrodes exhibited a broad absorption feature in the visible spectrum allowing the harvesting of more energy from photons. This can be attributed to the dye absorption (λ_{max} at 502 and 504 nm for bodipy dyes 1 and 2, respectively). Compared to the bodipy solutions, sensitized TiO_2 photoelectrodes exhibited a bathochromic shift that occurred as a result of the interaction, between the dye and TiO_2 nanoparticles (suppression of H-aggregation). [35] Moreover, to investigate the molecular packing of dye molecules [36] X-Ray diffraction (XRD) was also performed. Supplementary Materials Figure S4 shows the XRD diffraction patterns for unsensitized and sensitized TiO_2 photoelectrodes where no difference between the samples was observed.

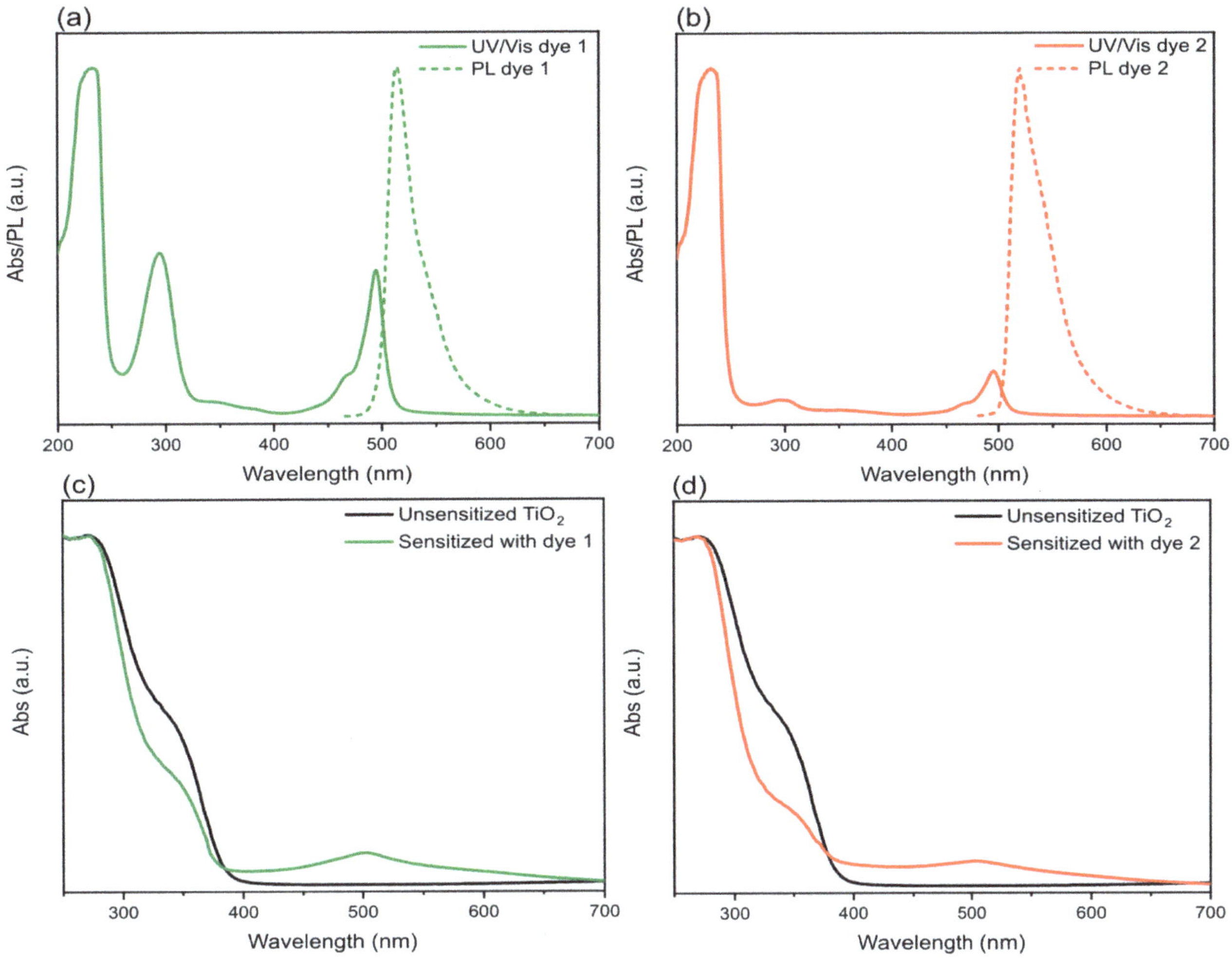

Figure 1. UV/Vis absorption (solid line) and photoluminescence (PL) emission (dotted lines) spectra of bodipy (**a**) dye 1 and (**b**) dye 2 solutions. (Excitation wavelength for PL emission was 400 nm) UV/Vis absorption spectra of TiO_2 sensitized photoelectrodes of bodipy (**c**) dye 1 and (**d**) dye 2, respectively. Black lines represent untreated TiO_2 photoelectrodes where green (dye 1) and red (dye 2) lines correspond to TiO_2 sensitized photoelectrodes.

Table 1. Summary of spectral properties of bodipy dye 1 and 2 in ethanol solution (0.5 mM) and on photoelectrodes. The band gap values (E_g) were determined following the procedure reported elsewhere [37] and also from the Tauc plots (Supplementary Materials Figure S3) where the obtained values were comparable.

Sensitizer	Solution Abs λ_{max} (nm)	Solution PL λ_{max} (nm)	Photoelectrode Abs λ_{max} (nm)	E_g (eV)	ε (M^{-1} cm^{-1})
Dye 1	232, 295, 495	514	502	2.39	6124 (at 495 nm)
Dye 2	232, 295, 496	520	504	2.38	1350 (at 496 nm)

4.3. Spectroscopy Characterization

FTIR spectroscopy was used to obtain infrared spectra (% transmittance) of synthesized bodipy dyes which can be seen in Figure 2. A broad band in the region of around 3500–2500 cm^{-1} can be observed (Figure 2a) which is a typical characteristic for compounds containing a carboxyl group (O–H stretch) [38]. Since this band is in the same region as the C–H stretching bands, a slightly distorted absorption pattern is observed where the former band is superimposed on the sharp C–H stretching bands. Furthermore, a strong peak appeared around 1700 cm^{-1} for both dyes which was assigned to

C=O stretching of the carbonyl group [35]. Additional peaks that were identified for both spectra were located in the regions of around 1400, 1200, and 910 cm^{-1} (see Supplementary Materials Table S1 for the exact values) and were assigned to O–H bend, C–O stretch, and O–H bend, respectively.

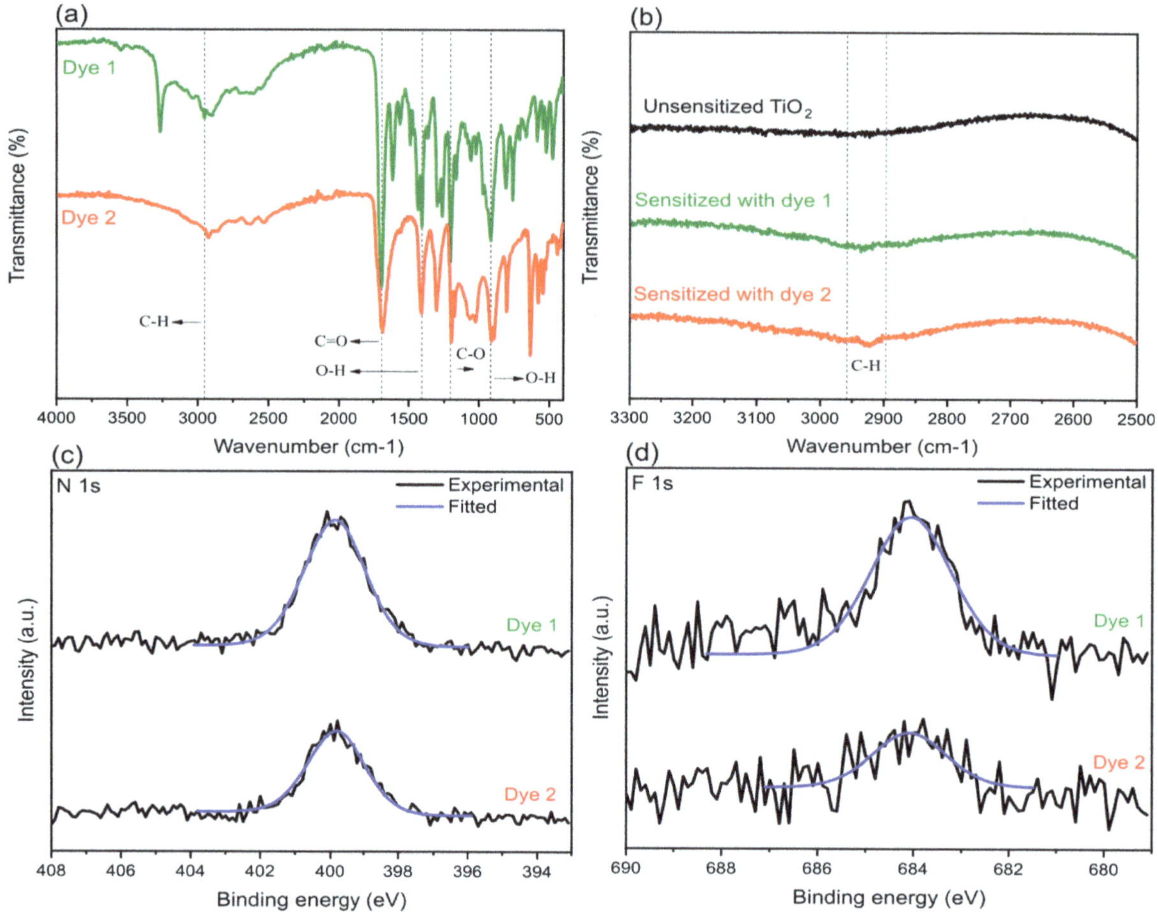

Figure 2. Fourier-transform infrared (FTIR) spectra of bodipy (**a**) dye 1 (green curve) and dye 2 (red curve) in a powder form along with the inset (**b**) of untreated (black curve) and sensitized TiO_2 photoelectrodes on glass substrate. The whole range for the FTIR spectra of TiO_2 photoelectrodes can be seen in Supplementary Materials Figure S5. The high-resolution (**c**) N 1s and (**d**) F 1s X-ray photoelectron spectroscopy (XPS) spectra for sensitized TiO_2 photoelectrodes.

FTIR measurements were also performed with the intention to investigate interaction of individual dyes on sensitized TiO_2 electrodes. However, our results showed that the substrates used for TiO_2 deposition strongly absorbed below 2500 cm^{-1} which prevented characterization in this region (Supplementary Materials Figure S5). Despite this, by comparing regions of around 3000 cm^{-1} between sensitized and untreated TiO_2 photoelectrodes, a slight increase in absorbance was observed for both sensitized TiO_2 photoelectrodes. (Figure 2b) These bands were assigned to C–H stretching of the dye compounds.

X-Ray photoelectron spectroscopy (XPS) was carried out to identify relevant elements present on the surface of untreated and sensitized TiO_2 photoelectrodes. Elements (N and F) associated to the compounds of the dyes were identified on both sensitized TiO_2 photoelectrodes (Figure 2c) while only noise was observed in the same binding energy regions for the untreated TiO_2 photoelectrode. (Figure S6). It is noteworthy that efforts related to the detection of boron resulted to be negative. However, such results are not surprising since it is notoriously difficult to detect B1's peak due to its very low sensitivity.

4.4. Photoelectrochemical Properties

Photoelectrochemical cells (PECs) were fabricated using bodipy dyes 1 and 2 as photosensitizers for nanocrystalline TiO_2 photoelectrodes. Figure 3a shows the photocurrent–voltage (J–V) curves of PECs and the photovoltaic parameters are summarized in Table 2. Compared to control (unsensitized), PECs sensitized with bodipy dyes showed a significant increase in J_{SC} values (≈2.5-fold increase for dye 1 and ≈ 2-fold increase for dye 2). Such a phenomenon is primarily attributed to the improved light harvesting due to the presence of the bodipy dyes. Moreover, the fill factor (FF) also increased for sensitized PECs while V_{OC} decreased. The latter can be related to the changes in the band gap energy (E_g) where an E_g decrease results in higher J_{SC} and lower V_{OC}. Nevertheless, the overall power conversion efficiency (PCE) of bodipy sensitized PECs doubled compared to the control. On the other hand, the results indicate that the proximity of the carboxylic acid group to the bodipy core notably influenced the photovoltaic parameters for PECs. Specifically, PECs sensitized with a bodipy dye 1 (distance of three carbons) were shown to exhibit better J_{SC} as opposed to the bodipy dye 2 (distance of two carbons) sensitized PECs that had higher V_{OC} and FF. Nevertheless, the overall PCE was not influenced by such changes (Table 2). It is noteworthy that the average values of the photovoltaic parameters for different batches of devices, shown in Figure S7, did not show a change in the trends, as discussed herein. Furthermore, the values of PV parameters for bodipy sensitized PECs were low where one of the reasons together with the nature of the absorption onset of the dyes can be related to their aggregation on the TiO_2 photoelectrodes, thus increasing charge recombination and reducing electron injection to the n-type layer of the photoelectrode [39,40].

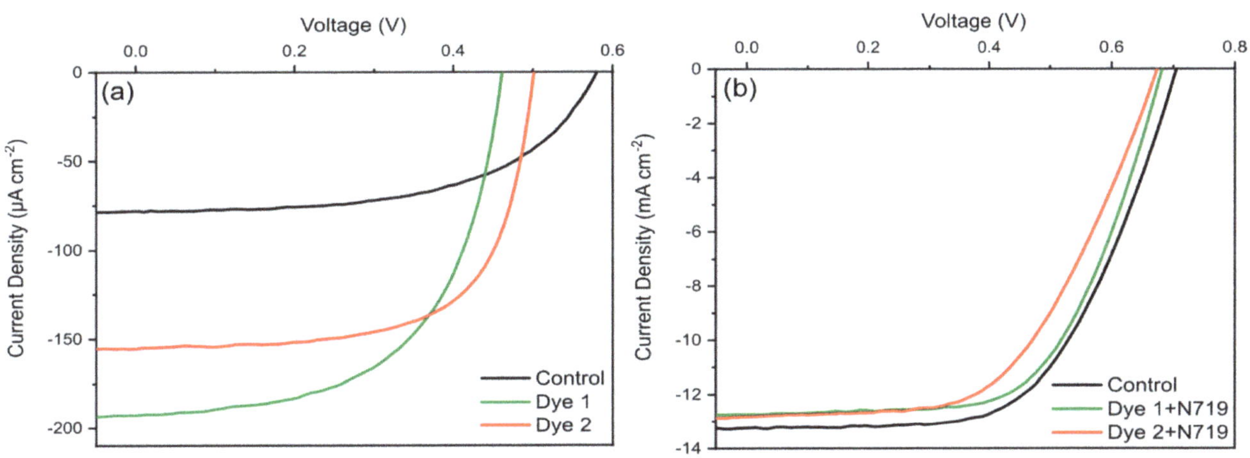

Figure 3. Photocurrent–voltage (J–V) characteristics of photoelectrochemical cells (PECs) sensitized only with (**a**) dye 1 and 2 including a control (unsensitized) together with (**b**) N719 co-sensitized. The control PEC in (**b**) was sensitized only with a dye N719.

Table 2. Summary of photovoltaic parameters for different PECs sensitized with bodipy dye 1 or 2 and co-sensitized with dye 1 + N719 or dye 2 + N719 including the corresponding control (unsensitized and N719 sensitized) PECs.

PEC	J_{SC} (μA cm^{-2})	V_{oc} (mV)	FF (%)	PCE (%)
Control (unsensitized)	78.2	579.6	56.3	0.025
Dye 1	192.6	461.3	58.0	0.051
Dye 2	154.8	501.0	66.4	0.051
Control (N719 sensitized)	13.2	704.7	59.2	5.5
Dye 1 + N719	12.7	681.3	61.6	5.3
Dye 2 + N719	12.8	673.1	55.5	4.8

Previous reports have shown that the performance of DSSCs containing co-adsorbents, bodipy and N719 can be improved compared to a single sensitizer [35]. Therefore, in an attempt to improve PV parameters bodipy dyes in this study were used and co-sensitized with N719 to investigate if the same conclusions can also be extended to the other combinations of sensitizers. In comparison with the control (sensitized only with N719), both J_{SC} and V_{OC} diminished for bodipy (dye 1 and 2) co-sensitized (5 vol%) [35] PECs (Figure 3b and Table 2). A particularly noticeable decrease was observed for V_{OC} where the parameter is influenced by the recombination rate and adsorption mode of the sensitizer [41] that when compared to the control, negatively influenced the co-sensitized PECs. Therefore, the bodipy dyes used in this study were shown to have a negative effect when implemented together with a N719 dye. We postulate that such an outcome was as a result of an increase in recombination centers which can be attributed to either a mismatch in energy levels or unsuitable molar absorption coefficient. Despite this, the bodipy dyes were shown to successfully sensitize TiO_2 photoelectrodes, and further work into modifying the functional groups could lead to improved single sensitized, and co-sensitized PV performance. Specifically, efforts should be made into replacing electronegative elements, for example fluorine, with other functional groups that do not have such tendency to attract a bonding pair of electrons since this characteristic negatively affects the PV performance. In addition, other methods that aim to extend the absorption onset include attachments of an electron donor or acceptor to the C2 and C6 positions [29].

5. Conclusions

Two bodipy dyes with different carboxylic acids on the *meso*-position of the bodipy core have been synthesized. The bodipy dyes were used to sensitize TiO_2 photoelectrodes which were characterized using spectroscopic techniques (UV/Vis absorption, FTIR and XPS). On the basis of these results, the TiO_2 photoelectrodes were used to fabricate PECs. Their PV parameters were analyzed, and the results showed superior light harvesting and thus higher power conversion efficiencies compared to the control (unsensitized) PECs. Furthermore, the bodipy dyes were co-sensitized (5 vol%) with a N719 dye to investigate their interactions in a PEC. The results showed a decrease in efficiencies indicating co-sensitization had a negative effect on the photovoltaic performance parameters of co-sensitized PECs. Further improvements in efficiencies are possible by designing/modifying bodipy dyes with different light absorbing groups within the organic framework to tune the absorption spectral range with high molar extinction coefficients and use them as co-sensitizers for solar cell devices.

Supplementary Materials

Figure S1: ^{1}H-NMR of bodipy dye 1, Figure S2: ^{1}H-NMR of bodipy dye 2, Figure S3: Tauc plots for solutions of bodipy (a) dye 1 and (b) dye 2, Figure S4: XRD patterns for TiO_2 photoelectrodes, Figure S5: FTIR spectra of untreated and sensitized TiO2 photoelectrodes, Figure S6: The high-resolution (a) N 1s and (b) F 1s XPS spectra for control (unsensitized) TiO2 photoelectrode, Figure S7: Average (a) J_{SC}, (b) V_{OC}, (c) FF and (d) power conversion efficiency (PCE) of the fabricated PECs sensitized only with bodipy dye 1 or 2 including a control (unsensitized), Table S1: FTIR peak wavenumbers and assignments for bodipy dye 1 and 2.

Author Contributions: Experiments were designed by T.J.M and I.P.P with contribution from J.D.W. All dyes were synthesized by J.L.D. Materials characterization and PECs fabrication was performed by F.A. The manuscript was written by F.A. with contributions of J.L.D, T.J.M and I.P.P. All authors have given approval to the final version of the manuscript.

References

1. Nayak, P.K.; Mahesh, S.; Snaith, H.J.; Cahen, D. Photovoltaic solar cell technologies: Analysing the state of the art. *Nat. Rev. Mater.* **2019**, *4*, 269. [CrossRef]
2. Khatri, I.; Shudo, K.; Matsuura, J.; Sugiyama, M.; Nakada, T. Impact of heat-light soaking on potassium fluoride treated CIGS solar cells with CdS buffer layer. *Prog. Photovolt. Res. Appl.* **2018**, *26*, 171–178. [CrossRef]

3. Munkhbayar, B.; Nine Md, J.; Jeoun, J.; Ji, M.; Jeong, H.; Chung, H. Synthesis of a graphene–tungsten composite with improved dispersibility of graphene in an ethanol solution and its use as a counter electrode for dye-sensitised solar cells. *J. Power Sources* **2013**, *230*, 207–217. [CrossRef]
4. Batmunkh, M.; Shrestha, A.; Bat-Erdene, M.; Nine, M.J.; Shearer, C.J.; Gibson, C.T.; Slattery, A.D.; Tawfik, S.A.; Ford, M.J.; Dai, S.; et al. Electrocatalytic activity of a 2D phosphorene-based heteroelectrocatalyst for photoelectrochemical cells. *Angew. Chem. Int. Ed.* **2018**, *57*, 2644–2647. [CrossRef] [PubMed]
5. Stroyuk, O.; Raevskaya, A.; Gaponik, N. Solar light harvesting with multinary metal chalcogenide nanocrystals. *Chem. Soc. Rev.* **2018**, *47*, 5354–5422. [CrossRef] [PubMed]
6. Shrestha, A.; Batmunkh, M.; Tricoli, A.; Qiao, S.Z.; Dai, S. Near-infrared active lead chalcogenide quantum dots: Preparation, post-synthesis ligand exchange, and applications in solar cells. *Angew. Chem. Int. Ed.* **2019**, *58*, 5202–5224. [CrossRef] [PubMed]
7. Ye, L.; Xiong, Y.; Chen, Z.; Zhang, Q.; Fei, Z.; Henry, R.; Heeney, M.; O'Connor, B.T.; You, W.; Ade, H. Sequential deposition of organic films with eco-compatible solvents improves performance and enables over 12%-efficiency nonfullerene solar cells. *Adv. Mater.* **2019**, *31*, 1808153. [CrossRef] [PubMed]
8. Zhang, J.; Tan, H.S.; Guo, X.; Facchetti, A.; Yan, H. Material insights and challenges for non-fullerene organic solar cells based on small molecular acceptors. *Nat. Energy* **2018**, *3*, 720–731. [CrossRef]
9. Ye, L.; Hu, H.; Ghasemi, M.; Wang, T.; Collins, B.A.; Kim, J.-H.; Jiang, K.; Carpenter, J.H.; Li, H.; Li, Z.; et al. Quantitative relations between interaction parameter, miscibility and function in organic solar cells. *Nat. Mater.* **2018**, *17*, 253–260. [CrossRef]
10. Jeon, N.J.; Na, H.; Jung, E.H.; Yang, T.-Y.; Lee, Y.G.; Kim, G.; Shin, H.-W.; Seok, S.I.; Lee, J.; Seo, J. A fluorene-terminated hole-transporting material for highly efficient and stable perovskite solar cells. *Nat. Energy* **2018**, *3*, 682. [CrossRef]
11. Macdonald, T.J.; Batmunkh, M.; Lin, C.-T.; Kim, J.; Tune, D.D.; Ambroz, F.; Li, X.; Xu, S.; Sol, C.; Papakonstantinou, I.; et al. Origin of performance enhancement in TiO_2-carbon nanotube composite perovskite solar cells. *Small Methods* **2019**, 1900164. [CrossRef]
12. Rong, Y.; Hu, Y.; Mei, A.; Tan, H.; Saidaminov, M.I.; Seok, S.I.; McGehee, M.D.; Sargent, E.H.; Han, H. Challenges for commercializing perovskite solar cells. *Science* **2018**, *361*, eaat8235. [CrossRef] [PubMed]
13. Ren, Y.; Sun, D.; Cao, Y.; Tsao, H.N.; Yuan, Y.; Zakeeruddin, S.M.; Wang, P.; Grätzel, M. A stable blue photosensitizer for color palette of dye-sensitized solar cells reaching 12.6% efficiency. *J. Am. Chem. Soc.* **2018**, *140*, 2405–2408. [CrossRef] [PubMed]
14. Ambroz, F.; Sathasivam, S.; Lee, R.; Gadipelli, S.; Lin, C.-T.; Xu, S.; Poduval, R.K.; Mclachlan, M.A.; Papakonstantinou, I.; Parkin, I.P.; et al. Influence of lithium and lanthanum treatment on TiO_2 nanofibers and their application in n-i-p solar cells. *ChemElectroChem* **2019**, *6*, 3590–3598. [CrossRef]
15. Macdonald, T.J.; Ambroz, F.; Batmunkh, M.; Li, Y.; Kim, D.; Contini, C.; Poduval, R.; Liu, H.; Shapter, J.G.; Papakonstantinou, I.; et al. TiO_2 nanofiber photoelectrochemical cells loaded with sub-12 nm AuNPs: Size dependent performance evaluation. *Mater. Today Energy* **2018**, *9*, 254–263. [CrossRef]
16. Nazeeruddin, M.K.; Klein, C.; Liska, P.; Grätzel, M. Synthesis of novel ruthenium sensitizers and their application in dye-sensitized solar cells. *Coord. Chem. Rev.* **2005**, *249*, 1460–1467. [CrossRef]
17. Nazeeruddin, M.K.; Kay, A.; Rodicio, I.; Humphry-Baker, R.; Mueller, E.; Liska, P.; Vlachopoulos, N.; Graetzel, M. Conversion of light to electricity by cis-X2bis(2,2′-bipyridyl-4,4′-dicarboxylate)ruthenium(II) charge-transfer sensitizers (X = Cl-, Br-, I-, CN-, and SCN-) on nanocrystalline titanium dioxide electrodes. *J. Am. Chem. Soc.* **1993**, *115*, 6382–6390. [CrossRef]
18. Kolemen, S.; Bozdemir, O.A.; Cakmak, Y.; Barin, G.; Erten-Ela, S.; Marszalek, M.; Yum, J.-H.; Zakeeruddin, S.M.; Nazeeruddin, M.K.; Grätzel, M.; et al. Optimization of distyryl-Bodipy chromophores for efficient panchromatic sensitization in dye sensitized solar cells. *Chem. Sci.* **2011**, *2*, 949–954. [CrossRef]
19. Ji, J.-M.; Zhou, H.; Kim, H.K. Rational design criteria for D–π–A structured organic and porphyrin sensitizers for highly efficient dye-sensitized solar cells. *J. Mater. Chem. A* **2018**, *6*, 14518–14545. [CrossRef]
20. Song, H.; Liu, Q.; Xie, Y. Porphyrin-sensitized solar cells: systematic molecular optimization, coadsorption and cosensitization. *Chem. Commun.* **2018**, *54*, 1811–1824. [CrossRef]
21. Kuang, D.; Walter, P.; Nüesch, F.; Kim, S.; Ko, J.; Comte, P.; Zakeeruddin, S.M.; Nazeeruddin, M.K.; Grätzel, M. Co-sensitization of organic dyes for efficient ionic liquid electrolyte-based dye-sensitized solar cells. *Langmuir* **2007**, *23*, 10906–10909. [CrossRef] [PubMed]

22. Chen, Y.; Yang, L.; Wu, J.; Wang, G.; Huang, W.; Melkonyan, F.S.; Lu, Z.; Huang, Y.; Marks, T.J.; Facchetti, A. Performance, morphology, and charge recombination correlations in ternary squaraine solar cells. *Chem. Mater.* **2018**, *30*, 6810–6820. [CrossRef]
23. Erten-Ela, S.; Ueno, Y.; Asaba, T.; Kubo, Y. Synthesis of a dibenzo-Bodipy-incorporating phenothiazine dye as a panchromatic sensitizer for dye-sensitized solar cells. *New J. Chem.* **2017**, *41*, 10367–10375. [CrossRef]
24. Singh, S.P.; Gayathri, T. Evolution of BODIPY dyes as potential sensitizers for dye-sensitized solar cells. *Eur. J. Org. Chem.* **2014**, *2014*, 4689–4707. [CrossRef]
25. He, H.; Ji, S.; He, Y.; Zhu, A.; Zou, Y.; Deng, Y.; Ke, H.; Yang, H.; Zhao, Y.; Guo, Z.; et al. Photoconversion-tunable fluorophore vesicles for wavelength-dependent photoinduced cancer therapy. *Adv. Mater.* **2017**, *29*, 1606690. [CrossRef]
26. Turksoy, A.; Yildiz, D.; Akkaya, E.U. Photosensitization and controlled photosensitization with bodipy dyes. *Coord. Chem. Rev.* **2019**, *379*, 47–64. [CrossRef]
27. Lee, C.Y.; Hupp, J.T. Dye sensitized solar cells: TiO_2 sensitization with a bodipy-porphyrin antenna system. *Langmuir* **2010**, *26*, 3760–3765. [CrossRef]
28. Urbani, M.; Grätzel, M.; Nazeeruddin, M.K.; Torres, T. Meso-substituted porphyrins for dye-sensitized solar cells. *Chem. Rev.* **2014**, *114*, 12330–12396. [CrossRef]
29. Klfout, H.; Stewart, A.; Elkhalifa, M.; He, H. BODIPYs for dye-sensitized solar cells. *ACS Appl. Mater. Interfaces* **2017**, *9*, 39873–39889. [CrossRef]
30. Islam, A.; Chowdhury, T.H.; Qin, C.; Han, L.; Lee, J.-J.; Bedja, I.M.; Akhtaruzzaman, M.; Sopian, K.; Mirloup, A.; Leclerc, N. Panchromatic absorption of dye sensitized solar cells by co-Sensitization of triple organic dyes. *Sustain. Energy Fuels* **2017**, *2*, 209–214. [CrossRef]
31. Cheema, H.; Younts, R.; Gautam, B.; Gundogdu, K.; El-Shafei, A. Design and synthesis of BODIPY sensitizers with long alkyl chains tethered to N-carbazole and their application for dye sensitized solar cells. *Mater. Chem. Phys.* **2016**, *184*, 57–63. [CrossRef]
32. Qin, C.; Mirloup, A.; Leclerc, N.; Islam, A.; El-Shafei, A.; Han, L.; Ziessel, R. Molecular engineering of new thienyl-bodipy dyes for highly efficient panchromatic sensitized solar cells. *Adv. Energy Mater.* **2014**, *4*, 1400085. [CrossRef]
33. Li, Z.; Mintzer, E.; Bittman, R. First synthesis of free cholesterol–BODIPY conjugates. *J. Org. Chem.* **2006**, *71*, 1718–1721. [CrossRef] [PubMed]
34. Pakhomov, A.A.; Kononevich, Y.N.; Stukalova, M.V.; Svidchenko, E.A.; Surin, N.M.; Cherkaev, G.V.; Shchegolikhina, O.I.; Martynov, V.I.; Muzafarov, A.M. Synthesis and photophysical properties of a new BODIPY-based siloxane dye. *Tetrahedron Lett.* **2016**, *57*, 979–982. [CrossRef]
35. Wanwong, S.; Sangkhun, W.; Wootthikanokkhan, J. The effect of co-sensitization methods between N719 and boron dipyrromethene triads on dye-sensitized solar cell performance. *RSC Adv.* **2018**, *8*, 9202–9210. [CrossRef]
36. Krauss, T.N.; Barrena, E.; Zhang, X.N.; de Oteyza, D.G.; Major, J.; Dehm, V.; Würthner, F.; Cavalcanti, L.P.; Dosch, H. Three-dimensional molecular packing of thin organic films of PTCDI-C_8 determined by surface X-Ray diffraction. *Langmuir* **2008**, *24*, 12742–12744. [CrossRef] [PubMed]
37. Dharma, J.; Pisal, A. *Simple Method of Measuring the Band Gap Energy Value of TiO2 in the Powder Form Using a UV/Vis/NIR Spectrometer*; Perkin Elmer: Shelton, CT, USA, 2012.
38. Jiang, X.; Li, S.; Xiang, G.; Li, Q.; Fan, L.; He, L.; Gu, K. Determination of the acid values of edible oils via FTIR spectroscopy based on the OH stretching band. *Food Chem.* **2016**, *212*, 585–589. [CrossRef]
39. Mao, M.; Song, Q.-H. The structure-property relationships of D-π-A BODIPY dyes for dye-sensitized solar cells. *Chem. Rec.* **2016**, *16*, 719–733. [CrossRef]
40. Mishra, A.; Fischer, M.K.R.; Bäuerle, P. Metal-free organic dyes for dye-sensitized solar cells: From structure: Property relationships to design rules. *Angew. Chem. Int. Ed.* **2009**, *48*, 2474–2499. [CrossRef]
41. Fan, K.; Yu, J.; Ho, W. Improving photoanodes to obtain highly efficient dye-sensitized solar cells: A brief review. *Mater. Horiz.* **2017**, *4*, 319–344. [CrossRef]

6

Analysis of Contact Reaction Phenomenon between Aluminum–Silver and p+ Diffused Layer for n-Type c-Si Solar Cell Applications

Cheolmin Park [1], Sungyoon Chung [2], Nagarajan Balaji [1], Shihyun Ahn [3], Sunhwa Lee [2], Jinjoo Park [4] and Junsin Yi [2,*]

1 Department of Energy Science, Sungkyunkwan University, Suwon 16419, Korea; cmpark8311@gmail.com (C.P.); balaji@skku.edu (N.B.)

2 College of Information and Communication Engineering, Sungkyunkwan University, Suwon 16419, Korea; sungyoun89@skku.edu (S.C.); ssunholic@skku.edu (S.L.)

3 Department of Electrical and Computer Engineering, University of Seoul, 163, Seoulsiripdae-ro, Dongdaemun-gu, Seoul 02504, Korea; shahn@uos.ac.kr

4 Major of Energy and Applied Chemistry, Division of Energy & Optical Technology Convergence, Cheongju University 298, Daeseong-ro, Cheongwon-gu, Cheongju-si, Chungcheongbuk-do, Cheongju 28503, Korea; jwjh3516@cju.ac.kr

* Correspondence: junsin@skku.edu

Abstract: In this study, the contact mechanism between Ag–Al and Si and the change in contact resistance (Rc) were analyzed by varying the firing profile. The front electrode of an n-type c-Si solar cell was formed through a screen-printing process using Ag–Al paste. Rc was measured by varying the belt speed and peak temperature of the fast-firing furnace. Rc value of 6.98 mΩ-cm^{-2} was obtained for an optimal fast-firing profile with 865 °C peak temperature and 110 inches per min belt speed. The contact phenomenon and the influence of impurities between the front-electrode–Si interface and firing conditions were analyzed through scanning electron microscopy (SEM) and energy-dispersive X-ray spectroscopy (EDS). The EDS analysis revealed that the peak firing temperature at 865 °C exhibited a low atomic weight percentage of Al (0.72 and 0.36%) because Al was involved in the formation of alloy of Si with the front electrode. Based on the optimal results, a solar cell with a conversion efficiency of 19.46% was obtained.

Keywords: metallization; contact formation; Ag/Al paste; p+ emitter; N-type bifacial solar cells

---

1. Introduction

The optimization of the characteristics of the front electrodes of solar cells is important in two major aspects—improving conversion efficiency and reducing production costs, which are the main aims while fabricating solar cells. In the solar cell industry, screen printing technologies occupy most of the market share of the metallization of front and rear sides [1–4]. Based on the advantages of manufacturing cost and throughput, the technology of the screen-printing process has continuously improved, and the technological improvement in conversion efficiency can be largely divided into two aspects. The first is minimizing the shading loss by applying fine-line printing to absorb as much light as possible [5], and the second is minimizing the series resistance to reduce electrical loss when the carriers generated by the absorbed light are collected at the electrode [6]. For fine-line printing, techniques such as changing the emulsion thickness and the application of a knotless screens were applicable to both p- and n-type substrates regardless of the substrate type [7,8].

However, when decreasing the series resistance (specifically contact resistance (Rc)), changes in substrate type are significant factors. For p-type substrates, an emitter is formed using phosphorus, and phosphorus atoms penetrate the phosphorus silicate glass (PSG) layer and do not out-diffuse. As a result, the phosphorus atoms accumulate on the surface; thus, the carrier concentration on the surface is the highest [9,10]. In contrast, for n-type substrates, boron atoms penetrate the borosilicate glass (BSG) and diffuses out, thereby forming a "deficient layer" with a relatively low boron concentration on the surface [11]. When a deficient layer is formed, there are limitations to reducing the Rc when using Ag paste. This is because the Rc (resistivity) and emitter surface concentration have the following formula [12]:

$$\rho_c = \rho_{co} exp\left(\frac{2\Phi_B}{\hbar}\sqrt{\frac{\varepsilon_s m^*}{N}}\right) ohm - cm^2 \quad (1)$$

where ρ_{co} is a constant dependent on the metal and semiconductor. The specific contact resistivity (ρ_c) primarily depends on the metal-semiconductor work function (Φ_B), doping density (N), and effective mass of the carrier (m^*). Therefore, the paste of the n-type substrate solar cell contains a certain amount of Al to compensate for the deficient layer. It is similar to the formation of the local back surface field (LBSF) of the p-type passivated emitter and rear cell (PERC) structure, where Al reacts with Si and heat to form a p+ region when an alloy is formed. A small amount of Al in the front paste aids in improving the properties of the front contact.

In this study, the contact formation between p+ diffused Si layers and Ag–Al paste with respect to the firing conditions were qualitatively and quantitatively analyzed using scanning electron microscopy (SEM) and energy-dispersive X-ray spectroscopy (EDS). Based on the analysis, we aimed at describing the bonding pattern according to the temperature using a schematic diagram. Finally, the output characteristics of the n-type c-Si solar cell fabricated under optimized conditions were analyzed.

2. Experimental Details

Figure 1 shows the schematic of the solar cell structure and process flow used in this study. Here, 200-μm thick n-type solar-grade Czochralski (Cz-Si) grown wafers with a crystal orientation of (100) and resistivity of 1.5 ohm-cm were used as the starting material for the sample to be analyzed and fabricated into a solar cell. To reduce the reflectance of the Si wafer, the sample was textured using 2% NaOH and 8.75% isopropanol at a temperature range of 84 to 86 °C. The textured wafers were doped with boron using a thermal diffusion process in a furnace using BBr_3 as the dopant source at 910 °C, 5 min pre-deposition, and 1000 °C, 7.5 min drive-in step. After the diffusion process, BSG was removed by dipping the wafer in a diluted hydrofluoric acid (HF) solution, followed by DI-water rinsing and drying. The emitter-sheet resistance was changed to 80 ohm/sq using an etch-back method to remove the boron-rich layer (BRL) using an $HF–HNO_3–CH_3COOH$ mixture. The n+ back surface field was formed by diffusing phosphorus with phosphoryl chloride ($POCl_3$) diffusion via 810 °C, 7 min pre-deposition and 880 °C, 5 min drive-in step using the dopant source.

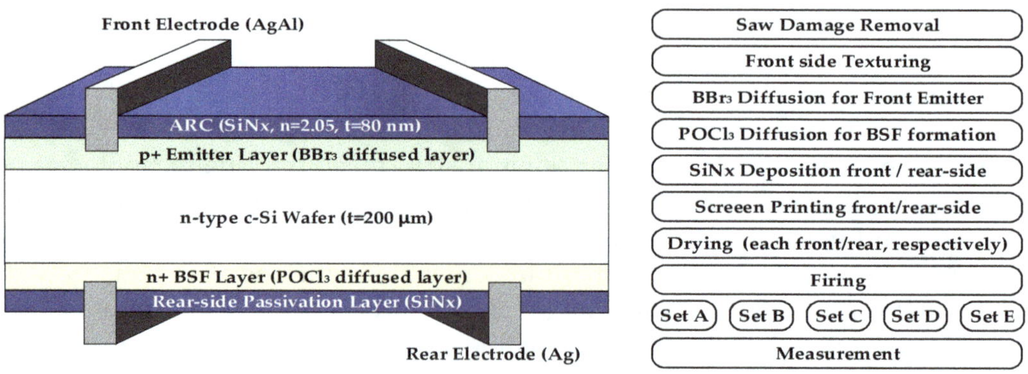

Figure 1. Schematic of the solar cell structure and process flow used in this study.

Then, 80-nm thick SiNx layers with a refractive index of 2.05 were deposited on the front side for both the passivation and anti-reflection coating layers using a plasma-enhanced chemical vapor deposition (PECVD) system with a plasma frequency of 13.56 MHz. In addition, a double-stacked SiNx layer with refractive indices 2.7 and 2.1 was deposited on the rear side for passivation and thermal stability characteristics. Metal electrodes were formed through screen printing using an Ag-Al mixture and Ag paste on the front and rear sides, respectively. After the screen printing process, the samples were dried at 150 °C for a duration of 4 min and subsequently co-fired in a four-zone infrared (IR) belt furnace at the test group temperatures with a peak temperature of 660–930 °C. Twenty-seven cells were produced for each group, and the size of each cell was 10.24 cm^2. The fast-firing profile designs of the experiment were grouped into five different sets, and the detailed characteristics of the co-firing groups are listed in Table 1.

Table 1. Fast-firing temperature profile groups of this study.

Co-Firing Temp Group	Temperature (°C)				Peak Temperature (°C)
	Zone 1	Zone 2	Zone 3	Zone 4	
Set A	545	670	570	400	593.4
Set B	545	735	570	400	677.8
Set C	545	800	570	400	719.6
Set D	545	865	570	400	775.0
Set E	545	930	570	400	849.5

For the analysis of the contact phenomenon, the atomic and weight percentages of Ag–Al–Si bonding were analyzed using SEM and EDS. When the fabrication of the solar cell was completed, the cells were characterized. The measurement was performed using illuminated current-voltage (LIV) characteristics (Pasan CT-801, Neuchâtel, Switzerland) under the global solar spectrum of AM 1.5 at 25 °C.

3. Results

Figure 2 shows the variation in R_C with respect to the variation in the peak firing temperature of the fast-firing process. A sample with the same structure as that of the completed cell was fabricated; subsequently, the transmission line (TLM) method pattern was printed, and the Rc was measured. As Figure 2a shows, the co-firing test group A had a very high R_C value, which can be attributed to the low peak firing temperature that was not sufficient to induce a reaction between the metal impurities in the screen-printed paste and the p+ diffused Si surface. Rc dramatically decreased from 181.82 $m\Omega \cdot cm^2$ to 6.98 $m\Omega \cdot cm^2$ for group D at a peak temperature of 865 °C. This revealed that an optimized co-firing temperature is required to initiate a reaction between metal impurities in screen-printing paste and Si atoms.

As the co-firing temperature for Set E was higher than that of Set D, R_C increased from 6.98 $m\Omega \cdot cm^2$ to 10.12 $m\Omega \cdot cm^2$. This increase in R_C may have been due to the variation in the reaction between metal impurities in the screen-printing paste and the Si surface caused by the over-fired electrode. Figure 2b shows the variation in Rc with respect to the variation in the co-firing belt speed. The highest Rc value of 39.19 $m\Omega \cdot cm^2$ was obtained at 60 ipm (inches per min). When the belt speed increased, the R_C value decreased to the lowest value of 6.98 $m\Omega \cdot cm^2$. The excessive co-firing process time effect on the output characteristics of the c-Si solar cell owing to the denaturalization of the metals and other compositions of the electrode. Additionally, the Rc value increased for an increase in belt speed of 110 ipm because of insufficient composition reactions between diffused layer and electrode over the optimal co-firing speed. The actual firing profile for Set D with the lowest Rc was measured using a thermal profiling system (Datapaq Solar Tracker Thermal Profiling System, Fluke, Everett, Washington, DC, USA), and the results are shown in Figure 2b. The measured peak temperature of set D was 775 °C.

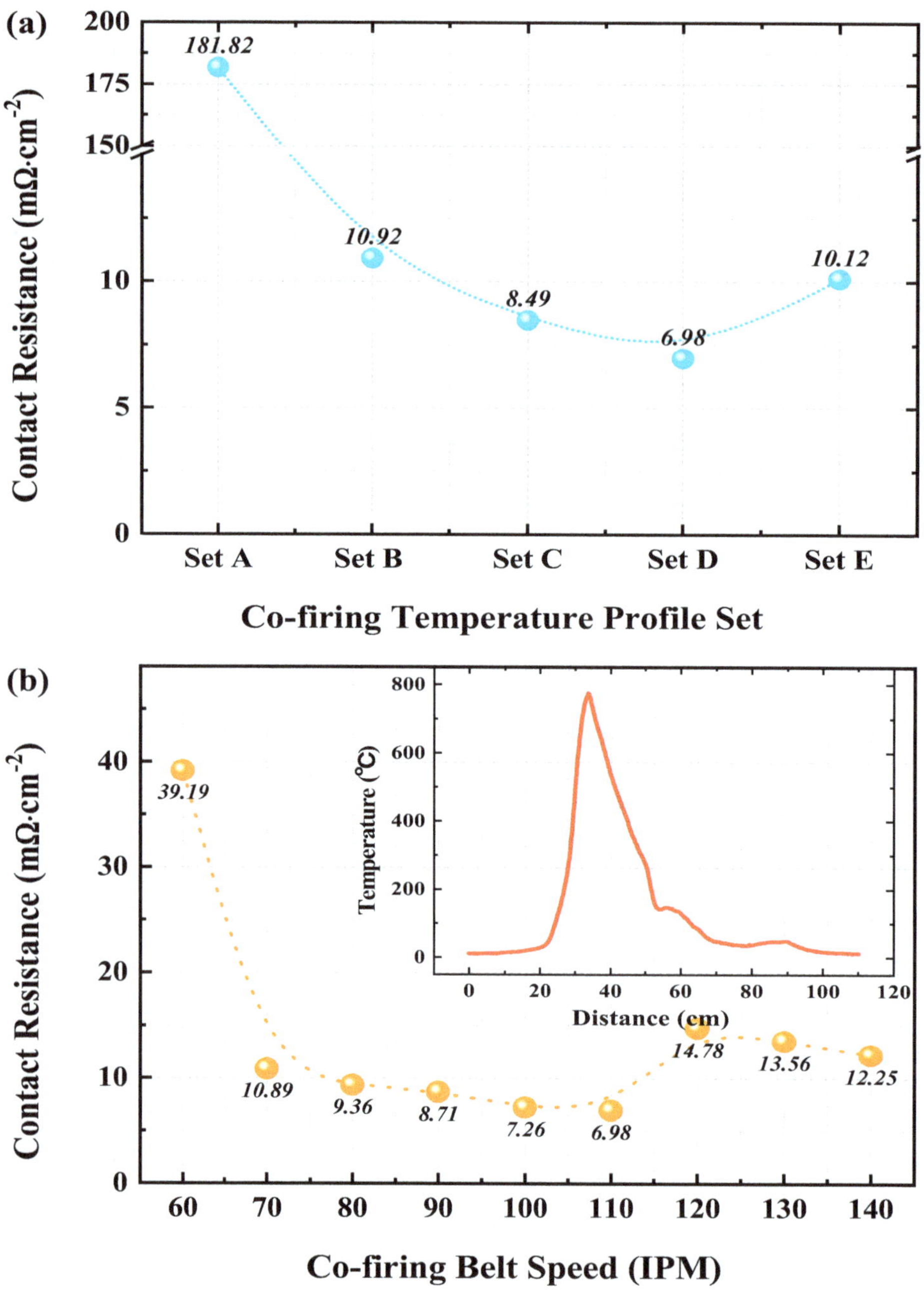

Figure 2. Analysis of R_C of front electrodes formed by the screen-printing process using Ag–Al paste with variation in the co-firing process condition: (**a**) peak temperature and (**b**) belt speed.

For further understanding of the characteristics of the contacts, the metal was etched, and the distribution of the crystallites was studied using SEM (Figure 3). To etch the metal contacts, the cells were dipped in mixed acid, and the glass layer was removed using HF. The removal of the interfacial glass layer made viewing the metal crystallites grown emitter surface easier. For sets A and B, the remaining metal paste was observed, which could be because of the low firing temperature profile. The firing temperature increased, and the color of the metal paste changed. This indicated that the reaction between the metal paste and Si was effective. For set E, the color of the metal paste was darker, which might have been due to over-firing between Si and the metal paste.

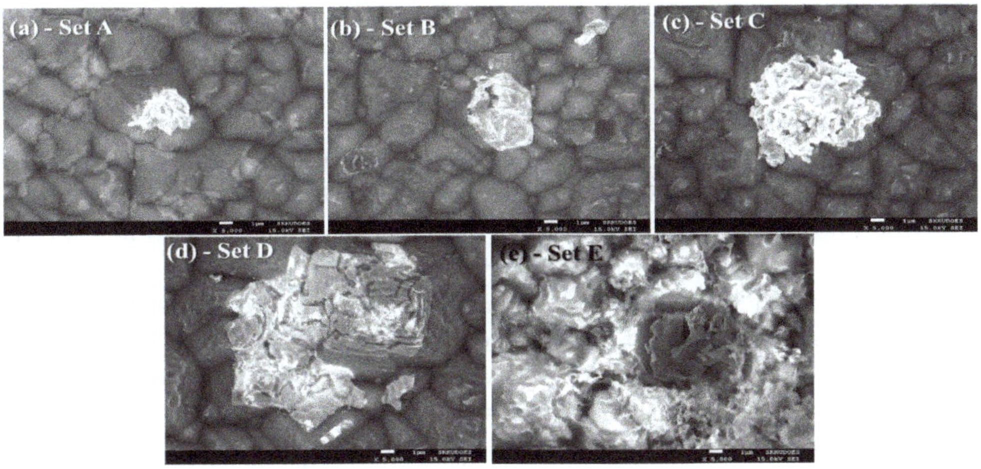

Figure 3. Scanning electron microscopy (SEM) images for the transition state of formed electrode between Si and the front-side electrode when the co-firing peak temperature profile set varied from (**a**) Set A to (**e**) Set E.

The at. wt % of the metals present on the surface were analyzed using EDS. By the electron beam of the SEM, each material emits its own specific X-ray during the sequence in which the electrons in the atom absorb and emit energy. The unique X-ray of each material inside the Ag–Al paste was collected by a detector and classified by intensity to perform a qualitative analysis (atomic/weight percentage) of the front electrode. Figure 4 shows the at. wt % of O, Al, Si, and Ag with variation in the co-firing peak temperature. The EDS results indicated that the Ag content varied dramatically with the co-firing temperature. The at. wt % Si decreased as the co-firing temperature increased, revealing that glass frit and other metal atoms in the metal paste reacted with Si. For set A, as a special scenario, because the co-firing temperature was not sufficient to react with the Si substrate and Ag–Al Paste, most of the paste components disappeared in the HF solution, resulting in a low percentage of Ag and Al and a high proportion of Si. In contrast, the atomic and weight percentages of Ag increased from 16.86% to 34.10% and 46.06% to 67.11%, respectively, as the co-firing temperature increased. Ag atoms were bound to the Si surface after reaction with the glass frit in the metal paste. This determined the binding quality and electrode formation quality between the screen-printed metal paste and the p+ diffused surface of c-Si solar cells.

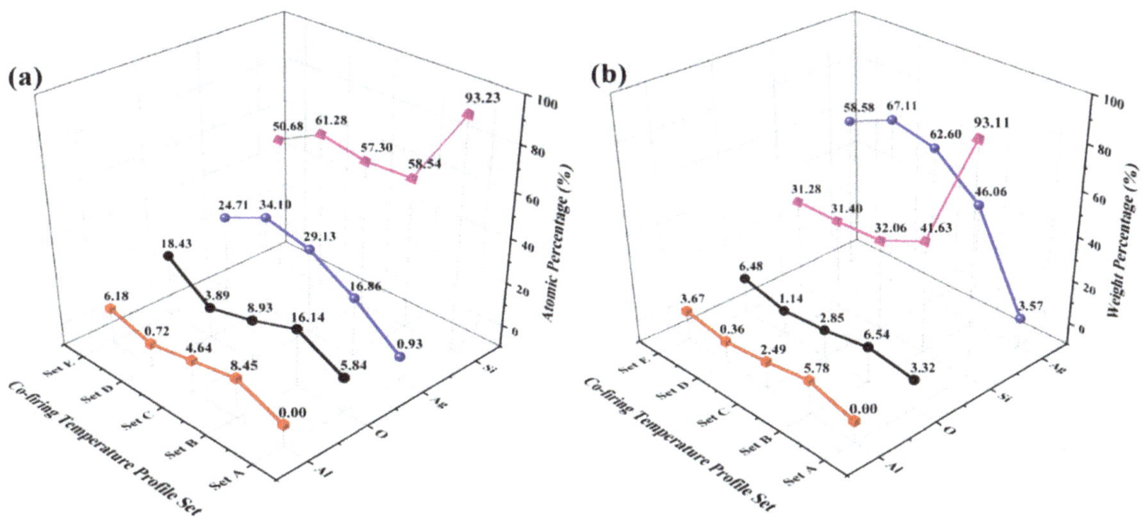

Figure 4. Energy-dispersive X-ray spectroscopy (EDS) analysis of the (**a**) atomic and (**b**) weight percentages of O, Al, Si, and Ag for the co-firing experimental sets A to E.

The at. wt % of Al exhibited a different trend compared with Ag and Si. This indicated that the Al content was bound to the p+ diffused surface with Ag in the Ag–Al metal paste as the co-firing temperature increased. When the co-firing temperature exceeded the optimized process window (>865 °C), Al remained in the electrode and surface because of over firing. The same behavior was observed in the weight percent analysis as the co-firing temperature varied. The difference in the percentage trend of Al in the EDS analysis results will be explained later in this section, but it can be considered as a difference in the contribution of Al to the formation of contact between Ag and Si. For set D, which had the best contact properties, the weight and atomic percentages of Al were the lowest (set A is excluded from this discussion). For set E, which had a higher temperature, Al melted rapidly in the eutectic phase and subsequently cooled, causing Al to escape and form additional voids. A similar pattern was observed in the Al-BSF of the PERC structure [13]. The four-element analysis of the variation in the atomic and weight percentages resulted in contact characteristics between the p+ surface and screen-printed Ag–Al paste determined by the binding ratio of the weight and atomic percentages of Si, Ag, and Al.

Figure 5 shows the SEM images of the alloy formation between the Ag–Al paste and the textured surface for the co-firing profile sets D and E. For the set E sample (Figure 5b), the region between the textured surface and the metal alloy electrode had more voids owing to the over-fired co-firing temperature. A larger void between the pyramid surface and metal electrode decreased the contact characteristics, narrowing down the current path to the electrode. In contrast, for set D, no void region (Figure 5a) occurred between the pyramid surface and metal alloy electrode region. This decreased the current loss.

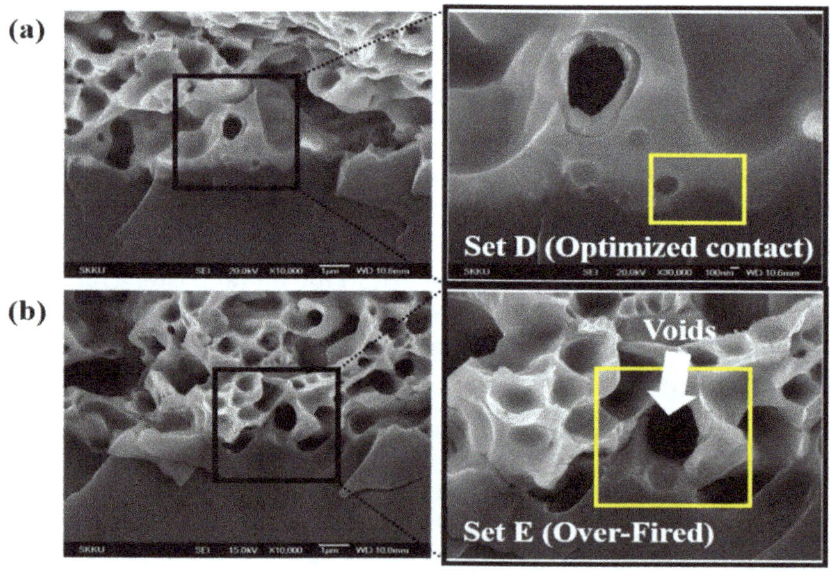

Figure 5. SEM images of the cross-sections of metal electrode formed on a pyramid structure for co-firing temperature profile for (**a**) set D and (**b**) set E.

Figure 6 depicts the reaction diagram of the metal electrode paste and p+ (boron) diffused layer on the Si substrate during the co-firing process after the front and rear-side electrodes were screen-printed. The glass etched the antireflection coating, reduced the melting point of the metal, and promoted adhesion to Si during the firing cycle. The melted glass frits dissolved both Ag and Si. Ag was observed to dissolve in the glass frit when firing occurred. As the glass cooled, the Si regrew epitaxially and Ag crystallites grew into the Si surface from the glass. As described above, the reaction mechanism of the Ag–Al paste had the same pattern until the initial reaction with the general Ag-paste reaction of the p-type substrate excluding Al role in reaction phenomenon. In each firing temperature scenario, Al reacted differently from the p+ diffused and dielectric layers. For Set B, the width of Al and the p+ diffused layer Si alloy (W_{p+}) was shallower than those in the other co-firing profiles because of insufficient reaction energy from thermal temperature. The relationship between W_{p+}, Al, and Si is expressed as [14–16]

$$W_{p+} = \frac{t_{Al} \cdot \rho_{Al}}{\rho_{Si}} \left[\frac{F(T)}{1 - F(T)} - \frac{F(T_o)}{1 - F(T_o)} \right] \tag{2}$$

where t_{Al} is the as-deposited Al thickness, ρ_{Al} and ρ_{Si} are the densities of Al and Si, respectively, $F(T)$ is the at. wt % of the molten phase of Si at the peak alloying temperature, and $F(T_o)$ is the at. wt % of Si at the eutectic temperature (constant). $F(T)$ and $F(T_o)$ were obtained from the Al–Si phase diagram. Thus, the Al–Si atomic percentage determined W_{p+}, and it had a role in the additional current path and locally electrical field similar to that of the Al–BSF layer in conventional p-type Si solar cells. Sets D and E exhibited optimized and over-fired contact phenomena after the co-firing process (Figure 6e). In the overfired scenario, an excessive-W_{p+} layer was formed by a continuous reaction between Al and the p+ diffused layer during the high-temperature co-firing process. In addition, some voids were created at the boundary of the metal paste and the p+ diffused surface by the high reaction temperature and continuous reaction with Ag, Al, Si, and other materials in the electrode metal paste. Consequently, the p–n junction was locally damaged by the excessive-W_{p+} layer, decreasing the device conversion efficiency [17].

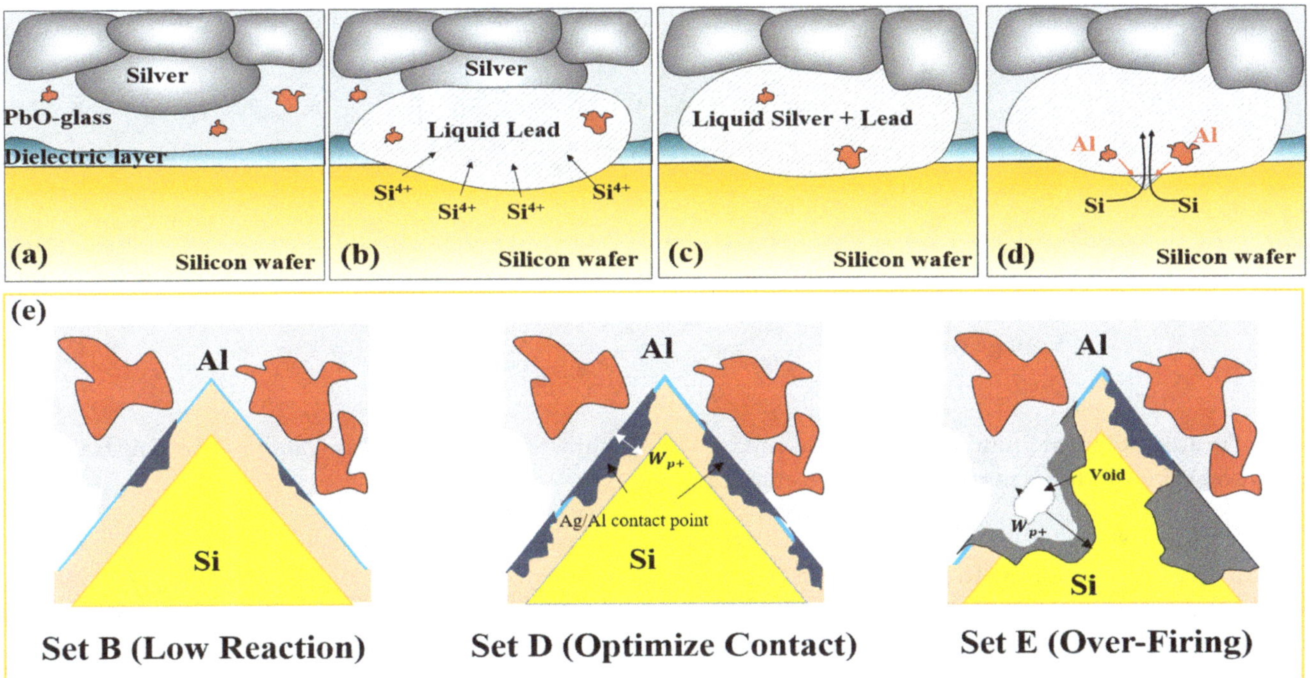

Figure 6. (**a–d**) Schematic of the reaction between the Ag–Al paste and the p+ diffused layer as the co-firing process progressed. (**e**) Surface reaction when firing temperature varied for sets B, D, and E.

Figure 7 shows the LIV results of the solar cell fabricated using a co-firing temperature profile set. For set A, the conversion efficiency of the solar cell is not measured because the contact between the front electrode and the p+ diffused layer was not established, as indicated in the SEM/EDS analysis of the electrical properties of the contact. Based on the analysis of the contact properties, the co-firing profile of the other sets were observed to form a sufficient contact characteristic between the p+ diffused layer and the front electrode, unlike that of set A. Although the output characteristics of the solar cell differed, unlike for set A, the conversion efficiency was not measured to be approximately zero. In Set D, which exhibited the best characteristics in terms of Rc, the current density (J_{SC}), open-circuit voltage (V_{OC}), and fill factor (FF) exhibited the best characteristics among the remaining groups, with a conversion efficiency of 19.46%. Moreover, for set E, the Rc value exhibited good properties compared to set B; hence, the fill factor was 60.03%, which was 12% better than that of set B (53.46%). However, owing to the local damage of the p–n junction (as described in Figure 5), an open-circuit voltage characteristic of 471 mV was observed, which was 17% lower than that of set B. The reason why it can be judged that the damage of the p–n junction by Al in the front paste affected the output characteristics of the solar cell, This is because the higher the peak temperature in the contact between the n+ diffused (phosphorus) layer on the back side and the Ag paste, the better it is in terms of contact resistance and fill factor [18].

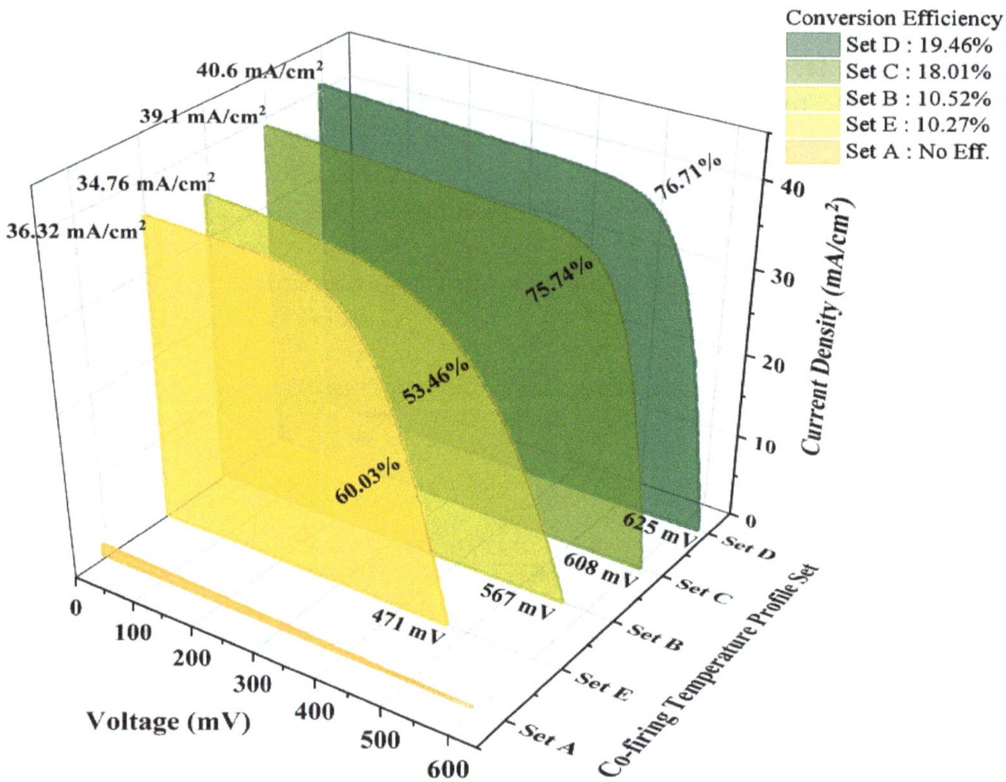

Figure 7. Output illumination current-voltage (LIV) characteristics of the fabricated n-type substrate solar cell for different co-firing temperature profile sets.

Figure 8 shows the loss analysis performed to determine which parameter decreased the output characteristics for the co-firing temperature profile sets B to E (set A was excluded from this analysis owing to low cell output properties in the LIV characteristics). The loss analysis was performed using the software Quokka 3 (March, Germany) technology computer-added design (TCAD) [19–21]. The Rc values obtained from the results shown in Figure 2 were input into the TCAD program. Table 2 lists the detailed TCAD simulation parameters.

In this TCAD simulation analysis, five parameters were selected for the loss analysis: the "front electrode recombination", which applied the characteristics of the front electrode loss such as the finger and busbar region; "lateral carrier transport", which was based on the quality of the interface trap density of the front and rear passivation layer; "current transport loss (resistive)", indicating the contact characteristics between the p+ diffused layer and front electrode; and the "SRH" and "Auger recombination" parameters, related to the recombination properties of the p+ diffused layer.

For set D, the current transport loss had a value of 1.464 mW/cm^2, while the other groups had a value from 1.65 (set C) to 5.52 mW/cm^2 (set B). Because the TCAD simulation was based on the same cell structure, most of the current transport loss could be estimated as the current transfer loss between the p+ diffused layer and the front electrode. In addition, the SRH and Auger recombinations, which represented recombination loss in the p+ diffused layer, exhibited the same loss percentage difference in all groups except set E. The ratio of current transport (resistive) loss accounted for approximately two-thirds of the total energy loss; therefore, it formed the major part of the key loss parameter.

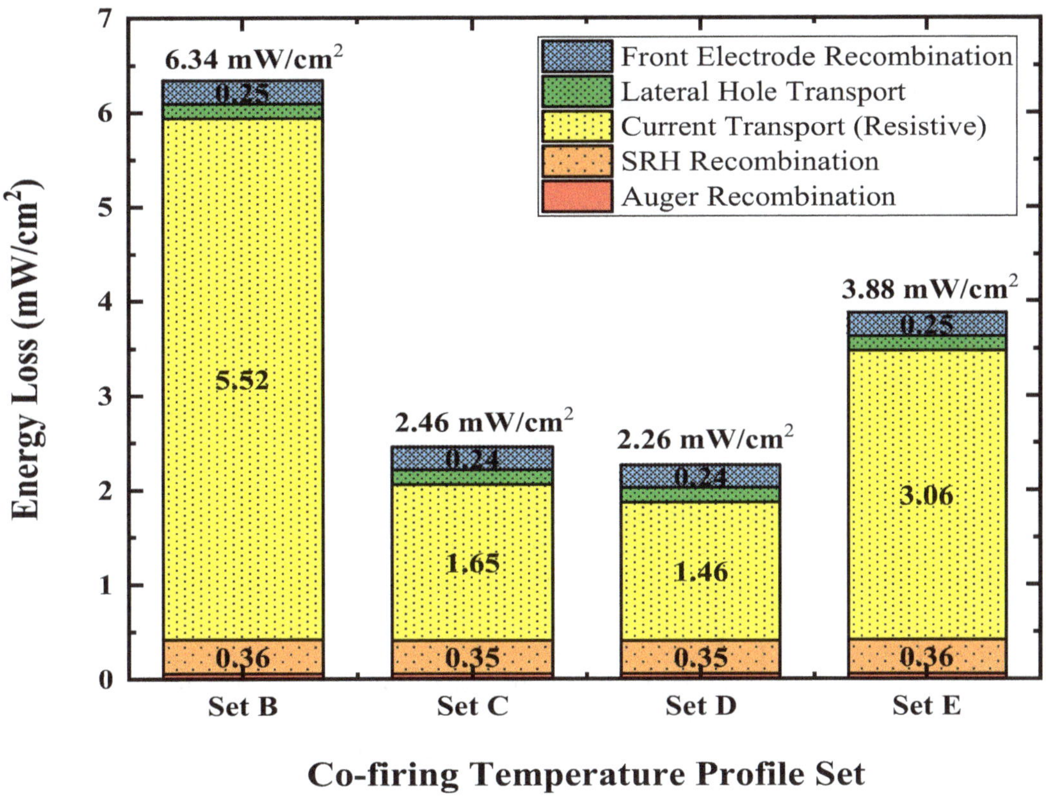

Figure 8. Energy loss analysis of recombination components for co-firing temperature profile sets using a TCAD tool.

Table 2. Detailed TCAD parameters for free energy loss analysis by co-firing temperature profile set.

TCAD Parameter	Set A	Set B	Set C	Set D	Set E
Wafer thickness (nm)			170		
Wafer type			n-type		
Wafer resistivity (Ω cm)			1.0		
Bulk mid-gap SRH lifetimes (μs)			1000 ($\tau_n = \tau_p$)		
p+ diffused emitter R_{sheet} (Ω cm^{-2})			80		
Front j_0: non-contact region(A/cm^2)			30.7×10^{-15}		
Front finger width (μm)			50		
Front contact resistivity (mΩ cm^{-2})	181.82	10.92	8.49	6.98	10.12
Rear local BSF R_{sheet} (Ω cm^{-2})			130		
Rear contact width (μm)			300		
Rear SRV: non-contact region (cm/s)			10		

4. Conclusions

In this study, the variation in the contact properties of the front electrode with the co-firing temperature was studied using SEM/EDS analysis. The proposed profile sets were applied to the production of solar cells to analyze changes in the output characteristics of the solar cells. For the temperature profile of 545, 865, 570, and 400 °C and 110 ipm, the optimal Rc value was 6.98 mΩ-cm^{-2}. This was confirmed by the Al alloy formed between the front electrode and the p+ diffused layer. When the firing temperature was low, the Al alloy did not react sufficiently with the p+ diffused layer. When the firing temperature exceeded the optimum point, it caused damage to the p+ diffused layer and created a void that affected the quality of the p–n junction as well as the current path.

Author Contributions: Conceptualization and writing—original draft preparation, C.P.; Data curation, S.C.; Writing—review and editing, N.B., S.A., S.L. and J.P.; Supervision, J.Y. All authors have read and agreed to the published version of the manuscript.

References

1. *International Technology Roadmap for Photovoltaic (ITRPV)*, 10th ed.; VDMA Photovoltaic Equipment: Frankfurt, Germany, 2019.
2. Lv, Y.; Zhuang, Y.F.; Wang, W.J.; Wei, W.W.; Sheng, J.; Zhang, S.; Shen, W.Z. Towards high-efficiency industrial p-type mono-like Si PERC solar cells. *Sol. Energy Mater. Sol. Cells* **2020**, *204*, 110202. [CrossRef]
3. Yin, H.P.; Tang, K.; Zhang, J.B.; Shan, W.; Huang, X.M.; Shen, X.D. Bifacial n-type silicon solar cells with selective front surface field and rear emitter. *Sol. Energy Mater. Sol. Cells* **2020**, *208*, 110345. [CrossRef]
4. Chen, D.; Chen, Y.; Wang, Z.; Gong, J.; Liu, C.; Zou, Y.; He, Y.; Wang, Y.; Yuan, L.; Lin, W.; et al. 24.58% total area efficiency of screen-printed, large area industrial silicon solar cells with the tunnel oxide passivated contacts (i-TOPCon) design. *Sol. Energy Mater. Sol. Cells* **2020**, *206*, 110258. [CrossRef]
5. Zhang, C.; Shen, H.; Sun, L.; Yang, J.; Wu, S.; Lu, Z. Bifacial p-Type PERC Solar Cell with Efficiency over 22% Using Laser Doped Selective Emitter. *Energies* **2020**, *13*, 1388. [CrossRef]
6. Musztyfaga-Staszuk, M.; Janicki, D.; Panek, P. Correlation of Different Electrical Parameters of Solar Cells with Silver Front Electrodes. *Materials* **2019**, *12*, 366. [CrossRef] [PubMed]
7. Tepner, S.; Wengenmeyr, N.; Ney, L.; Linse, M.; Pospischil, M.; Clement, F. Improving Wall Slip Behavior of Silver Pastes on Screen Emulsions for Fine Line Screen Printing. *Sol. Energy Mater. Sol. Cells* **2019**, *200*, 109969. [CrossRef]
8. Tepner, S.; Ney, L.; Linse, M.; Lorenz, A.; Pospischil, M.; Clement, F. Studying Knotless Screen Patterns for Fine-line Screen Printing of Si-Solar Cells. *IEEE J. Photovolt.* **2020**, *10*, 319–325. [CrossRef]
9. Wagner, H.; Dastgheib-Shirazi, A.; Min, B.; Morishige, A.E.; Steyer, M.; Hahn, G.; del Cañizo, C.; Buonassisi, T.; Altermatt, P.P. Optimizing phosphorus diffusion for photovoltaic applications: Peak doping, inactive phosphorus, gettering, and contact formation. *J. Appl. Phys.* **2016**, *119*, 185704. [CrossRef]
10. Li, H.; Ma, F.-J.; Hameiri, Z.; Wenham, S.; Abbott, M. An Advanced Qualitative Model Regarding the Role of Oxygen During POCl3 Diffusion in Silicon. *Phys. Status Solidi Rapid Res. Lett.* **2017**, *11*, 1700046. [CrossRef]
11. Tsutsui, K.; Matsuda, T.; Watanabe, M.; Jin, C.-G.; Sasaki, Y.; Mizuno, B.; Ikenaga, E.; Kakushima, K.; Ahmet, P.; Maruizumi, T.; et al. Activated boron and its concentration profiles in heavily doped Si studied by soft x-ray photoelectron spectroscopy and Hall measurements. *J. Appl. Phys.* **2008**, *104*, 93709. [CrossRef]
12. Scorzoni, A.; Finetti, M. Metal/Semiconductor contact resistivity and its determination from contact resistance measurements. *Mater. Sci. Rep.* **1988**, *3*, 79–137. [CrossRef]
13. Horbelt, R.; Hahn, G.; Job, R.; Terheiden, B. Void Formation on PERC Solar Cells and Their Impact on the Electrical Cell Parameters Verified by Luminescence and Scanning Acoustic Microscope Measurements. *Energy Procedia* **2015**, *84*, 47–55. [CrossRef]
14. Meemongkolkiat, V.; Hilali, M.; Rohatgi, A. Investigation of RTP and belt fired screen printed AL-BSF on textured and planar back surfaces of silicon solar cells. In Proceedings of the 3rd World Conference on Photovoltaic Energy Conversion, Osaka, Japan, 11–18 May 2003; Volume 1462, pp. 1467–1470.
15. Krause, J.; Woehl, R.; Rauer, M.; Schmiga, C.; Wilde, J.; Biro, D. Microstructural and electrical properties of different-sized aluminum-alloyed contacts and their layer system on silicon surfaces. *Sol. Energy Mater. Sol. Cells* **2011**, *95*, 2151–2160. [CrossRef]
16. Muller, J.; Bothe, K.; Gatz, S.; Plagwitz, H.; Schubert, G.; Brendel, R. Contact Formation and Recombination at Screen-Printed Local Aluminum-Alloyed Silicon Solar Cell Base Contacts. *IEEE Trans. Electron Dev.* **2011**, *58*, 3239–3245. [CrossRef]
17. Urban, T.; Krügel, K.; Heitmann, J. Formation of Ag-Al Alloy in context of PERC solar cell metallization. *Energy Procedia* **2017**, *124*, 930–935. [CrossRef]
18. Rudolph, D.; Olibet, S.; Hoornstra, J.; Weeber, A.; Cabrera, E.; Carr, A.; Koppes, M.; Kopecek, R. Replacement of Silver in Silicon Solar Cell Metallization Pastes Containing a Highly Reactive Glass Frit: Is it Possible? *Energy Procedia* **2013**, *43*, 44–53. [CrossRef]
19. Fell, A.; Schön, J.; Schubert, M.C.; Glunz, S.W. The concept of skins for silicon solar cell modeling. *Sol. Energy Mater. Sol. Cells* **2017**, *173*, 128–133. [CrossRef]

20. Fell, A.; Altermatt, P.P. A Detailed Full-Cell Model of a 2018 Commercial PERC Solar Cell in Quokka3. *IEEE J. Photovolt.* **2018**, *8*, 1443–1448. [CrossRef]
21. Fell, A.; Schon, J.; Muller, M.; Wohrle, N.; Schubert, M.C.; Glunz, S.W. Modeling Edge Recombination in Silicon Solar Cells. *IEEE J. Photovolt.* **2018**, *8*, 428–434. [CrossRef]

7

Improving Electron Extraction Ability and Device Stability of Perovskite Solar Cells using a Compatible PCBM/AZO Electron Transporting Bilayer

Hang Dong, Shangzheng Pang, Yi Zhang *, Dazheng Chen, Weidong Zhu, He Xi, Jingjing Chang, Jincheng Zhang, Chunfu Zhang * and Yue Hao

State Key Discipline Laboratory of Wide Band Gap Semiconductor Technology, School of Microelectronics, Xidian University, 2 South Taibai Road, Xi'an 710071, China; donghangxd@163.com (H.D.); 18717344558@163.com (S.P.); dzchen@xidian.edu.cn (D.C.); wdzhu@xidian.edu.cn (W.Z.); hxi@xidian.edu.cn (H.X.); jjingchang@xidian.edu.cn (J.C.); jchzhang@xidian.edu.cn (J.Z.); yhao@xidian.edu.cn (Y.H.)

* Correspondence: hyper_sys@163.com (Y.Z.); cfzhang@xidian.edu.cn (C.Z.)

Abstract: Due to the low temperature fabrication process and reduced hysteresis effect, inverted p-i-n structured perovskite solar cells (PSCs) with the PEDOT:PSS as the hole transporting layer and PCBM as the electron transporting layer have attracted considerable attention. However, the energy barrier at the interface between the PCBM layer and the metal electrode, which is due to an energy level mismatch, limits the electron extraction ability. In this work, an inorganic aluminum-doped zinc oxide (AZO) interlayer is inserted between the PCBM layer and the metal electrode so that electrons can be collected efficiently by the electrode. It is shown that with the help of the PCBM/AZO bilayer, the power conversion efficiency of PSCs is significantly improved, with negligible hysteresis and improved device stability. The UPS measurement shows that the AZO interlayer can effectively decrease the energy offset between PCBM and the metal electrode. The steady state photoluminescence, time-resolved photoluminescence, transient photocurrent, and transient photovoltage measurements show that the PSCs with the AZO interlayer have a longer radiative carrier recombination lifetime and more efficient charge extraction efficiency. Moreover, the introduction of the AZO interlayer could protect the underlying perovskite, and thus, greatly improve device stability.

Keywords: perovskite solar cells; aluminum-doped zinc oxide (AZO); electron transporting bilayer; stability

---

1. Introduction

Despite uncertainties regarding device stability and the usage of lead, metal halide perovskite solar cells (PSCs) have attracted increasing attention from both academia and industry due to their unprecedented properties, such as high absorption coefficients, long charge carrier diffusion lengths, and solution processing approach, since their first use as the active layer for photoelectron chemical cells in 2009 [1,2]. In the past few years, the power conversion efficiency (PCE) of PSCs has been rising dramatically, from 3.8% [3] to present values which are higher than 23% [4]. So far, two main device architectures have been used to fabricate PSCs. One is named the n-i-p structure, which usually involves depositing the perovskite material onto transparent substrates covered with a compact TiO_2 electron transport layer (ETL) and an optional mesoporous TiO_2 (or Al_2O_3) scaffold layer [5,6]. The other is named the p-i-n structure, which involves depositing the perovskite material onto transparent substrates which are covered with a hole transport layer (HTL), such as the poly(3,4-ethylenedioxythiophene):polystyrene sulfonic acid (PEDOT:PSS) [7,8]. High temperature annealing is usually necessary for the n-i-p structure to attaim high quality TiO_2 layers compared to the p-i-n structure, which could increase the production cost, and prevent its use in flexible substrates and multi-junction device architectures. As an alternative approach, the p-i-n structure is being used in this paper with PEDOT:PSS as the HTL and [6,6]-phenyl-C61-butyric acid methyl ester (PCBM) as the ETL, due to the low temperature fabrication process and reduced hysteresis effects [7,8].

In order to improve the PCE of p-i-n PSCs, many film growth methods such as the one step method [8], two step sequential deposition method [9–11], and the mixed-solvent-vapor annealing method [7] have been developed to achieve a highly uniform, dense, and pin-hole free perovskite films, which can yield more electron-hole pairs upon illumination with light, and thereby reduce the energy loss induced by recombination. With the continuous quality improvement of perovskite films, another issue in regard to interfaces in the PSCs is becoming more and more important, i.e., the p-i-n structure requires that the ETL be deposited on top of the perovskite layer. PCBM as a typical ETL is usually used in this structure. However, if the cathode electrode (Ag, Al) is evaporated onto the PCBM layer directly, there is always an energy barrier at the interface between the PCBM layer and the metal electrode due to the energy level mismatch [12]. Such a PCBM/metal electrode is not optimized for the electron extraction [12]. Therefore, in order to improve the device performance, a method of interface engineering between the PCBM layer and the metal electrode whereby additional layers are inserted has been proposed. In the past few years, some attempts have been proposed to improve the electron extraction properties between the PCBM layer and the metal electrode. For example, by inserting the interlayer of LiF between PCBM and electrode, Seo et al. achieved a PSC with a PCE of 14.1% for a unit cell and 8.7% for the module [13]. By using the polyethylenimine ethoxylated (PEIE) interlayer between the PCBM and electrode, Yang et al. demonstrated a high performance planar heterojunction PSC with a PCE of 14.82% [12]. By inserting the Ca between the PCBM and electrode, Chiang et al. achieved the best PCE, 16.31% [14]. By using the C_{60} and bathocuproine (BCP) interface modified layer, an optimal PCE of 17.9% was obtained [15]. By inserting PFN (poly [(9,9-bis(30-(N,N-dimethylamino)propyl)-2,7-fluorence)-alt-2,7-(9,9-dioct ylfluorene)]) interlayer between the PCBM and electrode, You et al. got an improved PCE of 17.1% [16]. All of these have shown that inserting the interlayer between the PCBM and electrode is an effective way to improve device performance. However, the interlayer materials such as the low workfunction metal (Ca) or organic materials are usually not stable enough. Thus, other interface materials are required. There are many n-type metal oxides such as zinc oxide (ZnO) [17,18], titanium oxide (TiO_x) [19,20], and tin oxide (SnO_2) [21–23], which has a low work function, and improves the electron extraction ability when used as the interlayer. However, in order to achieve good material quality, these metal oxides usually require a high temperature process, which is not suitable for the p-i-n structure in PSCs because the high process temperature will destroy the underlying perovskite materials. Thus, the means by which a high quality metal oxide may be obtained at a low temperature becomes important for the p-i-n structure in PSCs.

In this work, we adopt a room temperature solution processed Al-doped ZnO (AZO) as the interlayer between the PCBM and Ag electrode to improve device performance. AZO is a wide bandgap material with beneficial properties such as low workfunciton, high electron mobility, high optical transparency, and low-cost [24]. By using the AZO interlayer, the fabricated PSC shows an improved performance with low hysteresis and enhanced device stability. By comparing to devices without the AZO interlayer, it was found that the PSC with the AZO interlayer has a longer radiative carrier recombination lifetime. This may be attributed to the reduction of the energy mismatch between the PCBM layer and Ag electrode by introducing the AZO interlayer, which improved the electron extraction ability. By using the PCBM/AZO bilayer, a PSC with a short current density (Jsc) of 22.82 mA/cm^2, an open circuit voltage (Voc) of 0.99 V, a fill factor (FF) of 71.68%, and PCE of 16.18% was achieved. Moreover, the AZO interlayer can improve the stability of the device, with the PCE of the best device remaining 86.41% of its initial value after storing the device for over 720 h.

2. Materials and Methods

2.1. Materials

All solvents and reagents, Aluminum-doped zinc oxide nanoparticle ink (AZO, 2.5 wt.%, Sigma-Aldrich, Saint Louis, MI, USA), Methylammonium iodide (MAI, 99.8%, Dyesol, Queanbeyan, Australia), Formamidinium iodide (FAI, 99.8%, Dyesol, Queanbeyan, Australia), Lead iodide (PbI_2, 99.999%, Sigma-Aldrich, Saint Louis, MI, USA), Lead chloride ($PbCl_2$, 99.999%, Sigma-Aldrich, Saint Louis, MI, USA), Phenyl-C61-butyric acid methyl ester (PCBM, 98%, Nano-c, Westwood, MA, USA), Poly(3,4-ethy-lenedioxythiophene) Poly(styrenesulfonate) (PEDOT: PSS, Clevios PVP Al 4083, Hanau, Germany), N,N'-Dimethylformamide (DMF, 99.8%, Aladdin, Beijing, China), Chlorobenzene (CB, 99.8%, Sigma-Aldrich, Saint Louis, MI, USA), and Isopropanol (IPA, 99.5%, Sigma-Aldrich, Saint Louis, MI, USA), unless stated otherwise, are used as received without further purification.

2.2. Film Formation and Device Fabrication

The planer PSCs were fabricated on pre-patterned ITO glass substrates (10 Ω per square, around 2×2.5 cm^2 in size, Zhuhai Kaivo, Zhuhai, China). The patterned ITO glass substrates were sequentially ultrasonic cleaned with 5% decon-90 solution, de-ionized water, acetone, de-ionized water, alcohol at 50 °C for 20 min, respectively. Then the ITO substrates were dried with nitrogen and cleaned in a UV ozone oven for 30 min. A thin layer of PEDOT:PSS was spin-coated on the substrates at 7000 rpm for 45 s, and annealed at 150 °C for 15 min. After that, the substrates were transferred into a nitrogen-filled glovebox. To make a uniform perovskite layer, a perovskite precursor solution consisting of 1.36 M PbI_2 and 0.24 M $PbCl_2$ in the solvent of DMF (named PbX_2 solution) was stirred for 2 h at 75 °C, and 70 mg MAI and 30 mg FAI were dissolved in the solvent of IPA for the late use. Around 60 μL PbX_2 precursor solution pre-heated to 75 °C was transferred by pipettes to the PEDOT:PSS covered ITO substrates. Briefly, the spin coating process was programmed to run at 3000 rpm for 45 s. Then MAI and FAI mixed solution was spin-coated on top of the dried PbX_2 layer at room temperature at 3000 rpm for 45 s. All of the films were thermally annealed on the hotplate at 100 °C for 10 min. Next, a layer of PCBM (20 mg/mL in chlorobenzene) was spin-coated on the top of the perovskite layer at 2000 rpm for 45 s. After that, the AZO solution (8 mg/mL in IPA) was spin-coated on the top of the PCBM layer at 6000 rpm for 45 s, and the thickness of AZO film is about 90 nm measured by Stylus Profiler (Bruker Dektak XT, Bremen, Germany). The devices were finished by thermally evaporated 100 nm Ag. All the devices had an effective area of 7 mm^2.

2.3. Device Characterization

The morphologies of the perovskite layers were measured by scanning electron microscopy (SEM) (JSM-7800F, JEOL Ltd., Tokyo, Japan) and atomic force microscopy (AFM) (Agilent 5500, Santa Clara, CA, USA). X-ray diffraction (XRD) test was conducted on X' Pert Pro XRD (Bruker Optics, Ettlingen, Germany) and the samples were prepared as the same process of device fabrication. UPS measurements were acquired with a VG ESCA 220i-XL system (VG Instruments, Manchester, UK). The UV source was a He discharge lamp with a photon energy of 21.2 eV. The UV–visible absorption spectra were recorded with an UV–visible spectrophotometer (Perkin-Elmer Lambda 950, Waltham, MA, USA). Photovoltaic performances were measured by using a Keithley 2400 source meter (Tektronix, Inc., Beaverton, OR, USA) under simulated sunlight from XES-70S1 solar simulator (SEN-EI Electric. Co. Ltd, Osaka, Japan) matching the AM 1.5 G standard with an intensity of 100 mW/cm^2. The system was calibrated against a NREL certified reference solar cell. Incident photo-to-current conversion efficiencies (IPCEs) of PSCs were measured by the solar cell quantum efficiency measurement system (SCS10-X150, Zolix instrument. Co. Ltd, Beijing, China). Transient photocurrent measurement was performed with a system excited by a 532 nm (1000 Hz, 3.2 ns) pulse laser. Transient photovoltage measurement was performed with the same system excited by a 405 nm (50 Hz, 20 ms) pulse laser. A digital oscilloscope (Tektronix, D4105, Beaverton, OR, USA) was used to record the photocurrent or photovoltage decay process with a sampling resistor of 50 Ω or 1 MΩ, respectively. The thickness of AZO film was measured by Stylus Profiler (Bruker Dektak XT, Bremen, Germany). All measurements of the solar cells were performed under ambient atmosphere at room temperature without encapsulation.

3. Results and Discussion

The device structure in this work is shown in Figure 1a, where the PEDOT:PSS and PCBM act as the HTL and ETL, respectively. Here an inverted planar heterojunction structure is used instead of the conventional structure due to its advantages as mentioned earlier. AZO is inserted between the PCBM layer and Ag electrode to decrease the energy mismatch between them, so that the carrier collection may be efficiently improved. Figure 1b shows the schematic band diagram of this work with/without AZO. The energy level mismatch between PCBM and Ag electrode could lead to inefficient carrier transport. One of the methods to enhance the performance of the p-i-n PSCs is interface engineering. It is expected that the AZO interlayer with excellent photoelectric properties can reduce the work function of metal electrode [25,26], so that the carriers can be efficiently collected by the Ag electrode, as shown in Figure 1b.

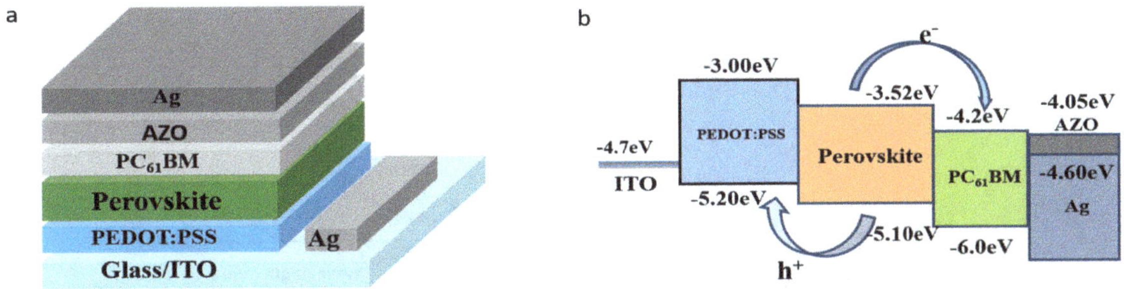

Figure 1. (**a**) Schematic structure of the $MA_{0.7}FA_{0.3}PbI_{3-x}Cl_x$ devices in this study: ITO/PEDOT:PSS/ $MA_{0.7}FA_{0.3}PbI_{3-x}Cl_x$ /PCBM/AZO/Ag. (**b**) The energy level diagram of the p-i-n solar cell. The energy levels of ITO, the bandgap of PEDOT:PSS, and the bandgap of PCBM were referenced from the previous report [12]. The workfunction of AZO was measured by UPS as stated in the following.

High quality perovskite material is the key to achieving a high performance PSC. In order to create a uniform and smooth perovskite film, a mixed-solvent-vapor annealing technique is used here, as in the previous report [7]. After spin-coated the MAI and FAI mixed solution, the perovskite films were put on top of a hot plate and covered by a glass petri dish. 40 μL of IPA:DMF (100:1 v/v) solvent was added in the petri dish around the substrates during the thermal annealing process so that the solvent vapor could react with the perovskite film. Figure S1 reveals the surface morphology of the fabricated perovskite film on the top of the ITO/PEDOT:PSS layer. It is obvious that the perovskite film is uniform and smooth with the grain size approximately 300–500 nm.

Figure 2a presents the champion current density-voltage (J-V) curves of devices with/without the AZO interlayer, and Table 1 shows the corresponding device parameters. The champion device with the AZO interlayer achieves a PCE of 16.19%, with a Jsc of 22.82 mA/cm^2, a Voc of 0.99 V and a FF of 71.68%. Compared with this device, the device without the AZO interlayer gives a PCE of 14.80%, with a Jsc of 22.18 mA/cm^2, a Voc of 0.94 V and a FF of 71.02%. The device performance with the AZO interlayer is much higher than that without the AZO layer. The higher Jsc, Voc and FF values for the device with the AZO interlayer are supposed to be mainly caused by the suitable work function and efficient charge extraction with the insertion of the AZO layer. Jsc of the devices is also checked by the IPCE spectra, as shown in Figure 2b. It can be seen that the Jsc of the devices integrated from the IPCE spectra matches well with those obtained from J-V measurements. The high IPCE value for the device with the AZO interlayer should be related with the excellent charge transport in the interface of PCBM/AZO bilayer. Figure S2 reveals the J-V curves of PSCs with/without the AZO interlayer measured via both forward and reverse bias sweeps. (Forward scan: from a negative bias −0.2 V to a positive bias 1.1 V, and reverse scan: from a positive bias 1.1 V to a negative bias −0.2 V). It was observed that the PSC with the AZO interlayer shows the neglect photocurrent hysteresis. On the other hand, the PSC without the AZO interlayer shows an obvious photocurrent hysteresis. The hysteresis effect in J-V measurements is one of the most challenging issues for PSCs. It has been shown that it is related to the accumulation of the interface [27]. The neglect photocurrent hysteresis for the device with the AZO interlayer means that there is negligible accumulation at the interface, and thus, better carrier extraction ability. Furthermore, the stabilized PCEs at the maximum power output point were examined, as shown in Figure 2c. The steady-state PCEs are measured to be 15.65% and 13.85% for the devices with/without the AZO interlayer, respectively. They are close to the values obtained from light J-V curves, indicating the reliability of the J-V curve measurement. Both the hysteresis measurement and the stead-state measurement show that the device with the AZO interlayer has the best performance, which is consistent with the PCE measurement shown in Figure 2a. This may be attributed to the fact that with the insertion of the AZO interlayer between the PCBM layer and Ag electrode, a bilayer with excellent charge transport efficiency will be formed.

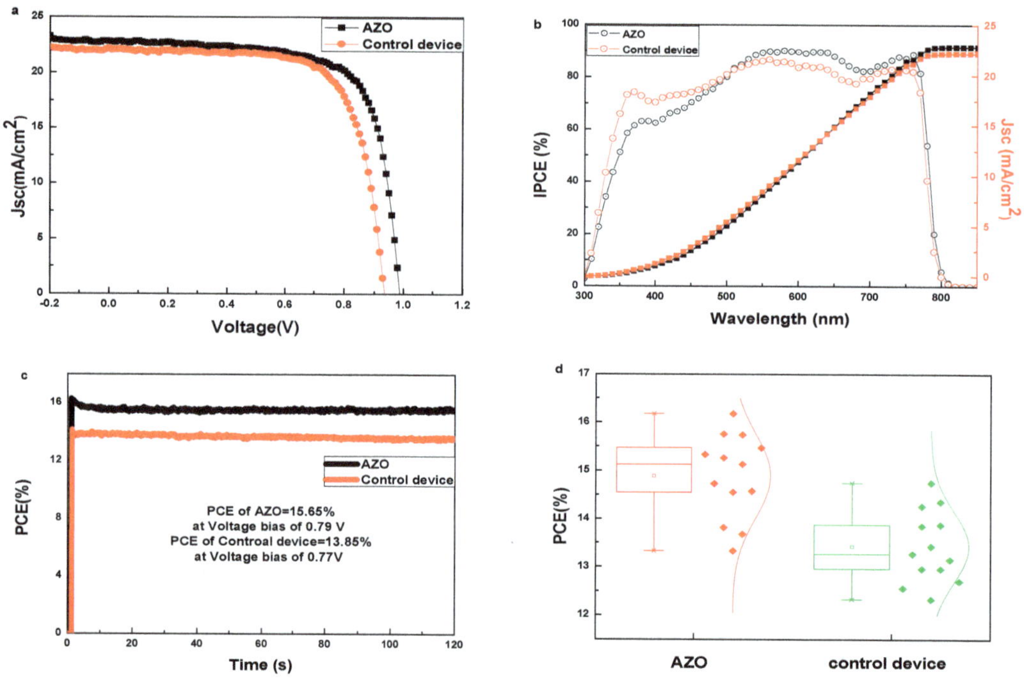

Figure 2. (**a**) Best performed J–V characteristics of PSCs with/without the AZO interlayer measured under 100 mW/cm^2 AM 1.5G illumination. (**b**) IPCE curves and integrated current density of PSCs with/without the AZO interlayer. (**c**) Steady output characteristics of PSCs with/without the AZO interlayer. (**d**) Statistics result of PCE for PSCs with/without the AZO interlayer.

Table 1. Photovoltaic performances of PSCs under AM 1.5G illumination (100 mW/cm^2).

	Jsc (mA/cm^2)	Voc (V)	FF (%)	PCE (%)
AZO	22.82	0.99	71.68	16.19
Control device	22.18	0.94	71.02	14.75

To confirm the above, the statistic results for the device parameters are summarized in Table S1, Figure 2d and Figure S3a–c. PSCs with the AZO interlayer exhibit a promising average PCE of 14.89% with an average Jsc of 20.99 mA/cm^2, an average Voc of 0.97 V and an average FF of 73.34%, outperforming the devices without the AZO interlayer, where an average PCE of 13.42% is achieved (the average Jsc, Voc and FF were 22.35 mA/cm^2, 0.95 V and 63.60% respectively). The parameters of

Voc and FF display significant enhancement, indicating the beneficial role of the AZO interlayer with enhanced charge extraction ability.

It is supposed that the PCBM/AZO bilayer could lower the energy barrier between the Ag electrode and the active layer; then, the electron extraction from the active layer to Ag electrode is facilitated. To verify this supposition, UPS measurement was carried out. Figure 3a shows the UPS spectra of Ag electrode, Ag/AZO and ITO/PEDOT:PSS/perovskite to investigate how the energy levels shift. The UPS measurement shows that the work function of Ag is 4.6 eV. After depositing AZO on Ag, there is an obvious shift toward lower work function of 4.05. Figure 3a reveals that the Fermi level (E_F) of the perovskite film on the ITO/PEDOT:PSS substrate is 3.2 eV, and the energy difference between the valance band maximum and E_F is 1.9 eV. We also measured the absorption of ITO/PEDOT:PSS/perovskite/PCBM (as shown in Figure 3b); this indicated that the energy gap of the perovskite is 1.58 eV. Therefore, we can obtain that the conduction band minimum of perovskite film in this work is 3.52 eV. The energy band of PCBM is referenced from the previous report [7] with the conduction band minimum of 4.20 eV and valance band maximum of 6.0 eV. It is obvious that the introduction of the AZO interface layer could decrease the energy level mismatch between PCBM layer and Ag electrode. Thus, electrons produced in the perovskite film could be collected by the Ag electrode more efficiently, which is consistent with the better performance of the device with the AZO interface layer.

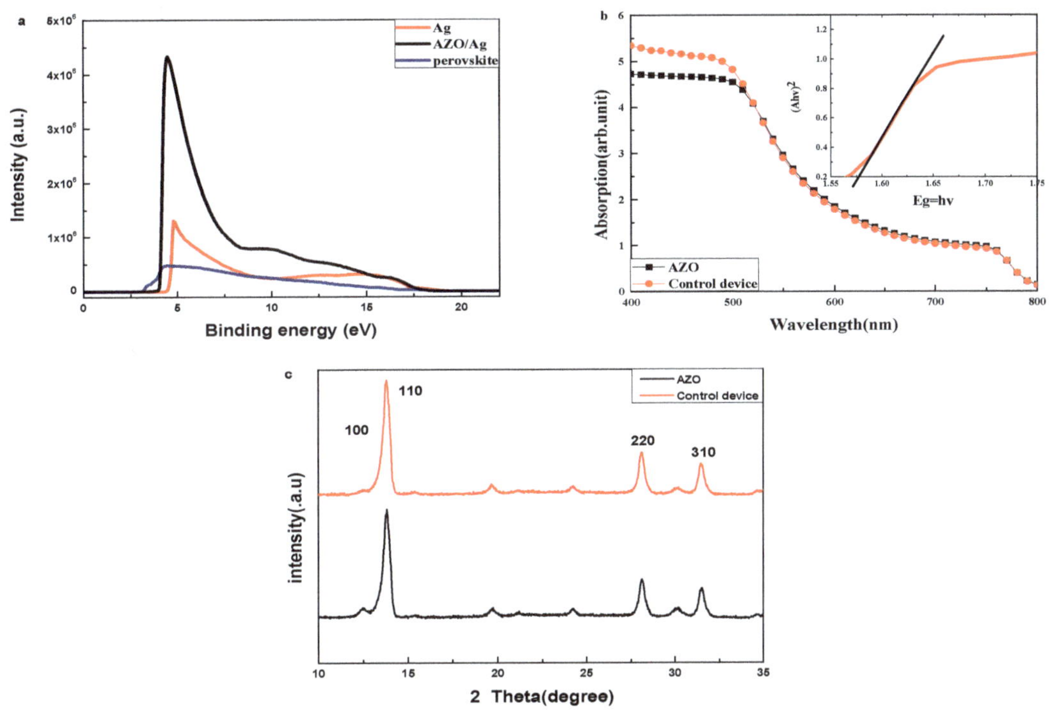

Figure 3. (**a**) UPS measurements of Ag electrode, Ag/AZO and ITO/PEDOT:PSS/perovskite. (**b**) The optical absorption spectra of ITO/PEDOT:PSS/perovskite/PCBM and ITO/PEDOT:PSS/perovskite/PCBM/AZO. inset: the corresponding bandgap of perovskite film. It is noted that the corresponding optical bandgap is about 1.58 eV. (**c**) XRD patterns of ITO/PEDOT:PSS/perovskite/PCBM and ITO/PEDOT:PSS/perovskite/PCBM/AZO.

Figure 3b shows the UV-vis spectra of ITO/PEDOT:PSS/perovskite/PCBM and ITO/PEDOT:PSS/perovskite/PCBM/AZO. It is obvious that at between 400 and 500 nm wavelengths, the absorbance of the device with AZO is lower than the device without AZO, which matches with the result of IPCE. The result of UV-vis measurement also indicates that the layers of AZO have minimal influence on the light absorption of the perovskite layer between 500 and 800 nm wavelengths. Because the absorbance of AZO layer is very weak and that of ITO/PEDOT:PSS/perovskite/PCBM is very strong, the absorbance of AZO has a little effect on the whole device. The XRD of ITO/PEDOT:PSS/perovskite/PCBM and ITO/PEDOT:PSS/perovskite/PCBM/AZO were also be measured, in order to observe the influence of AZO layer on the perovskite film. As shown in Figure 3c, diffraction peaks at the XRD patterns around 14.17°, 28.52°, and 31.98° are assigned to the (110), (220), and (310) lattice planes of the tetragonal perovskite structure, respectively. The peak at 12.70° is associated with the (100) crystal plane of PbI_2. The XRD patterns with/without AZO are similar; this demonstrates that the influence of AZO layer to perovskite layer is negligible. Thus, we can conclude that the improved performance of the device with the AZO interlayer is mainly induced by the improved electron extraction ability instead of the improved absorption or crystal quality.

The ITO/PEDOT:PSS/perovskite, ITO/PEDOT:PSS/perovskite/PCBM and ITO/PEDOT:PSS/perovskite/PCBM/AZO film morphologies and surface textures were also investigated by SEM. It is shown that with the method of mix-solvent-vapor annealing, a high uniform and smooth perovskite film was achieved as shown in Figure S1. Figure S4 shows the top-view SEM images of ITO/PEDOT:PSS/perovskite/PCBM and ITO/PEDOT:PSS/perovskite/PCBM/AZO. It is observed that with the PCBM or PCBM/AZO bilayer deposited on perovskite film their SEM images are breezing. However, the grain size and surface textures are unchanged. This demonstrates that the influence of AZO layer to perovskite layer is negligible, which is consistent with the result of the UV-vis and XRD which we measured above.

Figure 4a shows the steady-state photoluminescence (PL) spectra of the ITO/PEDOT:PSS/perovskite/PCBM layer with/without the AZO interlayer. The perovskite films were deposited on the glass/ITO/PEDOT:PSS substrates under the same experimental conditions. The excitation light wavelength is 700 nm and the PL emission is collected from the PCBM or PCBM/AZO side. To protect the samples during the measurement, the device was sealed by spin-coating a layer of insulating polymer (PMMA). When the sample has the AZO layer, it could be clearly observed that the peak intensity at 783 nm decreased, which demonstrates that the charge transfer from the perovskite to the ETL is more efficient. This indicates a suppressed non-radiative recombination with the optimized PCBM/AZO bilayer; this is consistent with the obtained UPS results and J-V characteristics discussed above. Figure 4b shows the TRPL behaviors for the sample with/without AZO layer. It is obvious that the sample with the AZO interlayer has a shorter lifespan, and that the sample without the AZO interlayer has a longer one. The lifespan for the sample without the AZO interlayer is around 239.08 ns; this is reduced to 201.84 ns when the AZO interlayer is inserted between the PCBM layer and the Ag electrode. This shows that the photogenerated charge carriers could be efficiently transported to the AZO layer and collected by the Ag electrode. All of these results indicate that efficient electron transfer occurs with the PCBM/AZO bilayer, which is necessary for efficient charge extraction and collection in PSCs, and corresponds with the higher performance and neglect photocurrent hysteresis for devices with the PCBM/AZO bilayer.

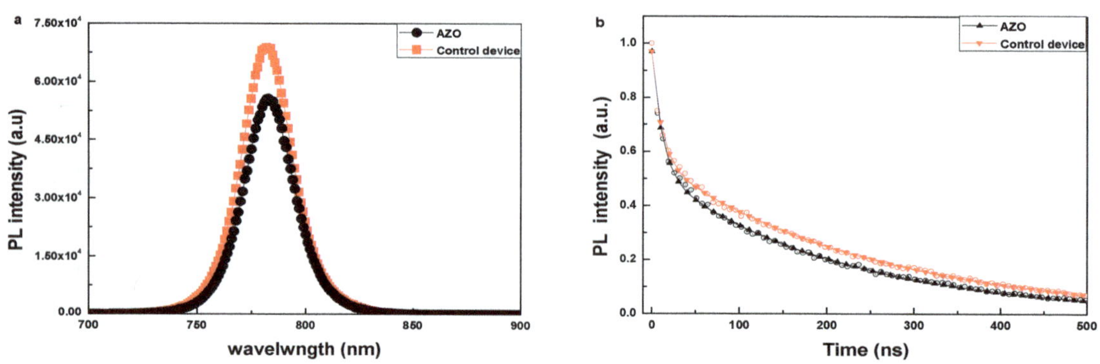

Figure 4. *Cont.*

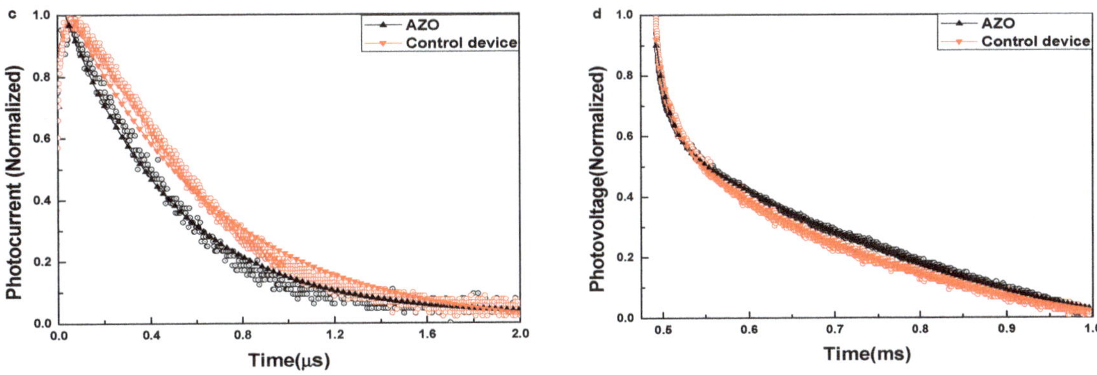

Figure 4. (**a**) Steady-state PL spectra and (**b**) Time-resolved PL spectra of perovskite/PCBM with/without the AZO layer. (**c**) Transient photocurrent and (**d**) transient photovoltage measurements of solar cells with/without the AZO interlayer.

The transient photocurrent and photovoltage measurements of PSCs were further performed to verify these results. Figure 4c shows the transient photocurrent of PSCs prepared with/without the AZO interlayer, measured at the short circuit condition. It was found that the device with the AZO interlayer has faster extraction (467.58 ns) than the device without the AZO interlayer (648.25 ns), indicating that the device former possesses much more efficient charge extraction and charge transport properties. Hence, the Voc and FF are enhanced. The transient photovoltage is used to determine the charge recombination lifetime, as shown in Figure 4d. The charge recombination lifetime of the device with the AZO interlayer increases to 0.41 ms, compared to that without the AZO interlayer (0.27 ms), indicating that the charge recombination is efficiently suppressed with the introduction of the AZO interlayer. These transient photocurrent and photovoltage measurement results are consistent with the PL measurement results.

Finally, we studied the stability of the device by storing it for 720 h. As shown in Figure 5 and Figure S5a–c, the device with the AZO interlayer exhibits remarkable PCE stability, retaining 86.41% of its initial value even after storing for over 720 h. The Jsc remained almost unchanged, with a slightly decreased FF value. The Voc has a small increase in the first 100 h, and stays around 1.00 V in the following measurement, while for the device without the AZO interlayer, the PCE decreases quickly to 60.14% of its initial value for the same storage condition and duration, which is mainly caused by the decreased FF and Jsc. These results suggest that the PSC with the PCBM/AZO bilayer exhibits better stability.

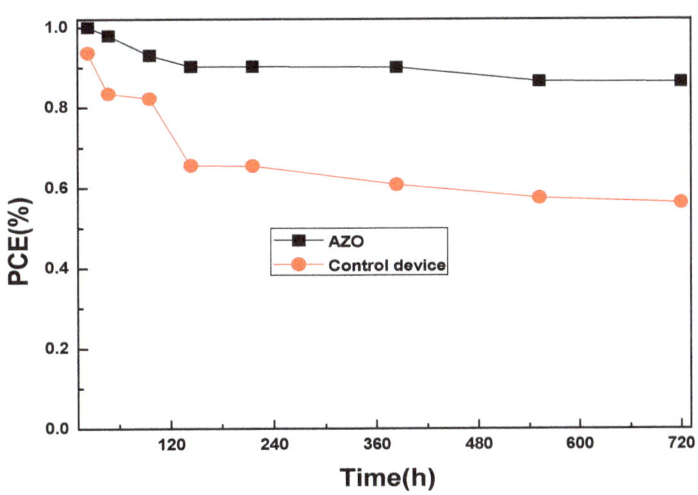

Figure 5. Stability of unencapsulated PSCs with/without the AZO interlayer.

4. Conclusions

In conclusion, we have demonstrated a stable and efficient p-i-n structure PSC with a PCBM/AZO electron transporting bilayer. The insertion of the AZO interlayer could greatly improve the performance of PSCs; the best-performing device exhibits a PCE of 16.18% with a Jsc of 22.82 mA/cm^2, a Voc of 0.99 V, and a FF 71.68%. The UPS measurement shows that the AZO interlayer can decrease the energy offset between PCBM and metal electrode. The measurements of PL, TRPL, transient photocurrent, and transient photovoltage show that the PSC with the AZO interlayer has a longer radiative carrier recombination lifetime, and more efficient charge extraction efficiency. In addition, the PSC with the AZO interlayer shows an improved stability.

Supplementary Materials
Figure S1: Top-view SEM image of the surface morphology of perovskite, Figure S2: (a) J–V curves in forward and reverse scans of PSCs with AZO, (b) J–V curves in forward and reverse scans of PSCs without AZO, Figure S3: Statistics results of Jsc (a) Voc (b) and FF (c) for PSCs with/without AZO, Figure S4: SEM graphs of (a) ITO/PEDOT:PSS/perovskite/PCBM and (b) ITO/PEDOT:PSS/perovskite/PCBM/AZO, Figure S5: Stability of (a) Jsc (b) Voc and (c) FF for unencapsulated devices with/without AZO, Table S1: Statistical results of perovskite solar cells under AM 1.5G illumination (100 mW/cm^2). The standard deviation results are derived from 12 perovskite solar cells.

Author Contributions: C.Z. conceived the idea. C.Z. and Y.Z. designed the experiment and guided the experiment. H.D. conducted most of the device fabrication and data collection and wrote the manuscript; C.Z. and J.Z. revised the manuscript; S.P., D.C., W.Z., H.X. helped the device measurement, J.C. helped the data analysis. Y.H. supervised the group. All authors read and approved the manuscript.

Acknowledgments: In this section you can acknowledge any support given which is not covered by the author contribution or funding sections. This may include administrative and technical support, or donations in kind (e.g. materials used for experiments).

References

1. Jeon, N.J.; Noh, J.H.; Yang, W.S.; Kim, Y.C.; Ryu, S.; Seo, J.; Seok, S. Compositional engineering of perovskite materials for high-performance solar cells. *Nature* **2015**, *517*, 476–480. [CrossRef] [PubMed]
2. Adnan, M.; Lee, J.K. All Sequential Dip-Coating Processed Perovskite Layers from an Aqueous Lead Precursor for High Efficiency Perovskite Solar Cells. *Sci. Rep.* **2018**, *8*, 1–10. [CrossRef] [PubMed]
3. Kojima, A.; Teshima, K.; Shirai, Y.; Miyasaka, T. Organometal halide perovskites as visible-light sensitizers for photovoltaic cells. *J. Am. Chem. Soc.* **2009**, *131*, 6050–6051. [CrossRef] [PubMed]
4. Best Research-Cell Efficiencies. Available online: https://www.nrel.gov/pv/assets/images/efficiency-chart-20180716.jpg (accessed on 16 July 2018).
5. Ma, J.; Chang, J.; Lin, Z.H.; Guo, X.; Zhou, L.; Liu, Z.; Xi, H.; Chen, D.; Zhang, C.; Hao, Y. Elucidating the Roles of $TiCl_4$ and PCBM Fullerene Treatment on TiO_2 Electron Transporting Layer for Highly Efficient Planar Perovskite Solar Cells. *J. Phys. Chem. C* **2017**, *122*, 1044–1053. [CrossRef]
6. Kim, J. The Stability Effect of Atomic Layer Deposition (ALD) of Al_2O_3 on $CH_3NH_3PbI_3$ Perovskite Solar Cell Fabricated by Vapor Deposition. *Key Eng. Mat.* **2017**, *753*, 156–162. [CrossRef]
7. Sun, X.; Zhang, C.; Chang, J.; Yang, H.; Xi, H.; Lu, G.; Chen, D.; Lin, Z.; Lu, X.; Zhang, J.; et al. Mixed-solvent-vapor annealing of perovskite for photovoltaic device efficiency enhancement. *Nano Energy* **2016**, *28*, 417–425. [CrossRef]
8. Pang, S.; Li, X.; Dong, H.; Chen, D.; Zhu, W.; Chang, J.; Lin, Z.; Xi, H.; Zhang, J.; Zhang, C.; et al. Efficient bifacial semitransparent perovskite solar cells using Ag /V_2O_5 as transparent anodes. *ACS Appl. Mater. Interfaces* **2018**, *10*, 12731–12739. [CrossRef] [PubMed]
9. Liu, Z.; Chang, J.; Lin, Z.; Zhou, L.; Yang, Z.; Chen, D.; Zhang, C.; Liu, S.; Hao, Y. High-Performance Planar Perovskite Solar Cells Using Low Temperature, Solution-Combustion-Based Nickel Oxide Hole Transporting Layer with Efficiency Exceeding 20%. *Adv. Energy Mater.* **2018**, *1703432*, 1–9. [CrossRef]
10. Mo, J.; Zhang, C.; Chang, J.; Yang, H.; Xi, H.; Chen, D.; Lin, Z.; Lu, G.; Zhang, J.; Hao, Y. Enhanced efficiency of planar perovskite solar cells via a two-step deposition using DMF as an additive to optimize the crystal growth behavior. *J. Mater. Chem. A* **2017**, *5*, 13032–13038. [CrossRef]

11. Shi, Z.; Ahalapitiya, H.J. Perovskites-Based Solar Cells: A Review of Recent Progress, Materials and Processing Methods. *Materials* **2018**, *11*, 729. [CrossRef] [PubMed]
12. Yang, H.; Zhang, J.; Zhang, C.; Chang, J.; Lin, Z.; Chen, D.; Sun, X.; Xi, H.; Han, G.; Hao, Y. Effect of polyelectrolyte interlayer on efficiency and stability of p-i-n perovskite solar cells. *Sol. Energy* **2016**, *139*, 190–198. [CrossRef]
13. Seo, J.; Park, S.; Kim, Y.C.; Jeon, N.J.; Noh, J.H.; Yoon, S.C.; Seok, S. Benefits of very thin PCBM and LiF layers for solution-processed p-i-n perovskite solar cells. *Energy Environ. Sci.* **2014**, *7*, 2642–2646. [CrossRef]
14. Chiang, C.H.; Tseng, Z.L.; Wu, C.G. Planar heterojunction perovskite/PC71BM solar cells with enhanced open-circuit voltage via a (2/1)-step spin-coating process. *J. Mater. Chem. A* **2014**, *2*, 15897–15903. [CrossRef]
15. Chen, C.; Zhang, S.; Wu, S.; Zhang, W.; Zhu, H.; Xiong, Z.; Zhang, Y.; Chen, W. Effect of BCP buffer layer on eliminating charge accumulation for high performance of inverted perovskite solar cells. *RSC Adv.* **2017**, *7*, 35819–35826. [CrossRef]
16. You, J.; Yang, Y.; Hong, Z.; Song, T.B.; Meng, L.; Liu, Y.; Jiang, C.; Zhou, H.; Chang, W.H.; Li, G.; et al. Moisture assisted perovskite film growth for high performance solar cells. *Appl. Phys. Lett.* **2014**, *105*, 945. [CrossRef]
17. Su, T.; Zheng, Y.; Ma, Z.; Cheng, L.; Xu, X.; Zhang, F.; Yu, G.; Sheng, Z. Band Engineering via Sn-doping of Zinc Oxide Electron Transport Materials for Perovskite Solar Cells. *ChemistrySelect* **2018**, *3*, 363–367. [CrossRef]
18. AitDads, H.; Bouzit, S.; Nkhaili, L.; Elkissani, A.; Outzourhit, A. Structural, optical and electrical properties of planar mixed perovskite halides/Al-doped Zinc oxide solar cells. *Sol. Energy Mater. Sol. Cells* **2016**, *148*, 30–33. [CrossRef]
19. Pérez-del-Rey, D.; Boix, P.P.; Sessolo, M.; Hadipour, A.; Bolink, H.J. Interfacial Modification for High-Efficiency Vapor-Phase-Deposited Perovskite Solar Cells Based on a Metal Oxide Buffer Layer. *J. Phys. Chem. Lett.* **2018**, 1041–1046. [CrossRef] [PubMed]
20. Li, Y.; Guo, Y.; Li, Y.; Zhou, X. Fabrication of Cd-Doped TiO_2 Nanorod Arrays and Photovoltaic Property in Perovskite Solar Cell. *Electrochim. Acta* **2016**, *200*, 29–36. [CrossRef]
21. Song, J.; Zheng, E.; Wang, X.F.; Tian, W.; Miyasaka, T. Low-temperature-processed ZnO–SnO_2 nanocomposite for efficient planar perovskite solar cells. *Sol. Energy Mater. Sol. Cells* **2016**, *144*, 623–630. [CrossRef]
22. Ren, X.; Yang, D.; Yang, Z.; Feng, J.; Zhu, X.; Niu, J.; Liu, Y.; Zhao, Y.; Liu, S. Solution-processed Nb:SnO_2 electron transport layer for efficient planar perovskite solar cells. *ACS Appl. Mater. Inter.* **2017**, *9*, 2421–2429. [CrossRef] [PubMed]
23. Li, W.; Jiang, Q.; Yang, J.; Luo, Y.; Li, X.; Hou, Y.; Zhou, S. Improvement of photovoltaic performance of perovskite solar cells with a ZnO/Zn_2SnO_4, composite compact layer. *Sol. Energy Mater. Sol. Cells* **2017**, *159*, 143–150. [CrossRef]
24. Zhang, Q.; Peng, R.; Zhang, C.; Chen, D.; Lin, Z.; Chang, J.; Zhang, J.; Hao, Y. Inverted Organic Solar Cells with Low-Temperature Al-Doped-ZnO Electron Transport Layer Processed from Aqueous Solution. *Polymers* **2018**, *10*, 127. [CrossRef]
25. Hwang, K.; Jung, Y.; Heo, Y.; Scholes, F.H.; Watkins, S.E.; Subbiah, J.; Jones, D.J.; Kim, D.Y.; Vak, D. Toward large scale roll-to-roll production of fully printed perovskite solar cells. *Adv. Mater.* **2015**, *27*, 1241–1247. [CrossRef] [PubMed]
26. Subbiah, J.; Mitchell, V.D.; Hui, N.K.C.; Jones, D.J.; Wong, W.W.H. A Green Route to Conjugated Polyelectrolyte Interlayers for High-Performance Solar Cells. *Angew. Chem. Int. Ed.* **2017**, *56*, 8431–8434. [CrossRef] [PubMed]
27. Zhou, L.; Chang, J.; Liu, Z.; Sun, X.; Lin, Z.; Chen, D.; Zhang, C.; Zhang, J.; Hao, Y. Enhanced planar perovskite solar cell efficiency and stability using a perovskite/PCBM heterojunction formed in one step. *Nanoscale* **2018**, *10*, 3053–3059. [CrossRef] [PubMed]

8

Elucidating the Effect of Etching Time Key-Parameter toward Optically and Electrically-Active Silicon Nanowires

Mariem Naffeti [1,2,3,*], Pablo Aitor Postigo [2], Radhouane Chtourou [1] and Mohamed Ali Zaïbi [1,3]

1 Laboratory of Nanomaterials and Systems for Renewable Energies (LaNSER), Research and Technology Center of Energy, Techno-Park Borj-Cedria, Bp 95, 2050 Hammam-Lif, Tunis, Tunisia; radhouane.chtourou@crten.rnrt.tn (R.C.); medali.zaibi@ensit.rnu.tn (M.A.Z.)

2 Instituto de Micro y Nanotecnología, IMN-CNM, CSIC (CEI UAM+CSIC) Isaac Newton, 8, E-28760 Tres Cantos, Madrid, Spain; pabloaitor.postigo@imn.cnm.csic.es

3 Tunis University—National High School of Engineering of Tunis, 5 Av Taha Hussein 1008 Tunis, Tunisia

* Correspondence: naffeti.mariam@gmail.com

Abstract: In this work, vertically aligned silicon nanowires (SiNWs) with relatively high crystallinity have been fabricated through a facile, reliable, and cost-effective metal assisted chemical etching method. After introducing an itemized elucidation of the fabrication process, the effect of varying etching time on morphological, structural, optical, and electrical properties of SiNWs was analysed. The NWs length increased with increasing etching time, whereas the wires filling ratio decreased. The broadband photoluminescence (PL) emission was originated from self-generated silicon nanocrystallites (SiNCs) and their size were derived through an analytical model. FTIR spectroscopy confirms that the PL deterioration for extended time is owing to the restriction of excitation volume and therefore reduction of effective light-emitting crystallites. These SiNWs are very effective in reducing the reflectance to 9–15% in comparison with Si wafer. I–V characteristics revealed that the rectifying behaviour and the diode parameters calculated from conventional thermionic emission and Cheung's model depend on the geometry of SiNWs. We deduce that judicious control of etching time or otherwise SiNWs' length is the key to ensure better optical and electrical properties of SiNWs. Our findings demonstrate that shorter SiNWs are much more optically and electrically active which is auspicious for the use in optoelectronic devices and solar cells applications.

Keywords: silicon nanowires; metal assisted chemical etching; etching time; optical properties; electrical properties

---

1. Introduction

In recent years, silicon nanowires (SiNWs) have aroused tremendous attention worldwide thanks to the following outstanding features: (1) Environment-friendly as second most earth-abundant materials; (2) unique dimensional structures (1 D); (3) interesting electrical and optical properties compared to bare silicon; (4) affordable fabrication; and (5) potential applications in several fields [1–5]. The various applications of these nanostructures may include lithium-ion batteries [6], biochemical sensors [7,8], electronics [9], catalysis [10], and solar cells [1,11].

Based upon the bottom-up and top down approaches, numerous methods have been used to fabricate SiNWs such as vapor–liquid–solid, thermal evaporation, molecular beam epitaxy, laser ablation, and lithography [12–16]. However, these techniques have some limitations as they generally require expensive and complex equipment, employ hazardous silicon precursors, and involve high vacuum and high temperature [17]. These features make the synthesis expensive and time-consuming and therefore hindered their applications for commercialized products. In contrast, an effective and promising synthetic method namely metal assisted chemical etching (MACE) has been proposed [2,4,18–20]. This technique is simple, rapid, low cost, and suitable for both industrial and laboratory scales. Moreover, MACE allows to obtain high crystalline SiNWs quality, as well as an easy control of the different parameters including orientation, doping type, length, and diameter.

The MACE method basically consists of two procedures, the formation of metal catalysts and the subsequent etching process which can be implemented either in a single step (1-MACE) [10,21] or in two steps (2-MACE) [17–20]. Moreover, the formation method, etching time, etching temperature, metal deposition time, and lastly the etchants' concentrations have a crucial influence on the morphology of SiNWs [2,5,17]. Ghosh et al. reported that SiNWs grown by MACE are usually covered with silicon nanocrystals due to the side wall etching and which are the origin of quantum confinement (QC) effects owing to their small dimensions [20].

Recently, several research groups have succeeded in the synthesis of optically-active SiNWs exhibiting a significant PL emission and a very low reflectance [3,4,17]. On the other hand, Qi et al. have demonstrated the fabrication of electrically-active SiNWs through heavily doped SiNWs with rough surface where a high Schottky barrier exists at the interface of SiNWs and the metal [22]. Nevertheless, further investigation is required to explore both optically and electrically active SiNWs. Indeed, a number of studies have investigated the optical and electrical properties of SiNWs [23–25]. It has been shown that SiNWs, obtained by 1-MACE in $AgNO_3$, HF, and H_2O_2 solution from a P^+ type starting Si wafer, possess such a remarkable low reflectance and the electronic properties are affected by SiNWs' homogeneity [23]. Otherwise, Hutagalung et al. also reported a low reflectance of less than 10% obtained by SiNW arrays synthesized via 1-MACE in an $AgNO_3$ and HF solution at 60 °C from an n-type Si, whereas its I–V characteristics show linear ohmic behavior [24]. Therefore, careful production of SiNWs, by tuning the different parameters such as the etching time, etchant composition and concentration, etching temperature, and the starting Si wafer characteristics, is important to ensure good physical properties of SiNWs [23–25]. However, these previous works have neglected the PL study, as well as a deeper insight and understanding of the electrical properties via determining the electrical parameters is missing.

In this work, we contribute our recent results on the fabrication of optically and electrically-active SiNWs which exhibit a strong PL emission, remarkable antireflections properties, and interesting electrical properties. First, a detailed explanation of the fabrication process is reported and then vertical aligned SiNWs with relatively high crystallinity were obtained through the 2-MACE method. The effect of etching time key parameter on SiNWs' filling ratio and length, optical, and electrical properties were investigated and evaluated. A broadband PL emission and low reflectance from these wires were obtained. Later on, the I–V characteristics and the electrical parameters were carefully determined and studied. Our findings could consider the optimized SiNWs as a promising candidate in PV and optoelectronic applications due to their unique structural, optical, and electrical properties.

2. Materials and Methods

2.1. Reagents and Materials

The chemical reagents used in this work for the cleaning or etching process such as acetone, ethanol, isopropanol, hydrofluoric acid (HF, 40%), silver nitrate ($AgNO_3$), hydrogen peroxide (H_2O_2, 35%), and nitric acid (HNO_3, 65%) were purchased from Sigma-Aldrich (Madrid, Spain). All of them were used without any purification. The single-side polished p-type silicon wafers were purchased from Siltronics (Archamps, France) and finally deionized water used in all the experiments was supplied by local sources.

2.2. Samples Preparation

SiNW arrays were synthesized by the Ag-assisted chemical etching method of (100) oriented p-type silicon wafers with the resistivity of 1–20 Ω cm. The fabrication process is as follows: (1) The silicon wafers were sequentially cleaned in acetone, ethanol, isopropanol, and deionized water (DI) in an ultrasonic bath for 15 min each. The cleaned wafers were then immersed in diluted HF for 3 min. (2) Silver nanoparticles were deposited onto the Si substrate using an aqueous solution composed of 4.8 M HF and 0.035 M $AgNO_3$ for 1 min. (3) Silicon wafers covered with AgNPs were dipped into the etching solution of 4.8 M HF and 0.5 M H_2O_2 at room temperature during 20 (sample S1), 40 (S2), 60 (S3), 80 (S4), 100 (S5), and 120 min (S6). (4) The resulting samples were rinsed with DI and then immersed in nitric acid for 15 min to remove the silver nanoparticles and dendrites. Finally, the as-formed homogeneous black SiNWs were washed again with DI and dried with nitrogen.

2.3. Characterizations

The samples were studied using several techniques. The morphologies of the synthesized SiNWs were characterized using scanning electron microscopy (SEM, FEI Verios 460, FEI Europe B.V., Eindhoven, Netherlands). The observations were performed in a top view, cross-section, and 30° tilt view. X-ray diffraction (XRD) measurements were performed using an automated Bruker D8 advance X-ray diffractometer (Bruker, Karlsruhe, Germany) with Cu Kα (λ = 1.54 Å) in 2θ ranging from 20 to 80°. Photoluminescence spectroscopic analyses of SiNWs were made with a 405 nm laser wavelength and all the measurements were done at room temperature (RT). Moreover, we have used an analytical model to deduce the SiNCs size through PL spectra of the fabricated samples. The FTIR analyses were taken on an absorbance mode using Bruker IFS66v/s FTIR spectrometer (Bruker, Karlsruhe, Germany) and investigated in the 400–4000 cm^{-1} range with a step of 4 cm^{-1}. Reflectance measurements were performed via Perkin Elmer Lambda 950 spectrophotometer (Perkin Elmer, Inc., Waltham, MA, USA). The current–voltage (I–V) measurements were measured via Keithley 2400 source meter (Keithley, Austin, TX, USA) in the dark and at room temperature.

3. Results and Discussion

3.1. Detailed Mechanism of SiNWs' Formation

Two-step MACE was introduced to prepare vertically aligned SiNWs which is a simple, reproducible, and inexpensive process using Ag catalysts in the HF/H_2O_2 etching agent. To better understand SiNWs' formation mechanism, the properties of Si, Ag^+, Ag, and H_2O_2 in HF solution should be understood. Figure 1 shows a scheme of the potential distribution of Si, Ag^+/Ag, and H_2O_2/H_2O redox pairs in HF solution when the energy is referred to the standard hydrogen electrode potential (SHE).

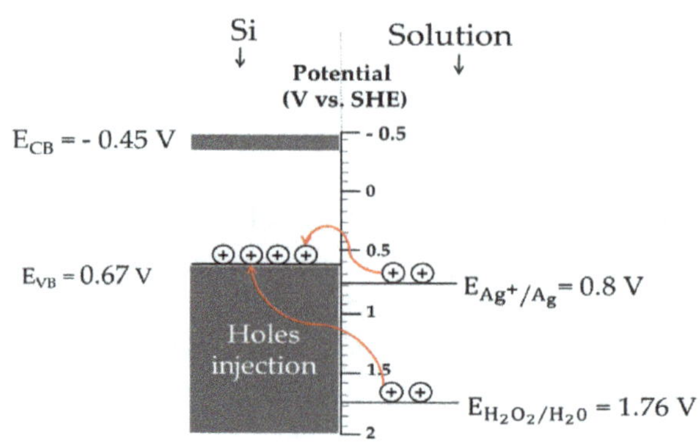

Figure 1. Schematic of the potential relationship between bands in a silicon (Si) substrate, Ag^+/Ag, and H_2O_2/H_2O redox pairs.

The energies of the valence band (VB) and conduction band (CB) of Si are 0.67 and −0.45 eV, respectively while the potential energies of the redox pairs of Ag^+/Ag and H_2O_2/H_2O are 0.8 and 1.76 V, respectively [26].

The growth mechanism of SiNW arrays was elucidated in this section and depicted in Figure 2.

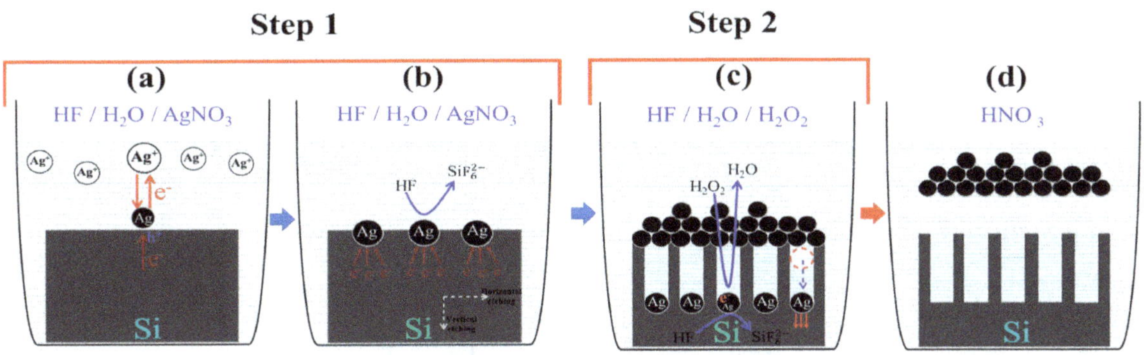

Figure 2. Schematic illustration of the formation mechanism of silicon nanowires (SiNWs) via two-step MACE. (**a**) Reduction of Ag^+ ions and formation of Ag nuclei at the Si surface. (**b**) Further silver nuclei growth, oxidative dissolution of Silicon atoms, and production of porous layer. (**c**) Vertical propagation of silver nanoparticles and faster etching leading to SiNWs formation. (**d**) Silver removal and vertical aligned SiNW arrays production.

After the cleaning stages, 2-MACE consists first of all in dipping the silicon wafers in a $HF/AgNO_3$ aqueous solution producing AgNPs deposition. Indeed, the $AgNO_3$ metallic salt come apart in the HF aqueous solution to yield metal ions Ag^+. The Ag^+ ions close to the Si surface extract electrons from the VB of Si resulting in a small silver nuclei on the surface of the silicon (Figure 2a). These electrons transfer continues to take place as Ag is more electronegative than Si leading to a large silver nuclei growth and then a formation of Ag-nanoclusters or a continuous silver layer. Furthermore, the excess of electrons extract forms an accumulation of holes beneath and around the catalyst.

This leads to silicon oxidation followed by HF dissolution into silicon hexafluoride ions (SiF_6^{2-}) while unveiling a newly exposed Si coming in contact with the Ag catalysts which will be further etched in the second step of the continuous MACE process. The synchronized mechanism of Ag^+ reduction (cathodic process) and silicon oxidation (anodic process) can be described by the following equations:

$$\text{cathode reaction} \quad Ag^+ + e^- \rightarrow Ag \quad E^0 = 0.79\ \text{V (vs. SHE)} \tag{1}$$

$$\text{anode reaction} \quad Si + 6HF \rightarrow SiF_6^{2-} + 4e^- + 6H^+ \tag{2}$$

$$\text{overall reaction} \quad Si + 6HF + 4Ag^+ \rightarrow SiF_6^{2-} + 4Ag + 6H^+ \tag{3}$$

At a certain while, the hole injection rate attenuates because almost all Ag^+ ions in the vicinity of the silicon surface are reduced into Ag. Therefore, silicon oxidation slows down causing a decrease in the Si etching rate which requires a complementary hole injection species to permit the continuation of the silicon nanostructuring. Subsequently, H_2O_2 is added to the etching solution because it is a strong oxidant allowing a hole injection instead of Ag ions. When the silicon wafer covered with the silver, dendritic layer was put into the HF/H_2O_2 aqueous solution, the second step in MACE mechanism which corresponds to the etching process starts to take place. Since the redox potential of H_2O_2 is more positive than that of Ag, H_2O_2 captures electrons from the previously nucleated Ag particles and reduces to H_2O. Hence, Ag is oxidized immediately to Ag^+, this oxidation allows the solution feeding with Ag^+ ions which enhances the hole injection. The silicon is then locally oxidized into SiO_2 and dissolved simultaneously by HF, so the AgNPs sink downwards vertically along the (100) direction in the pits thus formed leading to the generation of vertical SiNWs (Figure 2c). This is contrary to porous silicon (PS) which displays a porous surface morphology due to the absence of $AgNO_3$ in the etching solution and the slow etching rate. The redox reactions taking place during etching process can be described as follows [27,28]:

$$\text{cathode reaction} \quad H_2O_2 + 2e^- + 2H^+ \rightarrow 2H_2O \quad E^0 = 1.76\ \text{V (vs. SHE)} \tag{4}$$

$$\text{anode reaction} \quad Si + 2H_2O \rightarrow SiO_2 + 4H^+ + 4e^- \tag{5}$$

$$SiO_2 + 6HF \rightarrow SiF_6^{2-} + 2H_2O + 2H^+ \tag{6}$$

$$\text{Overall reaction} \quad Si + 2H_2O_2 + 6F^- + 4H^+ \rightarrow SiF_6^{2-} + 4H_2O \tag{7}$$

Using Reaction (7), the potential ΔE of etching process is as follows:

$$\Delta E = \Delta E^0 - \frac{0.059}{4}\left\{\log \frac{\left[5iF_6^{2-}\right]}{[H_2O_2]^2\left[H^+\right]^4[F^-]^6}\right\} \tag{8}$$

According to this equation, the increase of H_2O_2 and HF concentrations enhance the reaction potential with an increase in the etching rate. However, the etching rate may not increase infinitely and it is preferable to choose an appropriate range of etchant concentration to ensure good quality SiNWs formation. It is suggested that the self-grown SiNCs on SiNWs form due to sidewall etching such us the extra Ag^+ ions spread through the nanowires and trap electrons from the sidewalls leading to a lateral etching of SiNWs and consequently formation of porous SiNWs decorated with SiNCs [20]. Ultimately, the removal of AgNPs and dendrites by immersing the samples in an HNO_3 aqueous solution reveals highly oriented one-dimensional (1D) SiNWs. The corresponding reaction is expressed as:

$$3Ag + 4HNO_3 \rightarrow 3AgNO_3 + NO + 2H_2O \tag{9}$$

3.2. Etching Time Effect on Morphology of SiNW Arrays

In order to investigate the effect of etching time on SiNWs morphology and structure, the $AgNO_3$, HF, and H_2O_2 concentrations were fixed in both deposition and etching steps. SEM images (top view, cross-section, and tilt view at 30°) of the as-prepared SiNWs etched in different times (20–120 min), are shown in Figure 3. The SiNWs dependence on etching time was explored. Forest-like SiNW arrays can be noticed from the top view images (Figure 3a–f). The NWs etched during short etching time are isolated from each other. However, the tips of the nanowires congregate together with time increment to form bundles. This bundle-like structures distribute uniformly on the whole wafers and could be confirmed from the tilt view images. The possible reasons behind this conglomeration may be attributed to Van der Waals attraction between the nanowires [29,30] as well as to the increase in the length of SiNWs that likely causes them to bend to form bouquets under the action of gravity. Otherwise, it was also reported that some nanowires could remain unattached due to inhomogeneous etching induced by a random silver particle distribution [31]. The calculated values of average volume filling ratio (VFR) at the air/SiNW arrays interface versus etching time is plotted in Figure 4a and it showed the decreasing of VFR when the etching time increase. For example, VFR is 0.45 in dense SiNWs at 20 min decreases to 0.33 in convergent SiNW arrays at 120 min of etching time. This is due to the bundle-like structures of the wires in prolonged etching times. Chang et al. [32] reported a similar variation in VFR, however, as a function of $AgNO_3$ concentrations.

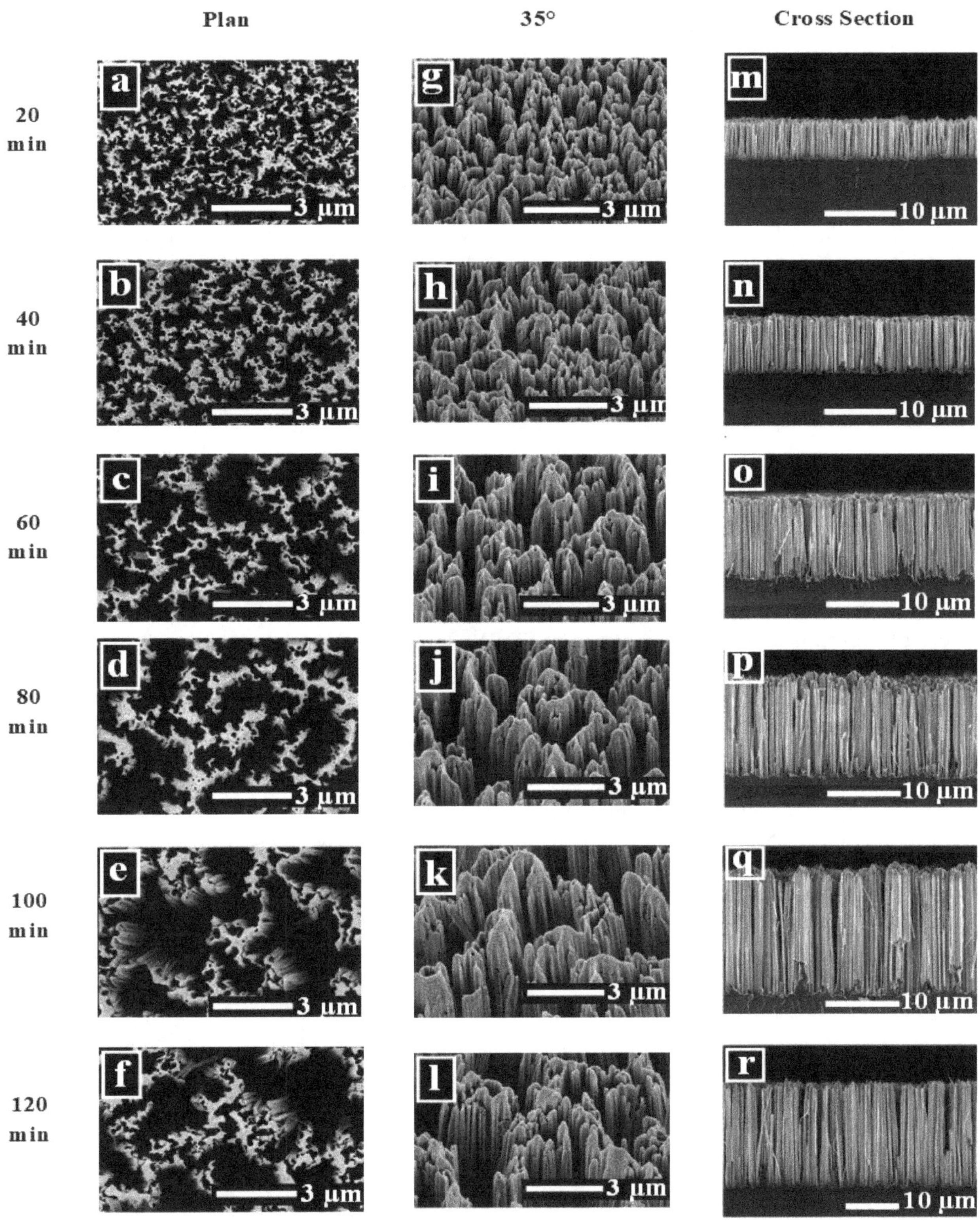

Figure 3. SEM images of SiNWs etched at different durations showing top view surface (**a**–**f**), tilt view 35° (**g**–**l**), and cross-sectional (**m**–**r**).

Vertically aligned SiNWs with good uniformity were clearly seen in the cross-section images (Figure 3m–r). The length of nanowires significantly increases from 5.48 to 20.84 µm with increasing the etching from 20 to 120 min whereas the wire diameter is approximately in the range of one hundred to a very few hundreds of nanometers.

Hence, the linear relationship of SiNWs length versus etching time plotted in Figure 4b gives an etching rate close to 0.152 μm/min.

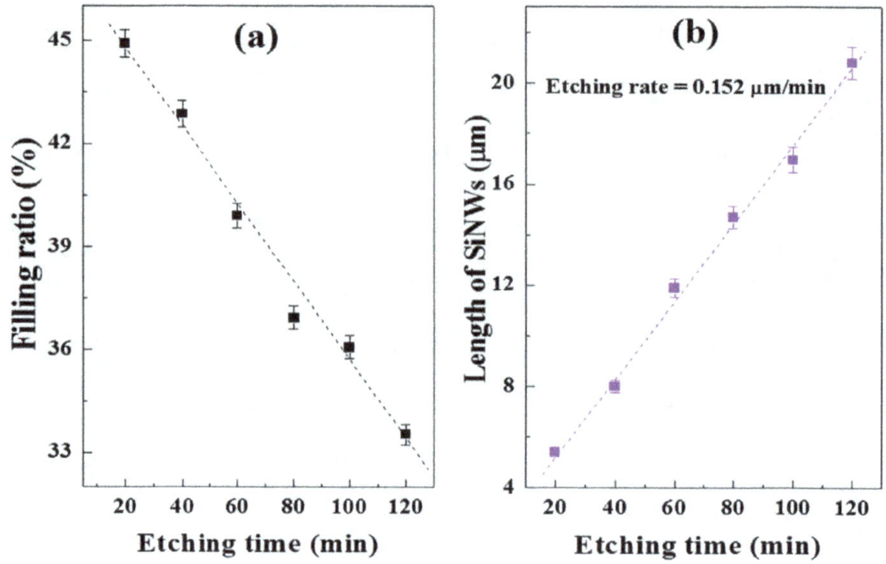

Figure 4. Variation of the (**a**) filling ratio and (**b**) length of SiNWs as a function of etching time.

This linear relation has been investigated in literature but with different etching rate values [2,18,33]. The possible reasons behind this increase of SiNWs length with etching time was ascribed to the fact that etching was given more time to proceed and the solution had enough oxidizing species to oxidize and dissolve the formed SiO_2. We retain that doubling the etching time would not necessarily double the length of SiNW [34,35]. From tilt view images, we can observe that SiNWs are covered by numerous porous structures especially on their tips due to the additional etching pathways via the re-nucleation of the AgNPs throughout the SiNWs.

These porous structures are the origin of quantum confinement effects owing to their small dimensions. SiNW arrays grown by MACE usually show these porous structures called silicon nanocrystals.

Figure 5 displays XRD patterns of untreated Si wafer and SiNWs. A unique sharp peak at 69° is observed for both samples, which is indexed to a (400) silicon plane. However, the silicon nanowires give a high peak intensity than that of Si wafer. This suggests high quality crystalline nanowires.

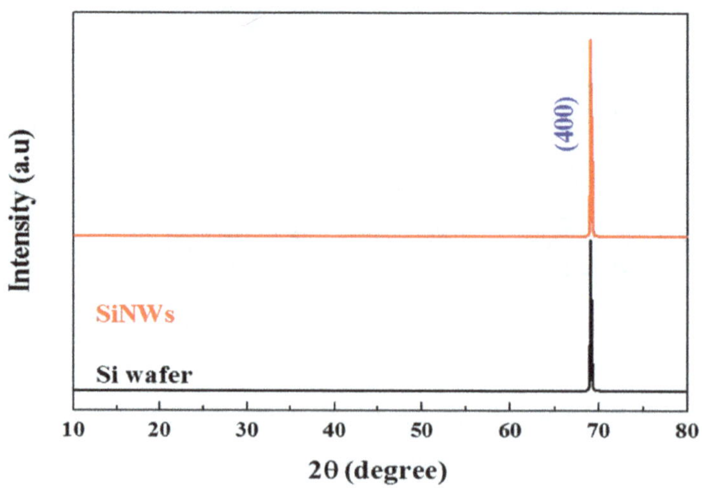

Figure 5. XRD patterns of silicon nanowires and untreated Si wafer.

On the other hand, no characteristic peaks of silver are observed in the XRD pattern which confirms the completely removal of silver nanoparticles by nitric acid.

3.3. Photoluminescence Spectroscopy

The room temperature PL measurements of the as-synthesized samples were carried out in order to study the etching time dependency on photoluminescence properties. The SiNWs PL spectra ranging from 1.4 to 2.3 eV are shown in Figure 6.

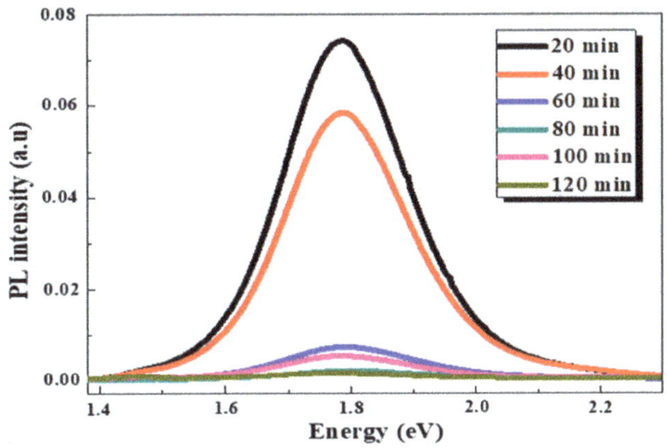

Figure 6. PL spectra of SiNWs synthesized at different etching times.

Broad PL emission bands were recorded with a maximum around 1.78 eV and a full width at half maximum (FWHM) of 0.24 eV. Their shape is a gaussian and did not present any significant shift. However, one can see a clear degradation in the PL intensity when increasing etching time.

The highest PL intensity was obtained for 20 and 40 min which are almost 26 times stronger compared to the prolonged etching time. Therefrom, the opposite trend between the integrated PL intensity and etching duration is plotted in Figure 7.

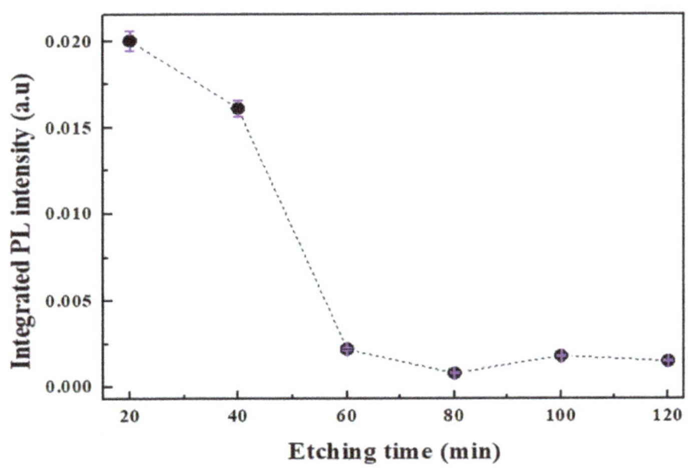

Figure 7. Variation of integrated PL intensities versus etching duration.

To explain the origin of this PL emission, several mechanisms have been proposed including the quantum confinement (QC) effects and the presence of defects in an SiO_x/Si interface and/or in the surface of the oxide related to the Si–O–Si bonds [36,37]. However, the QC effects, which enhance the radiative recombination of excitons, remain the most approved model which dictates that the crystallite Si size should be less than the Bohr radius of the free exciton of bulk silicon [17]. In the present work, the SiNW diameters are within one hundred to a very few hundred of nanometers, these values are much larger than the Bohr radius of excitons (5 nm) present in Si crystals. Taking into account the indirect band gap of SiNWs and their large diameters, the observed visible PL is unlikely to originate from the SiNWs and the origin of this light emission must be more complex. This is dissimilar to one earlier report that attribute the visible PL to the radiative recombination at the SiNWs itself [38]. On the other hand, it is widely reported that the SiNWs are decorated with self-grown silicon nanocrystals whose dimensions are smaller than the exciton Bohr diameter in silicon. These SiNCs cause a robust quantum confinement leading to the appearance of observed PL at room temperature [2–4,20]. In the current study, the nanocrystal size resulted from the spectral position of the PL peak by using the equation described below:

$$E = E_g + \frac{C}{d^{\alpha}} \tag{10}$$

where E is PL peak position (eV), E_g is the band gap of c-Si (1.12 eV), d is the SiNC size (nm), α and C are constants. Several works have reported different values of these constants from the SiNWs/NCs [19,39–41]. Comparison of SiNCs size values obtained by different parameters from earlier reports of the literature is investigated and the evaluated NC-sizes are summarized in Table 1. One can see clearly that all the estimated NCs diameter values are less than 5 nm and close to 2~4 nm regardless of the formula used (Table 1). According to these below values, we can assert that in these samples, the occurrence of QC in SiNCs is the origin of the recorded PL emission. Note that similar SiNCs size and PL peak position have been recently carried out [2,3,41] due to the known dependence of a SiNC band gap on its size.

Table 1. Comparison of the SiNCs sizes value estimated from Equation (10) for different reports of the literature.

Constants	Valenta et al. [39]	Yan et al. [40]	Gonchar et al. [41]	Bahera et al. [19]
C ($eV\ nm^{-2}$)	2.49	4	3.73	2.4
α	0.91	1.4	1.39	1.7
d (nm)	4.25	3.59	3.45	2.12

Interestingly, the compatibility between the samples (S1–S6) in terms of peak position, FWHM, NCs size, and the variation of the PL intensity indicate that these latter derive from different size distribution and densities of the inseparable set of SiNWs-NCs. This is confirmed by the reduction of SiNWs amount as seen above in the SEM images and more precisely by the reduction of the volume filling ratio over etching time. These latter indicate that the monotonous decrease of the PL intensity over etching duration is interpreted as restriction and diminution of excitation volume which in turn is proportional to the amount of effective light-emitting crystallites [17,41,42].

Therefore, an appropriate geometry is required to ensure PL and our findings demonstrate that shorter SiNWs are much more optically active. It is worth noting that the light emitted from the SiNW arrays upon laser irradiation is visible to the naked eye and appears orange obviously for the samples corresponding to 20 and 40 min etching time. This emission is attributed to the radiative recombination of excitons in small SiNCs present in the nanowire sidewalls in terms of a QC model as explained above.

3.4. FTIR Analysis

FTIR spectroscopy was used to study the surface composition of the as prepared SiNWs at different etching durations. An untreated silicon substrate was utilized as background to the measurements and the spectra were taken in absorption mode in the 400–4000 cm^{-1} spectral range as shown in Figure 8.

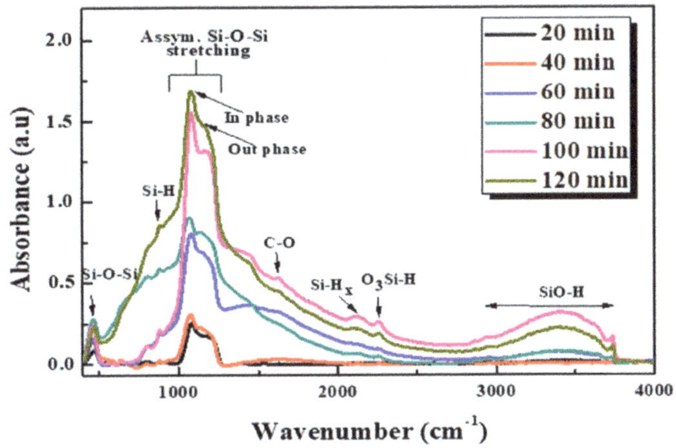

Figure 8. FTIR spectra of SiNWs obtained at different etching time.

All the spectra are closely similar with an increase in the absorbance bands intensity versus increasing etching time or otherwise with increasing SiNWs length. The most intense broad peak appears in the region between 1000–1300 cm^{-1} and is assigned to the Si–O–Si asymmetric stretching (AS) vibrations. This observed signal can be classified into a strong band at 1080 cm^{-1} attributed to AS_1 vibration mode and a shoulder at 1200 cm^{-1} related to AS_2 vibration mode. AS_1 results from the motion of the adjacent oxygen atoms moving in phase with one another while at AS_2 the oxygen atoms move 180° out of phase with one another. The appearance of this shoulder is a feature of the IR spectra of SiO_x material where x value equals to one or higher [43].

These Si–O–Si bonding, introduced during the electrochemical etching, were found to be more popular with increasing etching time. Hence, Figure 9 presents the evolution of peaks intensity of dominant bands existing in the spectra.

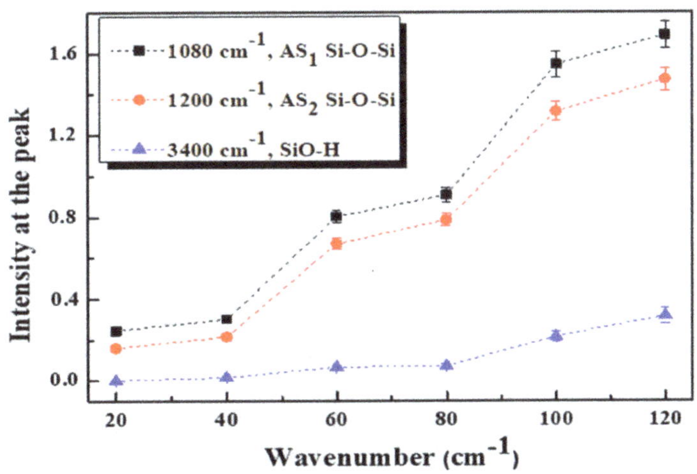

Figure 9. Evolution of peaks intensity of dominant FTIR bands versus etching time.

This increase of intensities indicates the continuity of chemical etching even during extended durations and then the formation of Si–O radicals becomes consistent when rising the immersion duration in HF/H_2O_2. Here, it involves that the charge transfer faces some difficulty in order to establish an electrical equilibrium leading to a decay in etching rate [44]. This behavior is in accordance with PL results and hence we could say that PL deterioration in higher etching time is related to the change in the distribution and density of the inseparable set of SiNWs-NCs owing to this etching rate decay leading to a restriction of excitation volume and therefore reduction of luminescent crystallites' density.

The broad peak at 465cm^{-1} was related to Si–O–Si, while the broad band within 3000–3700 cm^{-1} is attributed to the stretching of the O−H bonds in SiOH groups and surface adsorbed H_2O. Whereas the narrow band at 3747 cm^{-1}, is attributed to the stretching mode of surface free OH groups. The weak IR absorption peaks situated at around 900, 2125, and 2258 cm^{-1} corresponds to Si–H bending vibrations, Si–H_x bonds and to Si–H stretching mode in O_3–SiH, respectively. Finally, the tiny absorption band barely detectable at 1630 cm^{-1} is attributed to C–O bonds which mainly came from the SiNWs surface contamination from environment and can hardly be avoided [2,21,28,43]. We notice that the surface states assigned to Si–O and Si–H chemical bonds are popular in all samples.

3.5. Reflectance Analysis

Figure 10 shows the reflectance spectra of the untreated Si wafer and the synthesized SiNWs with various heights over the wavelength range of 250–700 nm which covers the main spectral irradiance of sunlight that is useful for Si solar cells.

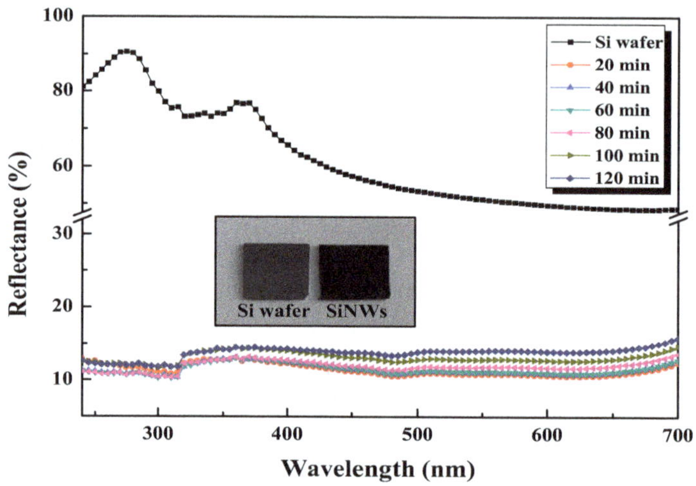

Figure 10. Reflectance spectra of SiNWs with various lengths etched during different times and corresponding Si wafer. The inset displays images of Si wafer and SiNWs.

When compared to the untreated bulk Si, all the etched samples show a significant lower reflectivity which is consistent with the colour changes from metallic grey to dark black as shown in the inset. The average reflectance for bulk Si is about 50%~90% in the UV–visible range while it reaches less than 15% for all the etched samples. Hence, such nanowires reduce reflection losses and further enhance the carrier collection which is extremely beneficial for developing high efficiency solar cells requiring good antireflective properties.

The sharp suppression in the reflectance is ascribed to three important factors: (i) The gradual variation towards the refractive index from air ($n \approx 1$) to SiNWs and to Si substrate ($n \approx 3.42$), (ii) the tapered morphology of silicon wires leading to light trapping due to the multiple reflections back and forth in the inner surface and (iii) the sub-wavelength structures for additional light trapping within the NWs [17,33,45]. The observed peaks of Si wafer at 275 and 367 nm come from the inter-band transitions of Si [32,45]. In addition, as readily seen in the figure, all the reflectance values of SiNW arrays with various lengths are almost close and low (9–15%) owing to the tapered NWs structure. The slightly higher reflectance observed for prolonged etching time is attributed to the coalescence of NWs at the apex and bundles formation as shown above in SEM images [33,45].

3.6. I–V Measurements

Electrical properties of SiNWs samples with various lengths of nanowires are studied through measurement of current–voltage (I–V) characteristics at room temperature and in the dark. First, Schottky contact was formed between metal and semiconductors (Al/Si and Ag/SiNWs). The I–V measurements of Ag/SiNWs/Si/Al structure from −4 to 4 V are depicted in Figure 11.

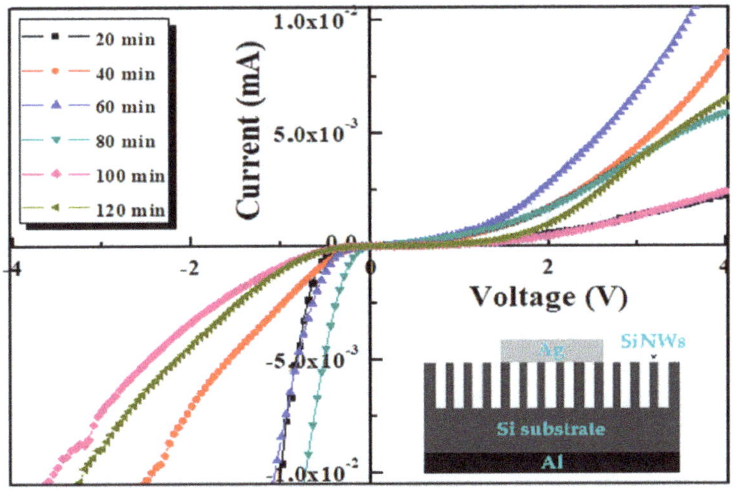

Figure 11. Curves of the as-prepared heterojunction based on SiNWs synthesised at various etching times. The inset displays a schematic illustration of Ag/SiNWs/Si/Al device.

All the curves corresponding to NWs show a diode-like nature with an obvious rectifying behaviour. Interestingly, the junctions based on these SiNWs etched at different etching time display considerably the same properties at different areas on the junction. This latter observation highlights the uniformity of the junctions and could be supported by the distribution of a uniform large area of SiNWs on the whole wafers as shown above in the SEM images. On the other hand, the fabrication of other devices with maintaining the same conditions pointed out an electrical reproducibility property of the junctions. One can note that SiNWs fabrication with good uniformity/reproducibility are a MACE feature [4,34]. It is worth noting that the observed rectifying behaviour of these SiNWs is similarly shown in the literature for porous silicon [42,46–48]. Importantly, the nonlinear I–V characteristics and the rectifying event are mainly controlled by the SiNW layers and ascribed to Ag/SiNWs:p-Si since the Si/Al interface was proved to be ohmic [47,48]. The electron flow into the SiNW layers from Ag and recombine with holes, leaving negatively charged electrons in SiNWs which leads to the N-type role of SiNWs with respect to Si. Thus, the formed depletion layer causes the rectifying behaviour. Another reason is that the quantum confinement coming from the SiNCs in the SiNWs increases the band gap in the SiNWs side which creates a potential barrier leading to the diodic behaviour [49,50]. It is clear to notice from the I–V curves the increase of rectification extent (current intensity) in the forward bias with etching time or in other words with SiNWs' length until reaching a threshold time (length) which corresponds to 60 min ($\approx$12 μm), then the current intensity decreases. Therefore, we can assume that the electrical transport properties of the device are governed by SiNWs' contribution. The current decrease for samples with higher etching times is attributed to the increase of SiNWs length. The wires, acting as carries trapped, will increase and form a high resistive region which lead to the decrease of the current flowing through the SiNW layers.

Therefore, a well-defined moderate length of SiNWs is required to ensure better electrical properties. Electrical parameters of the device are obtained using the conventional thermionic emission model (TE) as expressed:

$$I = I_s\left(\exp\left(\frac{q(v - IR_s)}{\eta kT}\right) - 1\right) \quad (11)$$

where q is the electronic charge, η is the ideality factor, R_s is the series resistance, k is the Boltzmann constant, T is the absolute temperature, and I_s is the reverse saturation current given by:

$$I_s = aA^*T^2 \exp\left(\frac{-q\varphi_b}{kT}\right) \quad (12)$$

Herein, a is the diode area, A* is Richardson constant (≈32 A cm^{-2} K^{-1} for p-type Si) and φ_b is barrier height. As eV >> ηKT at room temperature, Equation (11) can be rewritten as:

$$I = I_s \exp\left(\frac{q\,V}{\eta kT}\right) \tag{13}$$

The ideality factor η was estimated from the slope of the linear region of the plot of ln(I) vs. V (Figure 12) based on Equation (13). The saturation current was derived by extrapolating ln(I) vs. V plot to V = 0 while φ_b was calculated through Equation (12).

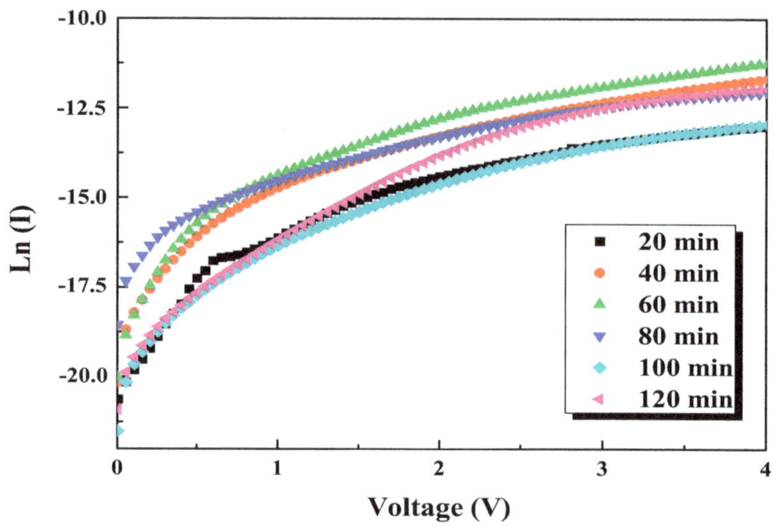

Figure 12. Logarithm of forward current versus voltage (Ln (I) vs. V) curves of SiNWs synthesized at different etching times.

In order to have more accurate ideality factor values owing to the non-linear part of ln(I) vs. V plots and to determine the series resistance (R_s), the Cheung's functions were utilized [51,52]. Thus,

$$\frac{dV}{d(LnI)} = R_s I + \frac{\eta kT}{q} \tag{14}$$

Figure 13 presents the experimental $\frac{dV}{d(LnI)}$ vs. I plots, the R_s and η were determined as the slope and y axis intercept, respectively.

The deduced values of Is, η, φ_b, and R_s for samples S1–S6 are summarized in Table 2 and show a notable dependence of SiNWs' length. According to the formula in Equation (11), high I require high Is and small η and R_s. The ideality factor values estimated from Cheung's functions are nearly matched with I–V method values. These relatively high values could be attributed to series resistance, barrier inhomogeneities, interfacial defects, or existence of an oxide layer spontaneously produced during the synthesis [51,52]. These latter give sense to the deviation from ideality and η deterioration especially for higher etching times. A similar trend is noticed for barrier height. The results point out that the electrical parameters depend upon the thicknesses of SiNWs and the use of shorter SiNWs not exceeding 12 μm (corresponding to 40 min etching time) can result in better electrical properties. One can note that these η values are lower compared to those reported on porous silicon [47]. This is suggested to be related to the unique properties of SiNWs compared to PS which is promising for optoelectronic devices application.

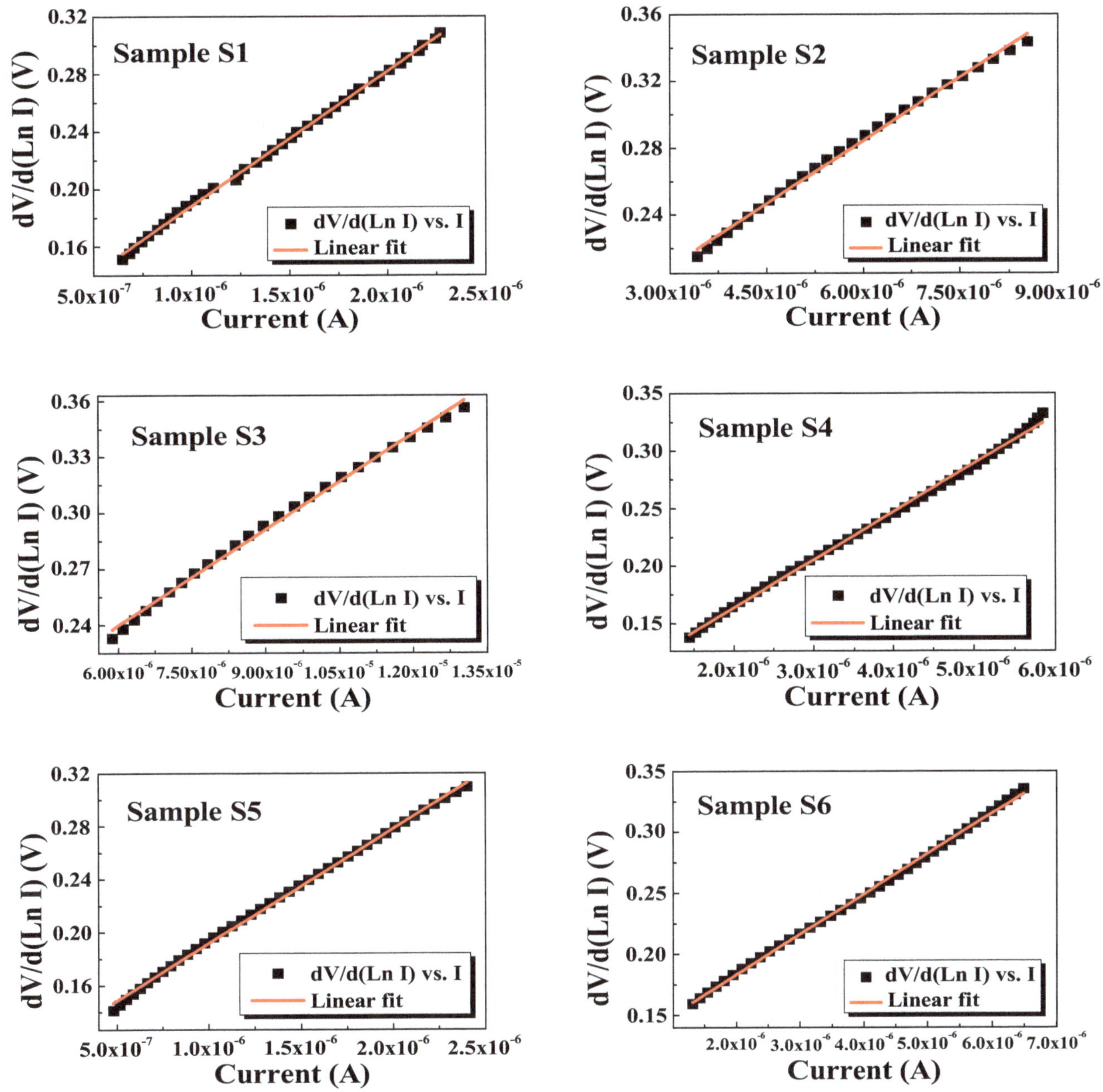

Figure 13. Plots of dV/d(LnI) vs. I of SiNWs samples (S1–S6).

Table 2. Various electrical parameters determined by conventional TE and Cheung's model.

	Ln (I) vs. V (TE model)			Cheung's Functions	
Etching Time	**I_s (μA)**	**η**	**φ_b(eV)**	**R_S (kΩ)**	**η**
20 min	0.211	5.91	0.806	93.65	3.65
40 min	0.495	5.31	0.790	32.10	3.56
60 min	0.695	4.38	0.771	25.06	2.60
80 min	0.566	6.17	0.768	41.75	3.10
100 min	0.239	6.05	0.801	86.33	4.08
120 min	0.224	6.41	0.804	33.01	4.53

4. Conclusions

In summary, optically and electrically-active SiNWs with strong PL emission, remarkable antireflections properties, and interesting electrical properties are successfully synthesized via 2-MACE method. First, an expanded explanation of the fabrication process is reported and then a systematic study is presented to visualize the effect of varying etching times on morphological, structural, optical, and electrical properties of SiNWs. The NWs length was increased by increasing etching time with a rate about 152 nm/min, whereas the wires filling ratio decreased. An analytical model is utilized to calculate the sizes of the self-grown silicon nanocrystallites which are the origin of the observed photoluminescence emission. FTIR spectroscopy supports PL results and confirms that the PL intensity degradation for expanded durations is owing to the restriction of excitation volume and therefore reduction of luminescent SiNCs. Therefore, appropriate SiNWs in terms of length and geometry is strongly required to ensure good PL. When compared to the silicon wafer, the formed SiNWs demonstrate a strong decrease of the reflectance to 9–15%. This strong reduction of reflectance certifies that SiNWs are an excellent candidate for photovoltaic cells. I–V characteristics revealed that the rectifying behaviour of the uniform-reproducible junctions and the diode parameters (Is, n, φ_b, and R_s) calculated from conventional thermionic emission and Cheung's model are found to depend significantly on the geometry of SiNWs. In brief, we deduce that etching time or otherwise SiNWs' length plays a key role on optical and electrical properties of SiNWs. Hence, judicious optimisation of these latter parameters is robustly required for better SiNWs' physical properties. Our findings demonstrate that shorter SiNWs are much more optically and electrically effective which pave the way for its application in the field of optoelectronic devices and solar cells.

Author Contributions: All authors have read and agree to the published version of the manuscript. M.N. performed the experiments, undertook the analysis, and wrote the manuscript. M.A.Z. and P.A.P. contributed to the analysis and supervised the progress of the research. P.A.P. and R.C. contributed reagents/materials/analysis tools.

Acknowledgments: The authors would like to acknowledge the Tunisian Ministry of Higher Education and Scientific Research as well as University of Tunis. P.A. Postigo acknowledge the service from the X-SEM Laboratory at IMN and funding from MINECO under project CSIC13-4E-1794 with support from EU (FEDER, FSE).

References

1. Sahoo, M.K.; Kale, P. Integration of silicon nanowires in solar cell structure for efficiency enhancement: A review. *J. Mater.* **2019**, *5*, 34–48. [CrossRef]
2. Cong, L.T.; Ngoc Lam, N.T.; Giang, N.T.; Kien, P.T.; Dung, N.D.; Ha, N.N. N-type silicon nanowires prepared by silver metal-assisted chemical etching: Fabrication and optical properties. *Mater. Sci. Semicond. Process.* **2019**, *90*, 198–204. [CrossRef]
3. Ashrafabadi, S.; Eshghi, H. Single-crystalline Si nanowires fabrication by one-step metal assisted chemical etching: The effect of etching time and resistivity of Si wafer. *Superlattices Microstruct.* **2018**, *120*, 517–524. [CrossRef]
4. Lajvardi, M.; Eshghi, H.; Izadifard, M.; Ghazi, M.E.; Goodarzi, A. Effects of silver and gold catalytic activities on the structural and optical properties of silicon nanowires. *Phys. E Low-Dimens. Syst. Nanostruct.* **2016**, *75*, 136–143. [CrossRef]
5. Amri, C.; Ouertani, R.; Hamdi, A.; Chtourou, R.; Ezzaouia, H. Effect of porous layer engineered with acid vapor etching on optical properties of solid silicon nanowire arrays. *Mater. Des.* **2016**, *111*, 394–404. [CrossRef]
6. Baek, S.-H.; Park, J.-S.; Jeong, Y.-M.; Kim, J.H. Facile synthesis of Ag-coated silicon nanowires as anode materials for high-performance rechargeable lithium battery. *J. Alloy. Compd.* **2016**, *660*, 387–391. [CrossRef]
7. Schwartz, M.; Nguyen, T.C.; Vu, X.T.; Wagner, P.; Thoelen, R.; Ingebrandt, S. Impedimetric Sensing of DNA with Silicon Nanowire Transistors as Alternative Transducer Principle. *Phys. Status Solidi A* **2018**, *215*, 1700740. [CrossRef]

8. Mirzaei, A.; Kang, S.Y.; Choi, S.-W.; Kwon, Y.J.; Choi, M.S.; Bang, J.H.; Kim, S.B.; Kim, H.W. Fabrication and gas sensing properties of vertically aligned Si nanowires. *Appl. Surf. Sci.* **2018**, *427*, 215–226. [CrossRef]
9. Borgne, B.L.; Salaün, A.C.; Pichon, L. Electrical properties of self-aligned gate-allaround polycrystalline silicon nanowires field-effect transistors. *Microelectron. Eng.* **2016**, *150*, 32–38. [CrossRef]
10. Amdouni, S.; Cherifi, Y.; Coffinier, Y.; Addad, A.; Zaïbi, M.A.; Oueslati, M.; Boukherroub, R. Gold nanoparticles coated silicon nanowires for efficient catalytic and photocatalytic applications. *Mater. Sci. Semicond. Process.* **2018**, *75*, 206–213. [CrossRef]
11. Yu, P.; Wu, J.; Liu, S.; Xiong, J.; Jagadish, C.; Wang, Z.M. Design and fabrication of silicon nanowires towards efficient solar cells. *Nanotoday* **2016**, *11*, 704–737. [CrossRef]
12. Latu-Romain, L.; Mouchet, C.; Cayron, C.; Rouviere, E.; Simonato, J.P. Growth parameters and shape specific synthesis of silicon nanowires by the VLS method. *J. Nanopart. Res.* **2008**, *10*, 1287–1291. [CrossRef]
13. Pan, H.; Lim, S.; Poh, C.; Sun, H.; Wu, X.; Feng, Y.; Lin, J. Growth of Si nanowires by thermal evaporation. *Nanotechnology* **2005**, *16*, 417–421. [CrossRef]
14. Fuhrmann, B.; Leipner, H.S.; Hoche, H.-R. Ordered arrays of silicon nanowires produced by nanosphere lithography and molecular beam epitaxy. *Nano Lett.* **2005**, *5*, 2524–2527. [CrossRef] [PubMed]
15. Kokai, F.; Inoue, S.; Hidaka, H.; Uchiyama, K.; Takahashi, Y.; Koshio, A. Catalyst-free growth of amorphous silicon nanowires by laser ablation. *Appl. Phys.* **2013**, *112*, 1–7. [CrossRef]
16. Hamdana, G.; Südkamp, T.; Descoins, M.; Mangelinck, D.; Caccamo, L.; Bertke, M.; Wasisto, H.S.; Bracht, H.; Peiner, E. Towards fabrication of 3D isotopically modulated vertical silicon nanowires in selective areas by nanosphere lithography. *Microelectron. Eng.* **2017**, *179*, 74–82. [CrossRef]
17. Lajvardi, M.; Eshghi, H.; Ghazi, M.E.; Izadifard, M.; Goodarzi, A. Structural and optical properties of silicon nanowires synthesized by Ag-assisted chemical etching. *Mater. Sci. Semicond. Process.* **2015**, *40*, 556–563. [CrossRef]
18. Vinzons, L.U.; Shu, L.; Yip, S.; Wong, C.-Y.; Chan, L.L.H.; Ho, J.C. Unraveling the Morphological Evolution and Etching Kinetics of Porous Silicon Nanowires During Metal-Assisted Chemical Etching. *Nanoscale Res. Lett.* **2017**, *12*, 1872. [CrossRef]
19. Behera, A.K.; Viswanath, R.N.; Lakshmanan, C.; Madapu, K.K.; Kamruddin, M.; Mathews, T. Synthesis, microstructure and visible luminescence properties of vertically aligned lightly doped porous silicon nanowalls. *Microporous Mesoporous Mater.* **2019**, *273*, 99–106. [CrossRef]
20. Ghosh, R.; Pal, A.; Giri, P.K. Quantitative analysis of the phonon confinement effect in arbitrarily shaped Si nanocrystals decorated on Si nanowires and its correlation with the photoluminescence spectrum. *J. Raman Spectrosc.* **2015**, *7*, 624–631. [CrossRef]
21. Leontis, I.; Othonos, A.; Nassiopoulou, A.G. Structure, morphology, and photoluminescence of porous Si nanowires: Effect of different chemical treatments. *Nanoscale Res. Lett.* **2013**, *1*, 383. [CrossRef] [PubMed]
22. Qi, Y.; Wang, Z.; Zhang, M.; Wang, X.; Ji, A.; Yang, F. Electron transport characteristics of silicon nanowires by metal-assisted chemical etching. *AIP Adv.* **2014**, *3*, 031307. [CrossRef]
23. Nafie, N.; Lachiheb, M.; Bouaicha, M. Effect of etching time on morphological, optical, and electronic properties of silicon nanowires. *Nanoscale Res. Lett.* **2012**, *1*, 393. [CrossRef]
24. Hutagalung, S.D.; Fadhali, M.M.; Areshi, R.A.; Tan, F.D. Optical and Electrical Characteristics of Silicon Nanowires Prepared by Electroless Etching. *Nanoscale Res. Lett.* **2017**, *12*, 425. [CrossRef]
25. Hasan, M.; Huq, M.F.; Mahmood, Z.H. A review on electronic and optical properties of silicon nanowire and its different growth techniques. *SpringerPlus* **2013**, *2*, 151. [CrossRef]
26. Huang, Z.; Geyer, N.; Werner, P.; de Boor, J.; Gösele, U. Metal-Assisted Chemical Etching of Silicon: A Review. *Adv. Mater.* **2010**, *2*, 285–308. [CrossRef] [PubMed]
27. Yang, C.; Wang, J.; Mei, L.; Wang, X. Enhanced Photocatalytic Degradation of Rhodamine B by Cu_2O Coated Silicon Nanowire Arrays in Presence of H_2O_2. *J. Mater. Sci. Technol.* **2014**, *11*, 1124–1129. [CrossRef]
28. Lin, L.; Guo, S.; Sun, X.; Feng, J.; Wang, Y. Synthesis and Photoluminescence Properties of Porous Silicon Nanowire Arrays. *Nanoscale Res. Lett.* **2010**, *11*, 1822–1828. [CrossRef]
29. Brahiti, N.; Hadjersi, T.; Amirouche, S.; Menari, H.; ElKechai, O. Photocatalytic degradation of cationic and anionic dyes in water using hydrogen-terminated silicon nanowires as catalyst. *J. Hydrog. Energy* **2018**, *43*, 11411–11421. [CrossRef]
30. Bai, F.; Li, M.; Song, D.; Yu, H.; gosh, B.; Li, Y. One-step synthesis of lightly doped porous silicon nanowires in $HF/AgNO_3/H_2O_2$ solution at room temperature. *J. Solid State Chem.* **2012**, *96*, 596–600. [CrossRef]

31. Qiu, T.; Wu, X.L.; Mei, Y.F.; Wan, G.J.; Chu, P.K.; Siu, G.G. From Si nanotubes to nanowires: Synthesis, characterization, and self-assembly. *J. Cryst. Growth* **2005**, *277*, 143–148. [CrossRef]
32. Chang, H.-C.; Lai, K.-Y.; Dai, Y.-A.; Wang, H.-H.; Lin, C.-A.; He, J.-H. Nanowire arrays with controlled structure profiles for maximizing optical collection efficiency. *Energy Environ. Sci.* **2011**, *4*, 2863. [CrossRef]
33. Li, S.; Ma, W.; Chen, X.; Xie, K.; Li, Y.; He, X.; Yang, X.; Lei, Y. Structure and antireflection properties of SiNWs arrays form mc-Si wafer through Ag-catalyzed chemical etching. *Appl. Surf. Sci.* **2016**, *369*, 232–240. [CrossRef]
34. Srivastava, S.K.; Kumar, D.; Schmitt, S.W.; Sood, K.N.; Christiansen, S.H.; Singh, P.K. Large area fabrication of vertical silicon nanowire arrays by silver-assisted single-step chemical etching and their formation kinetics. *Nanotechnology* **2014**, *25*, 175601. [CrossRef] [PubMed]
35. Ozdemir, B.; Kulakci, M.; Turan, R.; Unalan, H.E. Effect of electroless etching parameters on the growth and reflection properties of silicon nanowires. *Nanotechnology* **2011**, *22*, 155606. [CrossRef] [PubMed]
36. Najar, A.; Slimane, A.B.; Hedhili, M.N.; Anjum, D.; Sougrat, R. Effect of hydrofluoric acid concentration on the evolution of photoluminescence characteristics in porous silicon nanowires prepared by Ag-assisted electroless etching method. *J. Appl. Phys.* **2012**, *112*, 33502–33506. [CrossRef]
37. Vladimir, S.; Voigt, F.; Berger, A.; Bauer, G.; Christiansen, S. Roughness of silicon nanowire sidewalls and room temperature photoluminescence. *Phys. Rev. B* **2010**, *82*, 125446.
38. Chern, W.; Hsu, K.; Chun, I.S.; Azeredo BP, D.; Ahmed, N.; Kim, K.-H.; Zuo, J.-M.; Fang, N.; Ferreira, P.; Li, X. Nonlithographic Patterning and Metal-Assisted Chemical Etching for Manufacturing of Tunable Light-Emitting Silicon Nanowire Arrays. *Nano Lett.* **2010**, *10*, 1582–1588. [CrossRef]
39. Valenta, J.; Bruhn, B.; Linnros, J. Coexistence of 1D and Quasi-0D Photoluminescence from Single Silicon Nanowires, Coexistence of 1D and Quasi-0D. *Nano Lett.* **2011**, *7*, 3003–3009. [CrossRef]
40. Yan, J.-A.; Yang, L.; Chou, M.Y. Size and orientation dependence in the electronic properties of silicon nanowires. *Phys. Rev. B* **2007**, *76*, 115318. [CrossRef]
41. Gonchar, K.A.; Zubairova, A.A.; Schleusener, A.; Osminkina, L.A.; Sivakov, V. Optical Properties of Silicon Nanowires Fabricated by Environment-Friendly Chemistry. *Nanoscale Res. Lett.* **2016**, *1*, 357. [CrossRef] [PubMed]
42. Kulathuran, K.; Mohanraj, K.; Natarajan, B. Structural, optical and electrical characterization of nanostructured porous silicon: Effect of current density. *Spectrochim. Acta Part A Mol. Biomol. Spectrosc.* **2016**, *152*, 51–57. [CrossRef] [PubMed]
43. Zamchiy, A.O.; Baranov, E.A.; Khmel, S.Y.; Maximovskiy, E.A.; Gulyaev, D.V.; Zhuravlev, K.S. Deposition time dependence of the morphology and properties of tin-catalyzed silicon oxide nanowires synthesized by the gas-jet electron beam plasma chemical vapor deposition method. *Thin Solid Films* **2018**, *654*, 61–68. [CrossRef]
44. Moumni, B.; Jaballah, A.B. Correlation between oxidant concentrations, morphological aspects and etching kinetics of silicon nanowires during silver-assist electroless etching. *Appl. Surf. Sci.* **2017**, *425*, 1–7. [CrossRef]
45. Chaliyawala, H.A.; Ray, A.; Pati, R.K.; Mukhopadhyay, I. Strong light absorption capability directed by structured profile of vertical Si nanowires. *Optical Mater.* **2017**, *73*, 449–458. [CrossRef]
46. Garzon-Roman, A.; Cuate-Gomez, D.H. Graphene nanoflakes and carbon nanotubes on porous silicon layers by spin coating, for possible applications in optoelectronics. *Sens. Actuators A Phys.* **2019**, *292*, 121–128. [CrossRef]
47. Das, M.; Nath, P.; Sarkar, D. Influence of etching current density on microstructural, optical and electrical properties of porous silicon (PS):n-Si heterostructure. *Superlattices Microstruct.* **2016**, *90*, 77–86. [CrossRef]
48. Al Mortuza, A.; Hafijur, M.; Sinthia, R.; Mou, S.; Islam, M.J.; Ismail, A.B.M. Electrical and optical characteristics of porous silicon impregnated with LaF3 by a novel chemical bath technique. *Curr. Appl. Phys.* **2012**, *12*, 565–569. [CrossRef]
49. Dariani, R.S.; Zabihipour, M. Effect of electrical behavior of ZnO microparticles grown on porous silicon substrate. *Appl. Phys. A* **2016**, *122*, 1047. [CrossRef]
50. Haditale, M.; Zabihipour, A.; Koppelaar, H. A comparison of I-V characteristics of graphene silicon and graphene-porous silicon hybrid structures. *Superlattices Microstruct.* **2018**, *122*, 387–393. [CrossRef]

51. Somvanshi, D.; Jit, S. Analysis of I–V Characteristics of Pd/ZnO Thin Film/n-Si Schottky Diodes with Series Resistance. *J. Nanoelectron. Optoelectron.* **2014**, *9*, 1–6. [CrossRef]
52. Kumar, Y.; Kumr, H.; Rawat, G.; Kumar, C.; Pal, B.N.; Jit, S. Electrical and Optical Characteristics of Pd/ZnO Quantum Dots Based Schottky Photodiode on n-Si. In Proceedings of the 2016 IEEE International Symposium on Nanoelectronic and Information Systems (iNIS), Gwalior, India, 19–21 December 2016.

9

Complex Investigation of High Efficiency and Reliable Heterojunction Solar Cell based on an Improved Cu_2O Absorber Layer

Laurentiu Fara [1,2,*], Irinela Chilibon [3], Ørnulf Nordseth [4], Dan Craciunescu [1], Dan Savastru [3], Cristina Vasiliu [3], Laurentiu Baschir [3], Silvian Fara [1], Raj Kumar [5], Edouard Monakhov [5] and James P. Connolly [6]

1 Department of Physics, Faculty of Applied Sciences, Polytechnic University of Bucharest, 060042 Bucharest, Romania; dan.craciunescu@sdettib.pub.ro (D.C.); silvian.fara@gmail.com (S.F.)
2 Academy of Romanian Scientists, 050091 Bucharest, Romania
3 National Institute of Research and Development for Optoelectronics (INOE-2000), 077125 Bucharest, Romania; chilib@inoe.ro (I.C.); dsavas@inoe.ro (D.S.); icvasiliu@inoe.ro (C.V.); baschirlaurentiu@inoe.ro (L.B.)
4 Institute for Energy Technology (IFE), P.O. Box 40, NO-2027 Kjeller, Norway; Ornulf.Nordseth@ife.no
5 Department of Physics/Center for Materials Science and Nanotechnology (SMN), University of Oslo, P.O. Box 1048 Blindern, N-0316 Oslo, Norway; raj.kumar@smn.uio.no (R.K.); eduard.monakhov@fys.uio.no (E.M.)
6 GeePs (Group of electrical engineering—Paris), UMR CNRS 8507, CentraleSupélec, Univ. Paris-Sud, Université Paris-Saclay, Sorbonne Université, 11 rue Joliot-Curie, 91192 Plateau de Moulon, Gif-sur-Yvette CEDEX, France; jpgconnolly@gmail.com
* Correspondence: lfara@renerg.pub.ro

Abstract: This study is aimed at increasing the performance and reliability of silicon-based heterojunction solar cells with advanced methods. This is achieved by a numerical electro-optical modeling and reliability analysis for such solar cells correlated with experimental analysis of the Cu_2O absorber layer. It yields the optimization of a silicon tandem heterojunction solar cell based on a ZnO/Cu_2O subcell and a c-Si bottom subcell using electro-optical numerical modeling. The buffer layer affinity and mobility together with a low conduction band offset for the heterojunction are discussed, as well as spectral properties of the device model. Experimental research of N-doped Cu_2O thin films was dedicated to two main activities: (1) fabrication of specific samples by DC magnetron sputtering and (2) detailed characterization of the analyzed samples. This last investigation was based on advanced techniques: morphological (scanning electron microscopy—SEM and atomic force microscopy—AFM), structural (X-ray diffraction—XRD), and optical (spectroscopic ellipsometry—SE and Fourier-transform infrared spectroscopy—FTIR). This approach qualified the heterojunction solar cell based on cuprous oxide with nitrogen as an attractive candidate for high-performance solar devices. A reliability analysis based on Weibull statistical distribution establishes the degradation degree and failure rate of the studied solar cells under stress and under standard conditions.

Keywords: silicon tandem heterojunction solar cell; N-doped Cu_2O absorber layer; Al:ZnO (AZO); numerical electro-optical modeling; scanning electron microscopy (SEM); atomic force microscopy (AFM); X-ray diffraction (XRD); spectroscopic ellipsometry (SE); Fourier-transform infrared (FTIR) spectroscopy; degradation degree; failure rate

1. Introduction

Cuprous oxide has attracted much attention for thin film photovoltaic applications due to its high absorption coefficient in the visible region [1]. It is also a nontoxic, low cost material, which is abundant in the Earth's crust. Cu_2O crystallizes in a cubic structure with a lattice constant of 4.269 Å, where the copper (Cu) and oxygen (O) atoms are arranged in face-centered cubic (fcc) and body-centered cubic (bcc) sub-lattices, respectively. Cu_2O is a natural p-type semiconductor with a direct band gap of about 2.1 eV, with negatively charged Cu vacancies and O interstitials [2].

Concerning fabrication, cuprous oxide thin films have been prepared by a wide range of synthesis techniques including reactive sputtering, sol–gel technique, plasma evaporation, thermal oxidation, chemical vapor deposition, anodic oxidation, reactive sputtering, electrodeposition, and so on [3]. Wang et al. have demonstrated that the incorporation of nitrogen in thin films can significantly decrease the resistivity [4]. Nitrogen doping in Cu_2O has been shown to provide an effective approach to reducing the resistivity/enhancing the hole concentration, as the reported carrier concentrations in N-doped Cu_2O thin films are of the order of 10^{16}–10^{18} cm^{-3}.

Sberna et al. [5] analyzed the effects of nitrogen doping, introduced by ion implantation, on the parameters of Cu_2O films. They show that the optimization of structural, optical, and electrical properties of Cu_2O thin films synthesized by non-reactive magnetron sputtering can be achieved by post-deposition thermal annealing in oxygen atmosphere.

Foremost among the challenges in metal oxide solar cells development are the improvement of the performance and reliability of these devices. A potential candidate for high-efficiency metal oxide solar cells, such as the Cu_2O/ZnO/c-Si tandem heterojunction solar cell, could theoretically reach a conversion efficiency above 30% [6]. The key to achieving higher efficiencies for solar cells made from these materials is to tune the optical and electrical properties of absorber material in the solar cell, such as cuprous oxide (Cu_2O).

The conversion efficiency for the current ZnO/Cu_2O heterojunction solar cells is around 8%, while the potential is 14% conversion efficiency. Factors like low power conversion efficiency, high defect density, and large conduction band offset for the metal oxide heterojunction imply that there is room for research to further optimize the device. Probe-scanning instruments can provide valuable information towards the characterization and optimization of solar cell layers. However, interface defect states pose a challenge in reaching efficiencies close to the theoretical maximum. Currently, the conversion efficiency for cuprous oxide heterojunctions has the potential to reach 14% as reported by Takiguchi et al. [7] and a theoretical maximum efficiency of around 19%.

The present study is based on numerical electro-optical modeling and reliability analysis and is dedicated to increasing the performance and reliability of heterojunction solar cells [8–10]; it is correlated with the experimental investigation (morphological, structural, and optical) of an improved Cu_2O absorber layer. Its morphological and structural characterization is based on SEM, AFM, and XRD analyses [11]. The optical characterization of the Cu_2O layer uses FTIR spectroscopy and SE analyses [12]. Furthermore, a statistical reliability study is carried out, analyzing the lifetime, degradation degree, and failure rate of the studied solar cells.

2. Numerical Modeling of Cu_2O Heterojunction Solar Cell

2.1. Diagram of the Solar Cell Simulation Model

A schematic representation of the Cu_2O heterojunction solar cell simulation model is shown in Figure 1. The device structure includes a quartz glass superstrate, an aluminum-doped ZnO (AZO) n+ emitter layer, a Cu_2O p-type absorber layer, and a bottom Cu_2O:N layer.

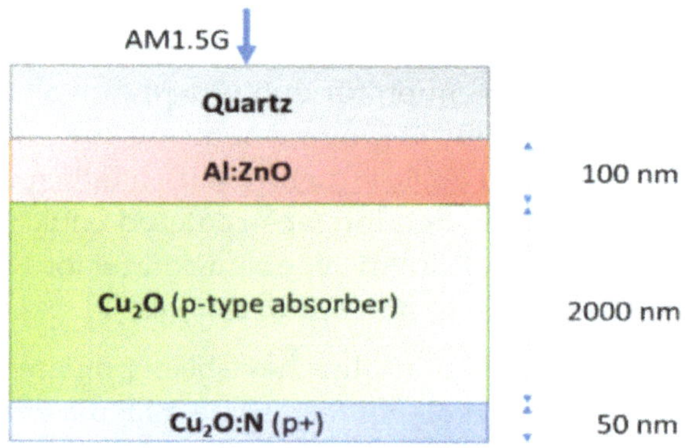

Figure 1. Simplified diagram of the simulated metal oxide solar cell.

2.2. Electrical Modeling and Simulation Results for the Two Subcells of the Tandem Heterojunction Solar Cell

Numerical modeling of the metal oxide heterojunction solar cell was developed based on three simulation software packages, namely Silvaco Atlas, MATLAB, and Quokka 2. The tandem heterojunction solar cell was constituted from two subcells: the metal oxide (Cu_2O) top subcell and c-Si bottom subcell. The electrical properties of the absorber (Cu_2O) and buffer layer (ZnO) from the structure of the top subcell were analyzed using Silvaco Atlas software. The implementation of defects, as well as the study of two efficient materials (ZnO and Ga_2O_3) for the buffer layer were investigated [13–17] and several analyses were considered using the Takiguchi model [7]. A transition interface defect layer (IDL) was introduced as well.

Several comparisons using the Silvaco numerical model were discussed. One of the most interesting buffer materials is ZnO generally, and in particular, the AZO layer (Al:ZnO). This material is very suitable from the point of view of the conduction bandgap offset in regards to Cu_2O. The dependence of the AZO thickness on top cell efficiency is plotted in the Figure 2; the optimum value for the AZO thickness looks to be 0.25 μm. Further analysis examined (1) the dependence of IDL thickness on top cell efficiency (Figure 3), (2) the impact of buffer electron mobility on subcell fill factor (FF) and efficiency (η) (Figure 4), and (3) the impact of buffer and IDL affinity on subcell efficiency (η) (Figure 5). We have remarked that a buffer electron mobility value under 100 $cm^2/V{\cdot}s$ would greatly affect the fill factor and efficiency. Defect densities are very important from the practical point of view in order to obtain required performance, especially for absorber and defect layers. We could compare different buffer and IDL materials from the point of view of their affinity. The optimal affinity interval (3.4–3.6 eV) for the buffer layer would determine the lower conduction band offset with the copper oxide absorber layer. The optimal affinity interval for IDL is larger (from 2.95 to 3.4 eV).

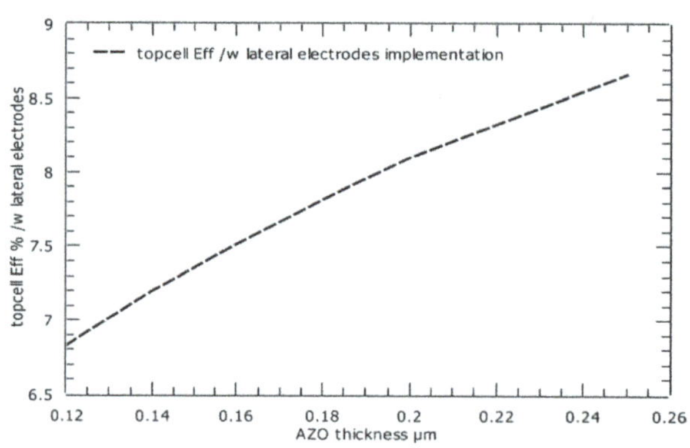

Figure 2. AZO (Al:ZnO) thickness vs. top cell efficiency.

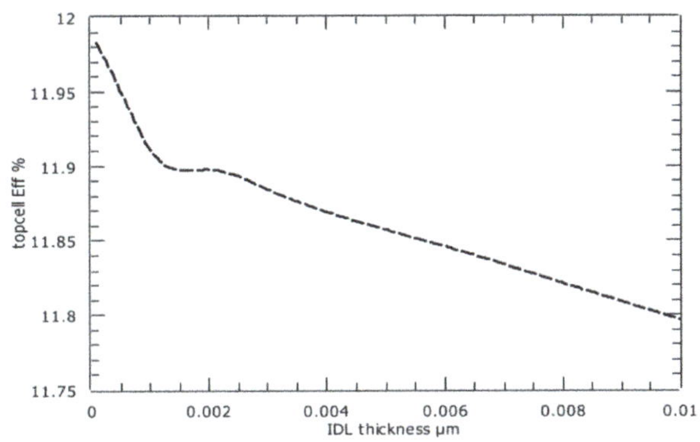

Figure 3. Interface defect layer (IDL) thickness vs. top cell efficiency.

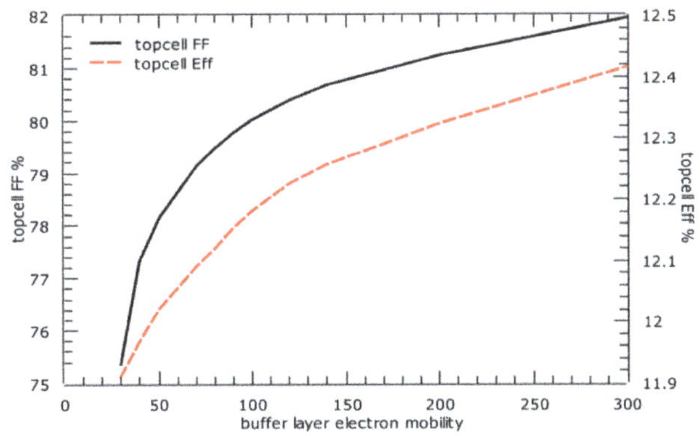

Figure 4. Buffer electron mobility ($cm^2/V{\cdot}s$) vs. top subcell fill factor and efficiency.

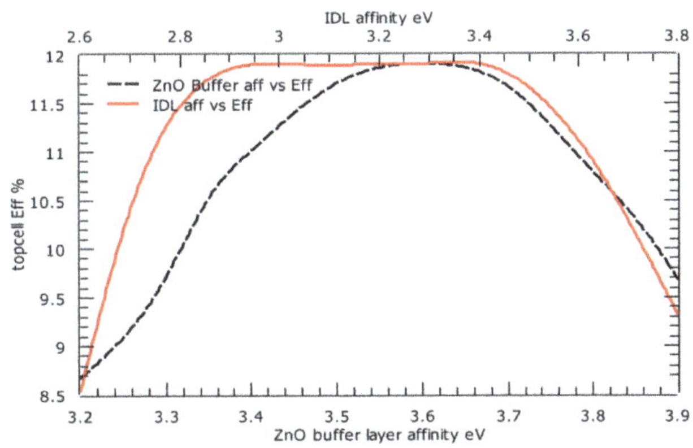

Figure 5. Buffer and IDL affinity vs. subcell efficiency.

A numerical model was adopted using the Quokka 2 software for the bottom subcell. The current-voltage (J-V) characteristics compared the experimental curve with the numerical modeled one (Figure 6).

The numerical model was used to fit to the experimental curve for control and prediction (Table 1) and to establish the essential electrical parameters for the silicone solar subcell.

A comparison between the experimental graph and Quokka modeled one for EQE (external quantum efficiency) is presented in Figure 7. Further improvement of the model would be required in the interval 750–1000 nm.

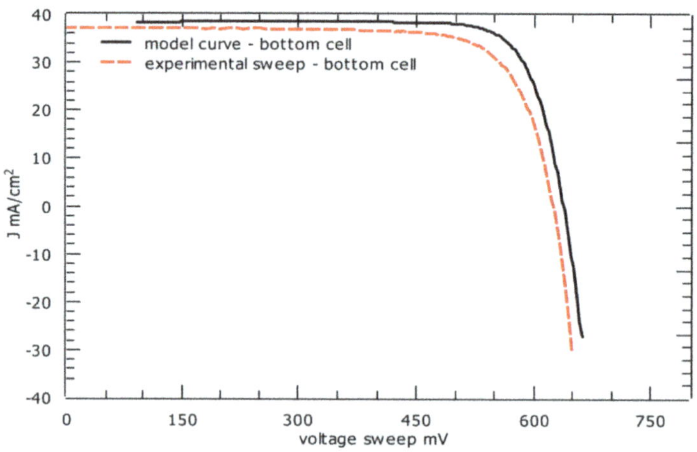

Figure 6. Experimental vs. Quokka 2 modeled J-V curve.

Table 1. Experimental J-V fit parameters for the c-Si subcell.

Parameter Name	Parameter Value
Open circuit voltage, V_{OC} (mV)	625.16
Fill Factor, FF (%)	76.83
Series resistance, R_S (ohm / cm^2):	0.196
Shunt resistance, R_{SH} (ohm / cm^2):	2382.8

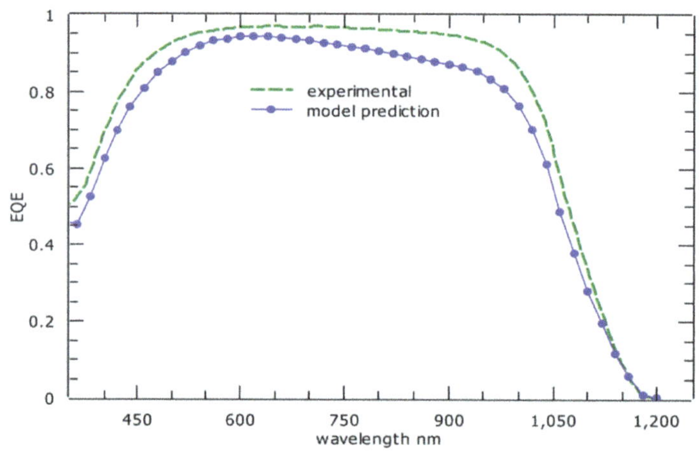

Figure 7. Experimental vs. Quokka modeled external quantum efficiency (EQE) curve.

The rather large current mismatch between the two subcells in tandem configuration, caused by the large band gap difference for the two absorbers, allows for individual optimization of the top and bottom subcells and favors the arrangement of a four-terminal configuration.

That is why it is possible to increase the conversion efficiency of the tandem heterojunction ZnO/Cu_2O silicone solar cell beyond the limit of conventional c-Si single-junction solar cell.

2.3. Optical Modeling and Simulation Results for the Top Subcell

In order to evaluate the optical characteristics of the device, we have used OPAL 2 software developed by PV (photovoltaic) Lighthouse [18]. OPAL 2 is a fast, reliable, and free online software tool that simulates the optics of advanced solar cells [19,20]. The optical modeling was performed for an AM1.5G solar spectrum and with experimentally determined complex refractive indices for the AZO and Cu_2O layers. The modeling was performed in the wavelength range from 300 nm to 800 nm.

Simulated optical functions were the normalized reflectance, absorptance, and transmittance as a function of wavelength for the studied device model, for the samples 1–3 of Cu_2O: N, to be discussed in the experimental section. The simulations are presented in Figure 8. For these simulations, a 100 nm thick AZO layer, a 2 μm thick Cu_2O layer, and a 50 nm thick Cu_2O:N layer were implemented in the model. A standard thickness of 50 nm was set for samples 1–3 of Cu_2O:N used in the solar cell structure.

The maximum reflectance in the 575–800 nm wavelength range for Sample 2 is a bit lower than the maximum values for Samples 1 and 3, as can be seen in Figure 8a. This result could prove useful for reflectance loss optimization.

The absorptance is high for all samples in the 300–525 nm wavelength range (with an average value of 90%). It decreases abruptly in the 525–600 nm range and tends to low values (<10%) in the 600–800 nm range, as seen in Figure 8b.

Transmittance was plotted concurrently for all three samples, as shown in Figure 8c. The transmittance suggests that Samples 1 and 3 yield a better transmission starting from the 560 nm wavelength. The cutoff wavelengths from the transmittance plot are 595 nm, 620 nm, 650 nm, 690 nm, 730 nm, 775 nm, and so on, a periodic spectral pattern being observed. The same approach can be taken for the reflectance plot. For the absorptance plot, the only significant cutoff wavelength is 590 nm. Based on the transmittance data, Tauc plots [20] for Samples 1–3 were calculated and presented in Figure 9.

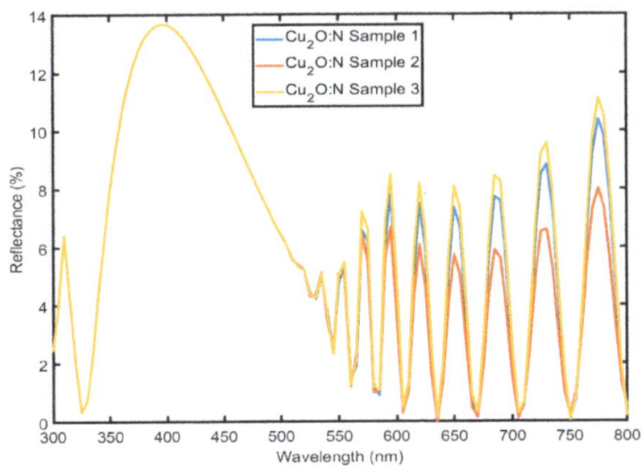

(a)

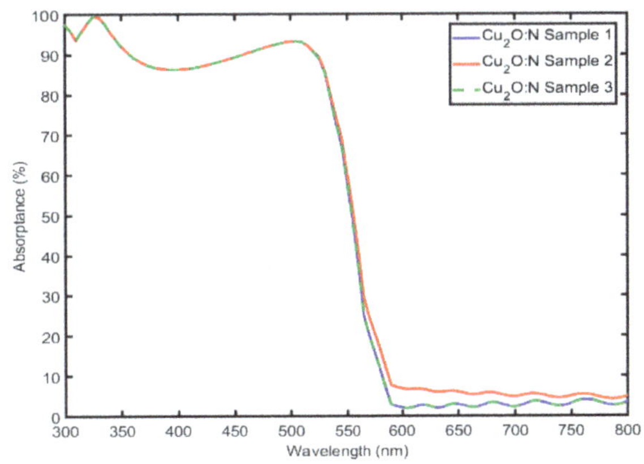

(b)

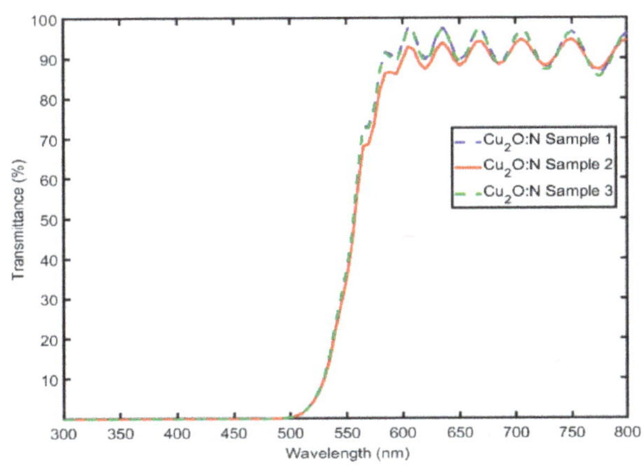

(c)

Figure 8. Percentage (**a**) reflectance, (**b**) absorptance, and (**c**) transmittance as a function of wavelength for the metal oxide structure.

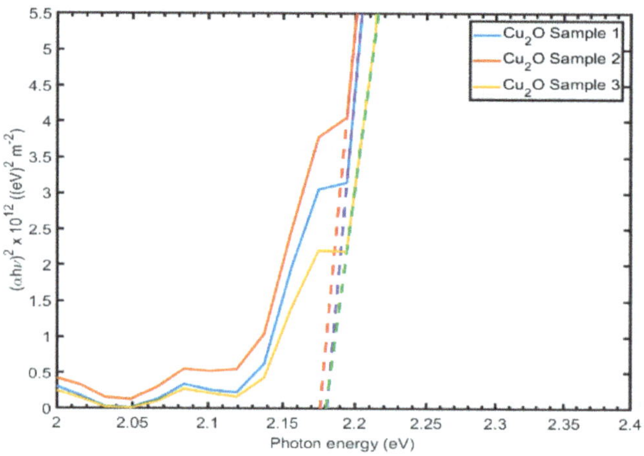

Figure 9. Tauc plots of Samples 1, 2, and 3 deposited on quartz substrate.

Tauc plots indicate an optical bandgap around 2.18 eV for Sample 1, 2.17 eV for Sample 2, and around 2.18 eV for Sample 3, which is in agreement with the ellipsometry measurements (see experimental section).

2.4. Relevance of the Best Simulation Tools for Numerical Modeling and Optimization of Metal Oxide Solar Devices

The simulation methodology could be used for design and material performance optimization. Two simulators were considered to be the best for the two subcells; Silvaco is the optimum tool for the top metal oxide solar subcell and absorbs the high energy part of sunlight, while PC1D is the most efficient tool for the bottom silicone solar subcell and absorbs the low energy part of sunlight. The simulated device with optimized parameters could be validated by comparing the numerical modeled results and performance with the experimental ones offered by an industrial manufactured solar cell with identical parameters.

3. Experimental Investigation of N-Doped Cu_2O Absorber Layer

3.1. Cu_2O Thin Film Fabrication and Overall Characterization

Cu_2O thin films were deposited on quartz substrates by direct current (DC) magnetron sputtering system (Semicore Triaxis) on 1 cm × 1 cm quartz substrates at 400 °C by varying the N_2 and Ar/O_2 flow [21]. Samples 1 and 2 were deposited at 400 °C with nitrogen flows 1 and 3 sccm (standard cubic centimeters per minute), respectively, while the Ar/O_2 flow was maintained constant. Sample 3 was deposited at 400 °C with Ar/N_2 flow of 1 sccm while the O_2 flow was fixed. The pre-sputtering time was 15 min, to avoid any contamination. The quartz substrates were cleaned in piranha solution and rinsed in deionized water, blown dry with nitrogen, and loaded into the deposition chamber. The same deposition time was used for all samples. The deposition power was fixed at 100 W. The base pressure before deposition was below 5×10^{-7} Torr while the sample stage was rotated at a constant speed of 12 rpm during deposition. The Hall effect measurements (LakeShore 7604) using the van der Pauw method were carried out to determine the hole mobility, resistivity, and hole carrier density.

The hole carrier density of as-deposited Sample 0 is ~10^{15} cm^{-3} with low mobility (~13 cm^2/Vs) and resistivity (~200 Ωcm), respectively. The hole carrier density for Samples 1 and 2 varied from ~10^{17}–10^{19} cm^{-3} with low carrier mobility (~0.1–1 cm^2/Vs) and resistivity (3–13 Ωcm), respectively, while Sample 3 has carrier density ~10^{16} cm^{-3} with resistivity ~54 Ωcm and carrier mobility ~1.5 cm^2/Vs. A reduced resistivity was achieved in the N-doped samples (1, 2, and 3) compared to the undoped Sample 0. This may reflect the presence of N in the DC Cu_2O film magnetron sputtering process influencing the electrical characteristics of solar cell as mentioned previously.

The process characteristics and layer structure of Cu_2O:N thin film samples, deposited by DC magnetron sputtering on quartz substrate, are presented in Table 2.

Table 2. Process characteristics and layer structure of Cu_2O:N thin films.

Sample Name	Gas Flow Conditions	Grains Size (nm)
Sample 0	Ar/O_2 was fixed at 42.5/7.5 sccm	20–90
Sample 1	N-doped 1 sccm (Ar/O_2 was fixed at 42.5/7.5 sccm)	20–48
Sample 2	N-doped 3 sccm (Ar/O_2 was fixed at 42.5/7.5 sccm)	30–54
Sample 3	N-doped 1 sccm (Ar was varied at 41.5 sccm, O_2 was fixed at 7.5 sccm)	40–88

3.2. *Morphological and Structural Characterization of the Cu_2O:N Films*

The surface morphology of the Cu_2O:N thin films was studied by SEM with the Quanta Inspect F 50 scanning electron microscope. Topographic AFM imaging was also performed by AFM with the Veeco Innova atomic force microscope, in the tapping mode, with 2.5 μm/s scanning speed. AFM images have a resolution of 512 × 512 pixels. The software used for image analysis was SPM Lab Analysis v.7.0 (Version 7.0, Veeco Innova, Bruker-Milton, ON, Canada, Santa Barbara-CA, USA.)

3.2.1. Scanning Electron Microscopy (SEM)

Experimental measurements for the determination of surface morphology were performed on Cu_2O:N thin films deposited by the DC magnetron sputtering on the quartz substrate under different gas flow conditions. Figure 10 shows the SEM images of four samples (see Table 2): one undoped Cu_2O film (Sample 0) and three Cu_2O:N thin film samples (Samples 1–3).

The grain size in the deposited layers is listed in Table 2 and compared with undoped Cu_2O film (Sample 0). With the increase of the N_2 flow, the films display a uniform layer surface with larger particle size (Samples 1 and 2). A varied Ar pressure induces a non-uniform particle size distribution (Samples 1 and 3). The particle size distribution of the undoped Cu_2O layer is non-uniformly distributed in the range 20–90 nm. As result, the presence of N_2 and an Ar/O_2 fixed flow improves the size uniformity of the deposited Cu_2O layers.

Figure 10. *Cont.*

Figure 10. SEM images of Cu_2O thin films deposited on quartz substrate: (**a**) Sample 0, (**b**) Sample 1, (**c**) Sample 2, and (**d**) Sample 3.

3.2.2. Atomic Force Microscopy (AFM)

The grain size, including surface roughness of Cu_2O:N thin film deposited on quartz substrate, was measured by topographic AFM imaging. The surface morphologies for each of the N-doped thin film samples compared with Sample 0 are shown in Figure 11 as 2D AFM images, along with their corresponding surface profile histograms, and in Figure 12 as 3D AFM images. The root-mean-square surface roughness (R_{RMS}) for each sample was extracted from their AFM images [22,23].

The histograms represent the statistical distribution of the height surface profile of the Cu_2O samples derived from the AFM images and show a higher distribution in the first half of the height range for Samples 1 and 2. However, the height profile distribution of Sample 3 shows less roughness. This is related to the deposition conditions and in accordance with the R_{RMS} parameter calculations in Table 3, which shows a lower R_{RMS} for this sample. According to the statistical distribution of the surface height profile, Samples 1, 2, and 3 have the highest frequency at a surface height of 26 nm, 25 nm, and 17 nm, respectively. For validation, R_{RMS} was derived from the histograms based on the 2D AFM images, processed in MATLAB, and presented in Table 3.

Table 3. R_{RMS} of Cu_2O thin films and height distribution standard deviation, derived from histograms of 2D AFM images.

Sample Deposited on Quartz	R_{RMS} (nm)	Height Distribution Standard Deviation (nm)
Sample 0	6.3	7.8
Sample 1	4.7	5.8
Sample 2	4.2	4.5
Sample 3	3.9	4.2

According to Table 3, the roughness deviation values follow the same trend as the R_{RMS} values, where Sample 1 has the highest deviation and Sample 3 presents the lowest one. Additionally, the undoped Cu_2O film (Sample 0) presents a greater roughness than the doped Samples 1–3.

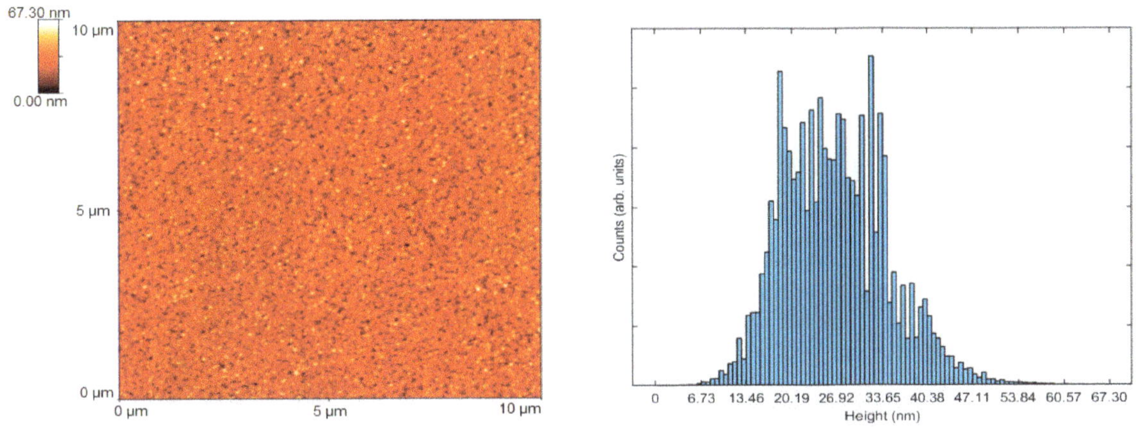

Sample 0

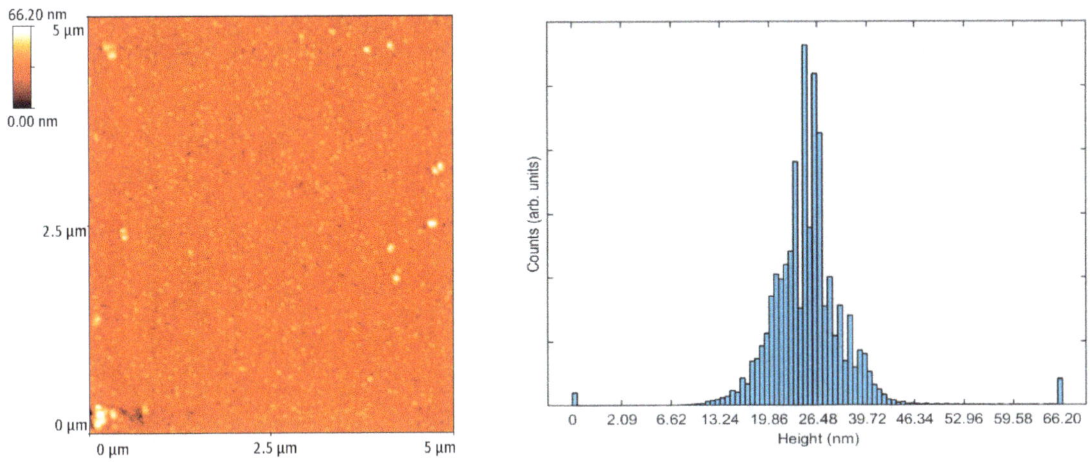

Sample 1

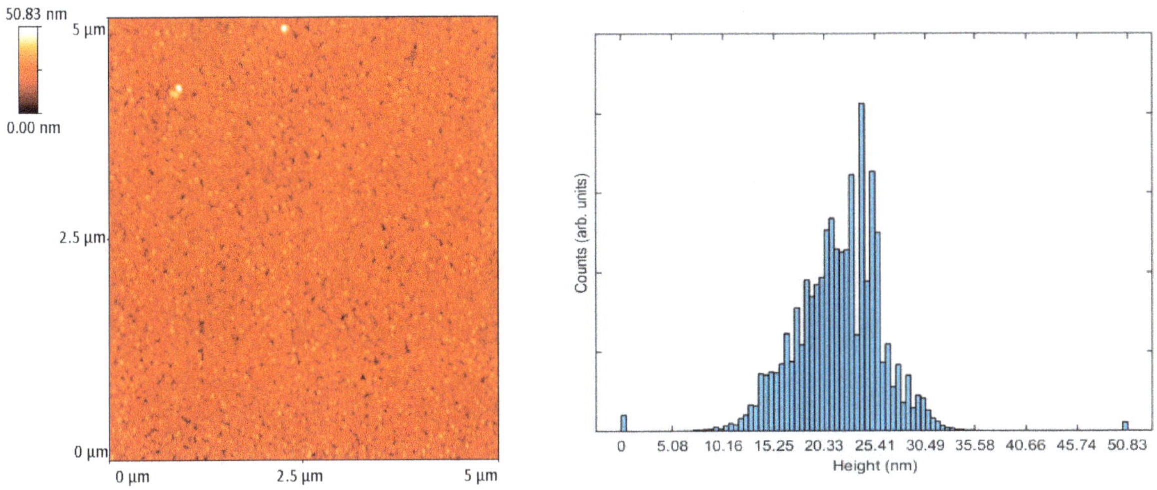

Sample 2

Figure 11. *Cont.*

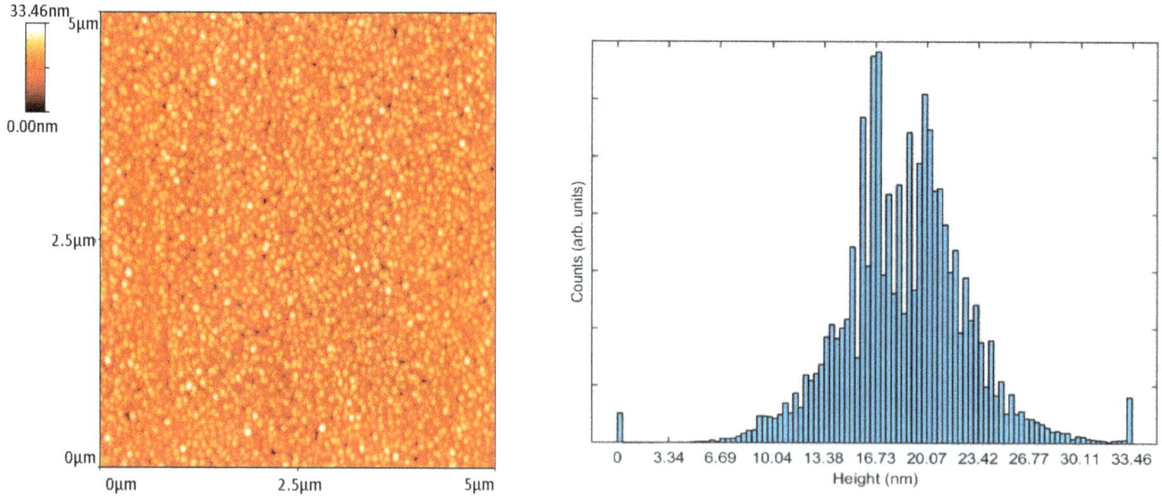

Sample 3

Figure 11. 2D AFM images of Samples 0–3 and their corresponding histograms.

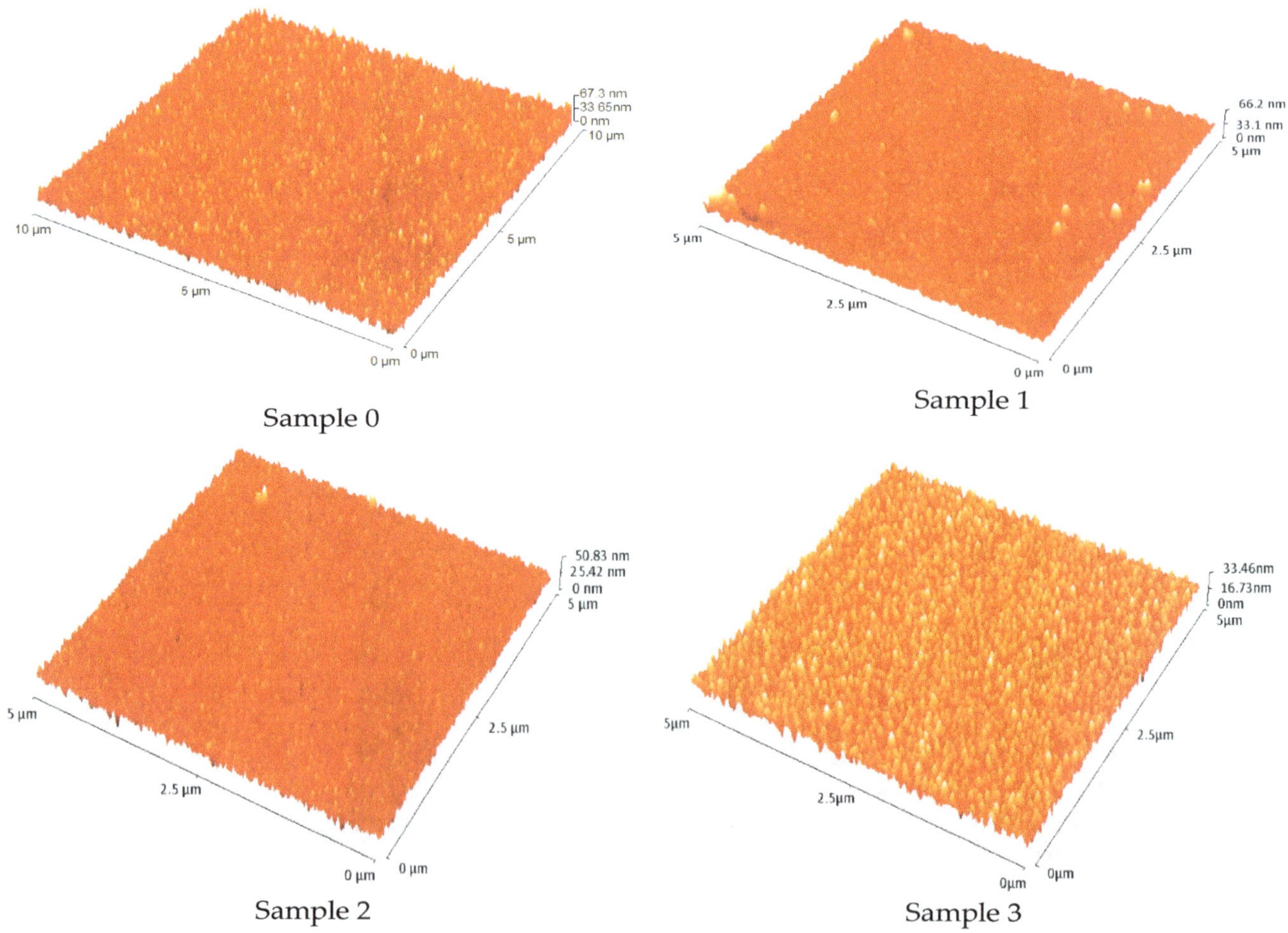

Figure 12. 3D AFM images of Samples 0–3.

3.2.3. Structural Characterization by XRD

The XRD pattern was obtained by a Bruker AXS D8 Discover Diffractometer, using Cu Kα-radiation and a Bragg–Brentano configuration. The background and Kα2 were stripped from the XRDs using the EVA software. The dominant XRD peaks were recorded in 2-theta range from 30–60°.

Figure 13 displays the XRD patterns of all samples, recorded in the range 30–60°. XRD patterns show that all films were dominated by a (200) reflection peak at 42.61° and a (111) reflection peak corresponding to 36.45°. The XRD pattern is not significantly influenced by nitrogen doping [24]. The Cu_2O peaks were determined by comparing experimental XRD peak patterns with the standard Powder Diffraction cards (ICDD patterns: 01-071-3645) [25].

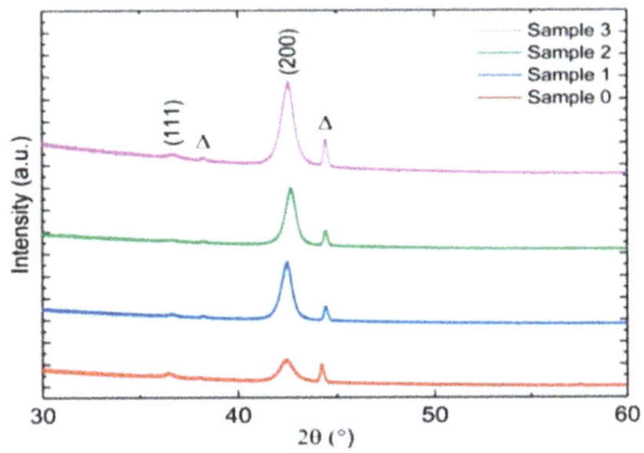

Figure 13. XRD patterns of Samples 0–3 in the range 30–60°, where Δ represents the background peak.

XRD patterns of the Cu_2O:N films prepared under various N_2 flow rates, together with the data of the Cu_2O reference films, show highly oriented (111) planes. For the Cu_2O:N samples, the peak intensity of the (111) plane decreases significantly with increasing of N_2 flow rate, and (200) planes show up in the heavy N-doping range [26].

3.3. *Optical Characterization*

Optical characterization involved two tools: FTIR spectroscopy and spectroscopic ellipsometry. FTIR spectroscopy analysis was carried out using a Perkin Elmer type Spectrum 100 spectrometer, with UATR accessory—PIKE GladiATR, in the wavenumber range 430–2000 cm^{-1} with a spectral resolution of 0.4 cm^{-1}. The refractive index and extinction coefficient of Cu_2O:N thin film were determined by spectroscopic ellipsometry, in the wavelength range from 190 to 2100 nm using an UVISEL spectroscopic ellipsometer from HORIBA Jobin Yvon. A model fit to the measured ellipsometry parameters was obtained using DeltaPsi 2.6 software (Version 2.6, HORIBA, Jobin Yvon, Palaiseau, France).

3.3.1. FTIR Spectroscopy

The FTIR transmission spectra obtained for Cu_2O:N thin films are shown in Figure 14 and the transmission peaks are listed in Table 4.

Table 4. Transmission peaks for Cu_2O:N thin films.

Sample Name	νCu_2O (cm^{-1})
Sample 1	608
Sample 2	616
Sample 3	615

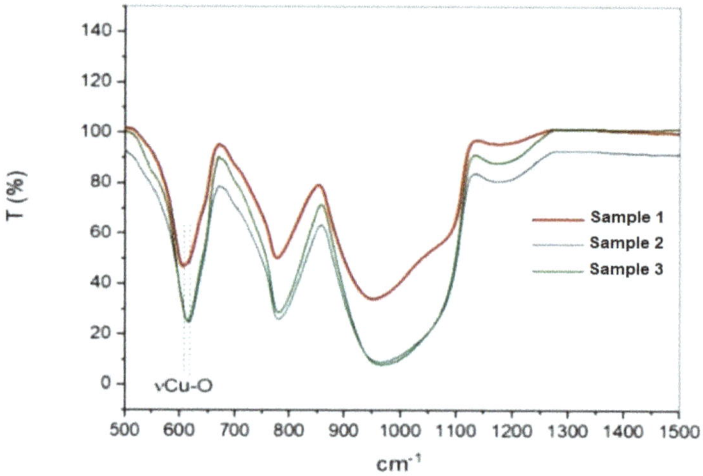

Figure 14. FTIR transmission spectra for Cu_2O:N thin film: Sample 1, Sample 2, and Sample 3.

The FTIR spectra of all samples deposited on quartz displayed the peaks corresponding to the vibrational modes of Si-O belonging to the quartz substrate (780 cm^{-1}, broadband centered at 980 cm^{-1} with a shoulder at 1080 cm^{-1}) [26,27]. The band located at 608–616 cm^{-1} corresponds to the stretching vibrational mode of Cu-O in Cu_2O phase. There is an increasing trend in the intensity of the Cu-O absorption band from Sample 1 to Sample 3, suggesting an increased concentration of N_2. This result is in agreement with the recent theoretical predictions that nitrogen mainly substitutes Cu in the molecular form, (N_2) Cu, in N-doped Cu_2O rather than as substitutional N on the O site, (NO) [5]. The decrease in resistivity, coherently with the increase in N doping concentration, was explained by annealing-driven substitution of oxygen atoms with implanted N atoms. The N doping may induce any definite variation in energy band structure of Cu_2O films [25].

In general, the photoluminescence (PL) spectra of Cu_2O films contain two main signals ranging from 450 to 650 nm, which are considered to be derived from free excitons and the bound excitonic region [28].

3.3.2. Spectroscopic Ellipsometry (SE)

The ellipsometry parameters of Cu_2O:N thin films deposited on quartz substrate were investigated with a resolution of 27 points in the range 1.5–4.1 eV (with step 0.1 eV). The investigated parameters were the band gap and the surface roughness [29,30] and were listed in Table 5.

Table 5. Investigated (SE) parameters of Cu_2O:N thin films.

Sample Name	Sample 1	Sample 2	Sample 3
Band gap (eV):	2.17	2.14	2.17
Surface roughness (nm):	15.9	19.6	17.2

By the Tauc–Lorentz oscillator methodology [31], modeling was performed to determine the band gap, surface roughness, refractive index (n), and the extinction coefficient (k) in the range 300–800 nm (from visible to near infrared) for Cu_2O thin films deposited on quartz substrates.

The model described a layer of Cu_2O deposited on the quartz substrate with a hard surface layer. The surface roughness was modeled by mixing the optical constants of the Cu_2O material surface and the air voids (on average, 90% Cu_2O and 10% air voids).

It is noticed that the surface roughness of Cu_2O film increases with the level of N_2 pressure (Sample 1 and Sample 3 vs. Sample 2), from 15.9 nm to 19.6 nm (Table 5).

The presence of N_2 during deposition of Cu_2O films on quartz substrate produces small changes over n and k in the range 300–600 nm for Samples 1–3 and 300–400 nm for Sample 2, as can be seen in Figure 15.

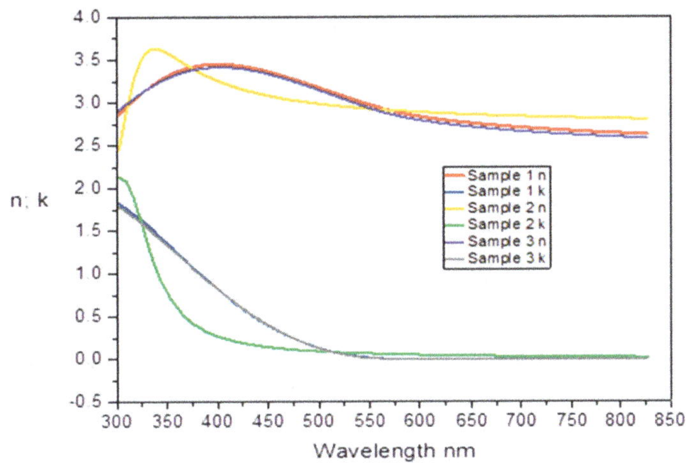

Figure 15. Graphs of n and k coefficients for Samples 1, 2, and 3 of Cu_2O:N thin films; λ is wavelength.

The changes of $n(\lambda)$ (yellow curve) and $k(\lambda)$ (green curve) coefficients observed in Figure 15 for Sample 2 (the highest N-doped sample) are associated with a structural change in Cu_2O induced by nitrogen doping. This can be a promising method for modifying the optical properties of the metal oxide [28,32] The crystal size decreases from approximately 23 nm to approximately 15 nm with increasing the N-doping concentration in the small N-doping range, where as it is almost unchanged (18–20 nm) in the heavy N-doping range [26].

For example, undoped Cu_2O films were deposited on quartz substrates by the RF magnetron sputtering system with a Cu_2O target (99.99%, purity); with the Ar flow rate at 20 sccm, the average transmittances in the visible range were 40–60%, and the band gap of Cu_2O films was approx. 18 eV [29].

3.4. General Comments on Experimental Investigation and Prospects of Heterojunction Cuprous Oxide Solar Cells with Nitrogen

The experimental investigation of N-doped Cu_2O thin films show they have good potential as an absorber layer for photovoltaic applications:

- The doped samples are characterized by higher whole carrier densities, lower carrier mobilities, and lower resistivities in comparison with the undoped sample;
- The SEM images showed the improvement in the size uniformity of the deposited Cu_2O layers due to N presence;
- The AFM measurements determined that the roughness is slightly influenced by the N doping, which could affect the electrical properties of solar cells. The histograms derived from the AFM images showed the surface profile height for the studied Cu_2O:N samples and emphasized their roughness deviation;
- The X-ray analysis of Cu_2O films put in evidence that all such films are polycrystalline and nitrogen doping does not show any significant change in the XRD pattern. A reduced resistivity was achieved in the N-doped samples compared to the undoped sample;
- The FTIR spectra of the samples displayed peaks corresponding to the vibrational modes of Si-O belonging to the quartz substrate;
- The SE analysis stressed the N influence over the optical properties of the Cu_2O thin films. The changes of refraction and extinction coefficients were given by the Cu_2O:N thin films associated with the structural change in Cu_2O induced by N-doping.

At the same time, it could be remarked that the N-doping can lead to the following on the heterojunction copper oxide solar cells with nitrogen:

- regulation of the band gap enabling increased transmittance;
- improved carrier lifetime and transport;
- increased conversion efficiency of the solar cell device.

All these results highlight the N-doped cuprous oxide heterojunction solar cell as a promising candidate for high efficiency solar devices.

4. Reliability Study

4.1. Methodological Tools

Definition of solar cell reliability: probability of device performance to fit to specific performance parameters for clear operational conditions without deviation outside determined tolerance limits.

There are different accelerated tests to be done regarding the reliability of a solar cell:

(1) Qualitative accelerated test: offers information regarding the solar cell malfunctions or failure ways;

(2) High accelerated life test (HALT) offers information regarding the solar cell failure mechanisms and average lifetime;

(3) High accelerated stress screening (HASS) offers information regarding the solar cell operating time and failure rate within the operating temperature range

In order to perform an accurate reliability analysis of the solar cell, the SYNTHESIS simulation tools developed by ReliaSoft company [33] were used, namely: (1) *the Weibull++ tools*, which performs life data analysis using multiple lifetime repartitions with a reliability-oriented interface; (2) *the ALTA tool*, which allows a quantitative analysis of accelerated life tests.

4.2. Solar Cells Thermal Stability Characteristics

Three main parameters can be studied using the Weibull statistical repartition: (1) *the solar cell lifetime* in different conditions: (2) *the degradation degree* based on the temperature influence; and (3) *number of defects (failures)*.

In order to make the most accurate predictions for the lifetime of the analyzed solar cells, the Weibull statistical repartition method (Figure 16a) of the Weibull++ tool was employed [33].

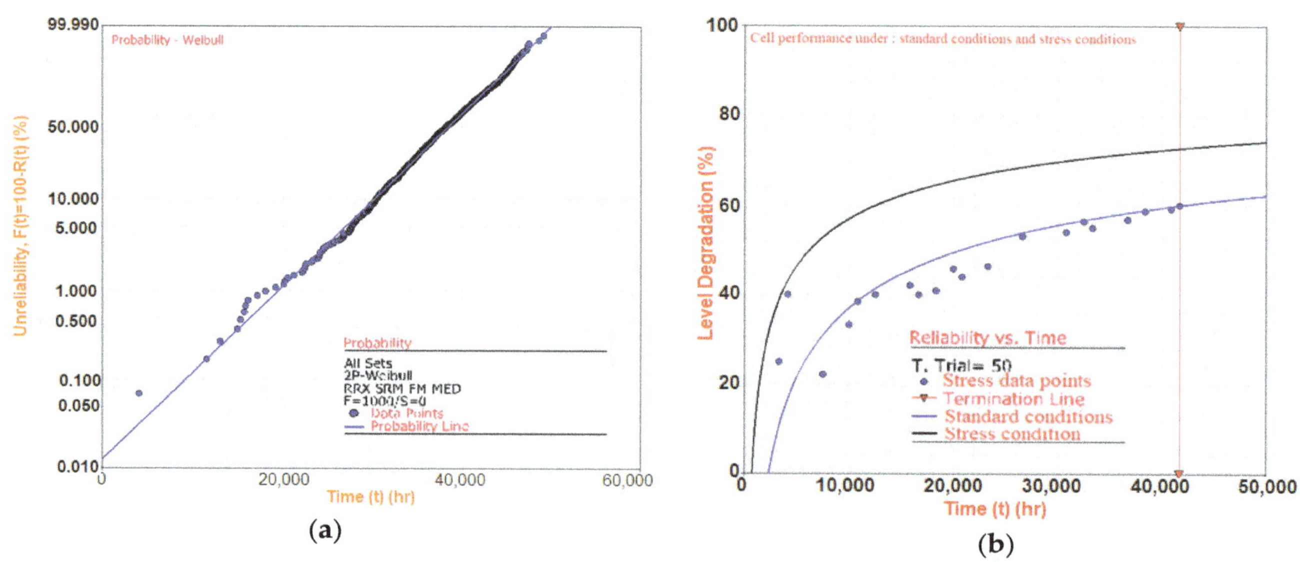

Figure 16. (**a**) Weibull distribution for the studied solar cell lifetime; (**b**) solar cell performance and degradation degree of the analyzed solar cell under standard and stress conditions.

The degradation degree of the analyzed solar cell can be characterized by the temperature influence on the device [34–36] and is obtained for two cases: (1) under stress conditions (black curve) and (2) under standard conditions (blue curve) (Figure 16b).

Based on the stress results of the HASS test presented in Figure 16b, it is possible to determine, by numerical simulation using *the ALTA tool*, the limiting value of specific solar cell operating time in normal conditions, as seen in Figure 17a. The HASS test also establishes the probable rate of solar cell failure within the operation temperature range of 50–100 °C (Figure 17b). It is remarked that the expected failures rate dependence increases strongly with temperature.

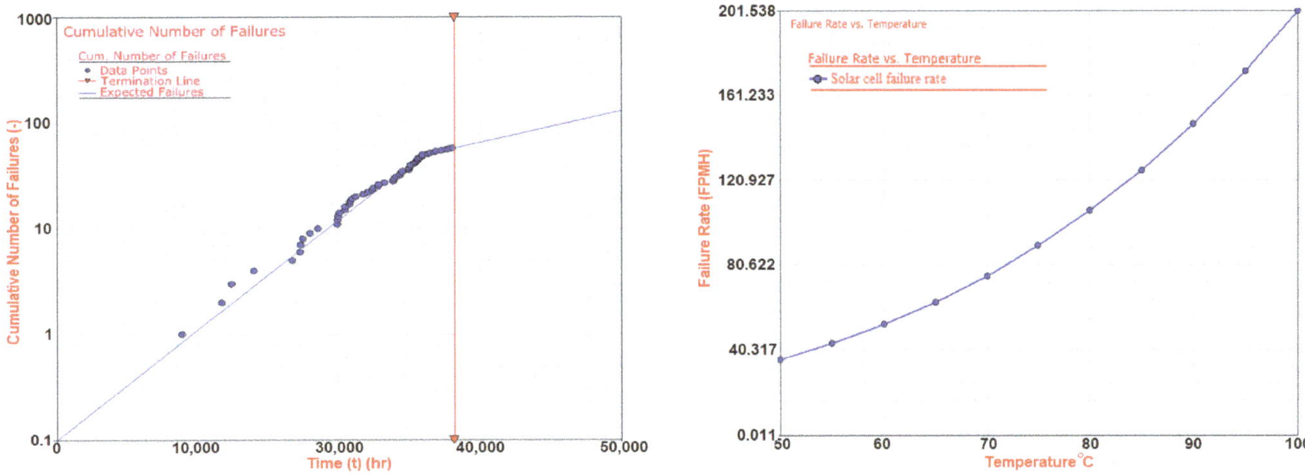

(a) (b)

Figure 17. (**a**) Cumulative number of failures as function of time (using ALTA module). (**b**) Baseline survival of solar cell simulation (using ALTA module) in temperature and stress conditions.

The simulated results obtained from this reliability analysis consist both in identifying the solar cell failure modes, as well as to predict the solar cell lifetime in normal operation conditions. At the same time, these simulations enable evaluation of the degradation time and number of defects for the analyzed solar cell. The number of defects increases alarmingly starting at 20,000 operating hours, and by the end of the accelerated HASS test, the solar device would fail (Figure 17a).

4.3. Further Work

The simulation tools used in this work, namely Weibull++ and ALTA, have limited capabilities for analyzing different types of advanced solar cells based on different materials. That is why new simulation tools must be considered for a further work dedicated to reliability studies and experimental analyses of metal oxide solar cells. The thermal stability of such devices is essential during solar cell operation. It could be noted that the Cu_2O:N layer is very stable because its annealing for 90 min does not degrade its nitrogen-doping, as reported by Ye F. et al. [37].

5. Conclusions

The optimization of a silicon tandem heterojunction solar cell with a Cu_2O absorber layer has been analyzed by numerical modeling based on its splitting into two subcells: a ZnO/Cu_2O top subcell and a c-Si bottom one. The buffer layer thickness and affinity are essential parameters for implementation of such devices defined by a low conduction band offset.

The optical characteristics of a Cu_2O heterojunction solar cell were obtained by numerical modeling, allowing the evaluation of optical losses and determination of maximum absorptance of the device.

A comprehensive characterization (morphological, structural, and optical) of the copper oxide layer with nitrogen doping enables us to propose the metal oxide heterojunction solar cell with nitrogen as a promising candidate for advanced high efficiency solar cells.

The main reliability parameters of metal oxide solar cells, namely the failure rate and degradation degree, were calculated and further work will stress the influence on electrical and optical properties of these devices from the thermal stability point of view.

Author Contributions: Conceptualization, L.F., I.C., and Ø.N.; methodology, L.F., I.C., R.K., and D.C.; formal analysis, L.F.; investigation, L.F., I.C., R.K., C.V., L.B., and D.S.; funding acquisition, L.F. and I.C.; software, S.F. and D.C.; validation, L.F., I.C., Ø.N., J.P.C., and D.C.; resources, I.C., R.K., and L.F.; data curation, S.F.; writing—original draft preparation, L.F., I.C., and S.F.; writing—review and editing, L.F., and I.C.; visualization, L.F. and I.C.; supervision, L.F., I.C., and J.P.C.; project administration, J.P.C., Ø.N., E.M., L.F., and I.C. All authors have read and agreed to the published version of the manuscript.

Acknowledgments: This work was supported by the project "Multiscale in modelling and validation for solar photovoltaics" (Multiscale Solar) MP1406, 2015–2019, for IPVF (Institute Photovoltaïque de l'Île de France)-GeePs, PUB (Polytechnic University of Bucharest) and INOE, funded by the European Commission through COST program. This research activity was conducted under the research project "High-performance tandem heterojunction solar cells for specific applications (SOLHET)", for IFE, UiO, PB and INOE funded by the Research Council of Norway (RCN), project no. 251789, and the Romanian Executive Agency for Higher Education, Research, Development and Innovation Funding (UEFISCDI), contracts no. 34/2016 and no. 35/2016, through the M-ERA.NET program, and project no. PN 19_18.01.01/2019, Ad. Act. 2020, for INOE, funded by Ministry of Research and Innovation (MCI), through the NUCLEU program and institutional project 19PFE/2018, Stage IV/2020.

References

1. Wang, Y.; Miska, P.; Pilloud, D.; Horwat, D.; Mücklich, F.; Pierson, J.F. Transmittance enhancement and optical band gap widening of Cu_2O thin films after air annealing. *J. Appl. Phys.* **2014**, *115*, 073505. [CrossRef]
2. Jung, Y.S.; Choi, H.W.; Kim, K.H. Properties of p-type N-doped Cu_2O thin films prepared by reactive sputtering. *Jpn. J. Appl. Phys.* **2014**, *53*, 11RA10. [CrossRef]
3. Siddiqui, H.; Parra, M.R.; Pandey, P.; Singh, N.; Qureshi, M.S.; Haque, F.Z. A review: Synthesis, characterization and cell performance of Cu_2O based material for solar cells. *Orient. J. Chem.* **2012**, *28*, 1533–1545. [CrossRef]
4. Wang, Y.; Ghanbaja, J.; Horwat, D.; Yu, L.; Pierson, J.F. Nitrogen chemical state in N-doped Cu_2O thin films. *Appl. Phys. Lett.* **2017**, *110*, 131902. [CrossRef]
5. Sberna, P.M.; Crupi, I.; Moscatelli, F.; Privitera, V.; Simone, F.; Miritello, M. Sputtered cuprous oxide thin films and nitrogen doping by ion implantation. *Thin Solid Film.* **2016**, *600*, 71–75. [CrossRef]
6. Mitroi, R.M.; Ninulescu, V.; Fara, L. Tandem solar cells based on Cu_2O and c-Si subcells in parallel configuration: Numerical simulation. *Int. J. Photoenergy* **2017**, *2017*, 1–6. [CrossRef]
7. Takiguchi, Y.; Miyajima, S. Device simulation of cuprous oxide heterojunction solar cells. *Jpn. J. Appl. Phys.* **2015**, *54*, 112303. [CrossRef]
8. Nordseth, O.; Fara, L.; Kumar, R.; Foss, S.E.; Dumitru, C.; Muscurel, V.F.; Drăgan, F.; Crăciunescu, D.; Bergum, K.; Haug, H.; et al. Electro-optical modeling of a ZnO/Cu_2O subcell in a silicon-based tandem heterojunction solar cell. In Proceedings of the 33rd European Photovoltaic Solar Energy Conference and Exhibition, Amsterdam, The Netherlands, 25–29 September 2017; pp. 172–177. [CrossRef]
9. Nordseth, O.; Haug, H.; Kumar, R.; Bergum, K.; Drăgan, F.; Craciunescu, D.; Fara, L.; Chilibon, I.; Monakhov, E.; Foss, S.E.; et al. Performance optimization of a four-terminal Cu_2O/c-Si Tandem heterojunction solar cell. In Proceedings of the 35th European Photovoltaic Solar Energy Conference and Exhibition, Bruxelles, Belgium, 24–28 September 2018; pp. 29–34. [CrossRef]
10. Dumitru, C.; Fara, L.; Nordseth, O.; Chilibon, I.; Kumar, R.; Svensson, B.G.; Dragan, F.; Muscurel, V.; Craciunescu, D.; Sterian, P. Electro-optical analysis and numerical modeling of Cu_2O as the absorber layer in advanced solar cells. In Proceedings of the 2018 International Conference on Photovoltaic Science and Technologies (PVCon), Ankara, Turkey, 4–6 July 2018; pp. 1–7. [CrossRef]

11. Chilibon, I.; Fara, L.; Nordseth, O.; Kumar, R.; Svensson, B.G.; Sterian, P.; Dumitru, C.; Dragan, F.; Muscurel, V.; Vasiliu, C. Structural and electrical analysis of Cu_2O layers for solar cell application. *Ann. Acad. Rom. Sci.* **2018**, *11*, 53–60.
12. Chilibon, I.; Fara, L.; Nordseth, O.; Kumar, R.; Svensson, B.G.; Sterian, P.; Dumitru, C.; Dragan, F.; Muscurel, V.; Vasiliu, C. Characterization of Cu_2O thin films used in solar cells by fluorescence and FTIR spectroscopy. *Ann. Acad. Rom. Sci. Ser. Sci.* **2018**, *11*, 61–68.
13. Lloyd, M.A.; Siah, S.C.; Brandt, R.E.; Serdy, J.; Johnson, S.W.; Lee, Y.S.; Buinassisi, T. Intrinsic defect engineering of cuprous oxide to enhance electrical transport properties for photovoltaic applications. In Proceedings of the IEEE 40th Photovoltaic Specialists Conference (PVSC), Denver, CO, USA, 8–13 June 2014; Volume 2016, pp. 3443–3445.
14. Tolstova, Y.; Omelchenko, S.T.; Blackwell, R.E.; Shing, A.M. Polycrystalline Cu_2O photovoltaic devices incorporating Zn (O, S) window layers. *Sol. Energy Mat. Sol. Cells* **2017**, *160*, 340–345. [CrossRef]
15. Minami, T.; Nishi, Y.; Miyata, T. High-efficiency Cu_2O-based heterojunction solar cells fabricated using a Ga_2O_3 thin film as n-type layer. *Appl. Phys. Express* **2013**, *6*, 044101. [CrossRef]
16. Robertson, J.; Falabretti, B. Band offsets of high K gate oxides on III-V semiconductors. *J. Appl. Phys.* **2006**, *100*, 014111. [CrossRef]
17. Brandt, R.E.; Young, M.; Hejin, H.P.; Dameron, A.; Chua, D.; Lee, S.Y.; Teeter, G.; Gordon, R.; Buonassisi, T. Band offsets of n-type electron-selective contacts on cuprous oxide (Cu_2O) for photovoltaics. *Appl. Phys. Lett.* **2014**, *105*, 263901. [CrossRef]
18. PV Lighthouse. Available online: https://www.pvlighthouse.com.au/ (accessed on 20 January 2020).
19. Baker-Finch, S.C.; McIntosh, K.R. OPAL 2: Rapid optical simulation of silicon solar cells. In Proceedings of the 38th IEEE Photovoltaic Specialists Conference, Austin, TX, USA, 3–8 June 2012; pp. 265–271. [CrossRef]
20. Dumitru, C.; Muscurel, V.; Nordseth, O.; Fara, L.; Sterian, P. Optimization of electro-optical performance and material parameters for a tandem metal oxide solar cell. In Proceedings of the International Conference on Computational Science and Its Applications–ICCSA, Melbourne, Australia, 2–5 July 2018; pp. 573–582. [CrossRef]
21. Murali, D.S.; Kumar, S.; Choudhary, R.J.; Wadikar, A.D.; Jain, M.K.; Subrahmanyam, A. Synthesis of Cu_2O from CuO thin films: Optical and electrical properties. *AIP Adv.* **2015**, *5*, 047143. [CrossRef]
22. Bennett, J.M.; Mattson, L. *Introduction to Surface Roughness and Scattering*; Optical Society of America: Washington, DC, USA, 1989.
23. Nordseth, O.; Kumar, R.; Bergum, K.; Chilibon, I.; Fara, L.; Foss, S.E.; Monakhov, E. Nitrogen-doped Cu_2O thin films for photovoltaic applications. *Materials* **2019**, *12*, 3038. [CrossRef] [PubMed]
24. Lee, S.W.; Lee, Y.S.; Heo, J.; Siah, S.C.; Chua, D.; Brandt, R.; Kim, S.B.; Mailoa, J.P.; Buonassisi, T.; Gordon, R.G. Improved Cu2O-based solar cells using atomic layer deposition to control the Cu oxidation state at the p-n junction. *Adv. Energy Mater.* **2014**, *4*, 1301916. [CrossRef]
25. *JCPDS 78-2076 for Cu_2O, JCPDS 03-0879 for Cu_4O_3, JCPDS 48-1548 for CuO, and JCPDS 85-1326 for Cu*; International Centre for Diffraction Data-JCPDS: Newtown Square, PA, USA, 1996.
26. Nakano, Y.; Saeki, S.; Morikawa, T. Optical bandgap widening of p-type Cu_2O films by nitrogen doping. *Appl. Phys. Lett.* **2009**, *94*, 022111. [CrossRef]
27. Reddy, M.H.P.; Pierson, J.F.; Uthanna, S. Structural, surface morphological, and optical properties of nanocrystalline Cu_2O and CuO films formed by RF magnetron sputtering: Oxygen partial pressure effect. *Phys. Status Solidi A* **2012**, *209*, 1279–1286. [CrossRef]
28. Ogwu, A.A.; Darma, T.H. A reactive magnetron sputtering route for attaining a controlled core-rim phase partitioning in Cu_2O/CuO thin films with resistive switching potential. *J. Appl. Phys.* **2013**, *113*, 183522. [CrossRef]
29. Su, J.; Niu, Q.; Sun, R.; An, X.; Zhang, Y. Investigation of Cu_2O films sputtered with ceramic target: Effect of RF power. *J. Optoelectron. Adv. M.* **2018**, *20*, 441–444.
30. Gong, J.B.; Dong, W.L.; Dai, R.C.; Wang, Z.P.; Zhang, Z.M.; Ding, Z.J. Thickness dependence of the optical constants of oxidized copper thin films based on ellipsometry and transmittance. *Chin. Phys. B* **2014**, *23*, 1–5. [CrossRef]
31. Gan, J.; Venkatachalapathy, V.; Svensson, B.G.; Monakhov, E.V. Influence of target power on properties of Cu_xO thin films prepared by reactive radio frequency magnetron sputtering. *Thin Solid Film.* **2015**, *594*, 250–255. [CrossRef]

32. Li, J.; Mei, Z.; Liu, L.; Liang, H.; Azarov, A.; Kuznetsov, A.; Liu, Y.; Ji, A.; Meng, Q.; Du, X. Probing defects in nitrogen-doped Cu_2O. *Sci. Rep.* **2014**, *4*, 7240. [CrossRef] [PubMed]
33. Reliability Engineering Software Products–ReliaSoft. Available online: https://www.reliasoft.com (accessed on 20 January 2019).
34. Sharma, V.; Chandel, S.S. Performance and degradation analysis for long term reliability of solar photovoltaic systems: A review. *Renew. Sustain. Energy Rev.* **2013**, *27*, 753–767. [CrossRef]
35. Guo, D.; Brinkman, D.; Shaik, A.R.; Ringhofer, C.; Vasileska, D. Metastability and reliability of CdTe solar cells. *J. Phys. D Appl. Phys.* **2018**, *51*, 15. [CrossRef]
36. Fara, L.; Craciunescu, D. Reliability analysis of photovoltaic systems for specific applications. In *Reliability and Ecological Aspects of Photovoltaic Modules*; Gok, A., Ed.; Intech Open: London, UK, 2020; pp. 79–92.
37. Ye, F.; Zeng, J.J.; Qiu, Y.B.; Cai, X.M.; Wang, B.; Wang, H.; Zhang, D.P.; Fan, P.; Roy, V.A.L. Deposition-rate controlled nitrogen-doping into cuprous oxide and its thermal stability. *Thin Solid Film.* **2019**, *674*, 44–51. [CrossRef]

10

Improving Two-Step Prepared $CH_3NH_3PbI_3$ Perovskite Solar Cells by Co-Doping Potassium Halide and Water in PbI_2 Layer

Hsuan-Ta Wu [1], Yu-Ting Cheng [1], Ching-Chich Leu [2,*], Shih-Hsiung Wu [3] and Chuan-Feng Shih [1,4,*]

1 Department of Electrical Engineering, National Cheng Kung University, Tainan 70101, Taiwan; n28004012@mail.ncku.edu.tw (H.-T.W.); n26041068@mail.ncku.edu.tw (Y.-T.C.)
2 Department of Chemical and Materials Engineering, National University of Kaohsiung, Kaohsiung 81148, Taiwan
3 Green Energy and Environment Research Laboratories, Industrial Technology Research Institute, Hsinchu 31040, Taiwan; shihhsiung@itri.org.tw
4 Hierarchical Green-Energy Materials (Hi-GEM) Research Center, National Cheng Kung University, Tainan 70101, Taiwan
* Correspondence: ccleu@nuk.edu.tw (C.-C.L.); cfshih@mail.ncku.edu.tw (C.-F.S.)

Abstract: Incorporating additives into organic halide perovskite solar cells is the typical approach to improve power conversion efficiency. In this paper, a methyl-ammonium lead iodide ($CH_3NH_3PbI_3$, $MAPbI_3$) organic perovskite film was fabricated using a two-step sequential process on top of the poly(3,4-ethylenedioxythiophene) polystyrene sulfonate (PEDOT:PSS) hole-transporting layer. Experimentally, water and potassium halides (KCl, KBr, and KI) were incorporated into the PbI_2 precursor solution. With only 2 vol% water, the cell efficiency was effectively improved. Without water, the addition of all of the three potassium halides unanimously degraded the performance of the solar cells, although the crystallinity was improved. Co-doping with KI and water showed a pronounced improvement in crystallinity and the elimination of carrier traps, yielding a power conversion efficiency (PCE) of 13.9%, which was approximately 60% higher than the pristine reference cell. The effect of metal halide and water co-doping in the PbI_2 layer on the performance of organic perovskite solar cells was studied. Raman and Fourier transform infrared spectroscopies indicated that a PbI_2-dimethylformamide-water related adduct was formed upon co-doping. Photoluminescence enhancement was observed due to the co-doping of KI and water, indicating the defect density was reduced. Finally, the co-doping process was recommended for developing high-performance organic halide perovskite solar cells.

Keywords: perovskite solar cells; water doping; potassium halide doping

---

1. Introduction

The first report on lead halide organic perovskites for photovoltaic applications was published in 2009 [1]. Kojima et al. used methylammonium lead iodide ($CH_3NH_3PbI_3$, $MAPbI_3$) to replace organic dyes in dye-sensitized solar cells (DSSCs), where mesoporous titanium oxide (TiO_2) and a liquid electrolyte were used, achieving a power conversion efficiency (PCE) of 3.8%. Recently, the best solar cell efficiency achieved was 22%, giving perovskites a reasonable chance to reach commercial competiveness [2]. PCE of the $MAPbI_3$-based solar cell has been close to 20% in both mesoporous structure devices [3] as well as in planar heterojunction architectures [4]. High temperature annealing (>400 °C) is required to crystallize the TiO_2 layers used in mesoporous-type solar cells. Compared to the high-temperature processing of mesoporous solar cells, the planar heterojunction perovskite photovoltaics has the advantage of a low-temperature (100 °C) solution process, and, therefore, can be adopted in the roll-to-roll production of flexible devices [5].

One key aspect that affects the performance of planar heterojunction perovskite solar cells is the quality of the organic perovskite film, which is determined by the thermodynamics and the growth kinetics of the film [6–8]. Solution-processed perovskite films usually have abundant defects when compared to single-crystal samples. The introduction of additives into the perovskite precursor solution was reported to be an effective way to prepare a high quality perovskite film with fewer defects, leading to enhanced device performance [4,9–11]. Potassium halides were wildly used as additives in the perovskite precursor solutions adopted for solar cell research [12–14]. Potassium halide-doped perovskite solar cells that have a record PCE of more than 20%, without I–V hysteresis, have been constructed [13,15]. Potassium halides were found to significantly facilitate the crystal growth of perovskite films and ameliorate the perovskite morphology, resulting in a reduced density in trap states and enhanced device performance [13,16]. However, it is not easy to form a homogeneous organic precursor by adding considerable amounts of potassium halide salts due to its restricted solubility in some organic solvent used for organic perovskite processing, which limits the application of potassium halides as additives. Water is a good solvent for potassium halide salts. Additionally, water additives have been reported to enhance the property of a two-step processed $MAPbI_3$ [17,18]. Water additives changed the characteristics of dimethylformamide (DMF), which is a general solvent for PbI_2, thus helping to make a homogeneous PbI_2 solution. A smooth and dense PbI_2 film was fabricated by adding a small amount of water into the PbI_2 precursor with DMF, and then a high-quality $MAPbI_3$ was obtained after the methylammonium iodide (MAI) conversion. Therefore, water was considered to be a suitable candidate for a co-doping additive for potassium halides during the $MAPbI_3$ solution process.

However, it is challenging to deposit high-quality organic perovskite films on PEDOT:PSS, which is wildly used as the hole-transporting material in planar heterojunction perovskite solar cells. To obtain a homogeneous film structure on top of an organic surface, such as the PEDOT:PSS film, through a simple-solution process is not easy for an ionic material (such as $MAPbI_3$) [17]. Compared to $MAPbI_3$, PbI_2 is less polar and, therefore, can easily form a continuous film on the PEDOT:PSS surface. Therefore, the two-step sequential process (PbI_2 layer + MAI conversion) is an appropriate way to fabricate a perovskite film when PEDOT:PSS is used as the hole-transporting layer [17]. However, very few studies have reported the effects of water or potassium halide additives on two-step processed perovskite solar cells [19]. According to this report, the alkali metal halides (KCl, NaCl, and LiCl) were incorporated with the PbI_2 layer and chelated with Pb^{2+} ions, enhancing the crystal growth of PbI_2 films that, in turn, improved the crystallinity of the perovskite films and their photovoltaic properties. However, to the best of our knowledge, no report has investigated the performance of a planar heterojunction perovskite device by considering their interactive effects by using both water and potassium halides.

In this work, we proposed an effective way to enhance the efficiency of the $MAPbI_3$-based perovskite device through the co-doping of water and potassium halides (KI, KBr, and KCl) during the PbI_2 deposition process. Systematic studies of the effects of water and potassium halides co-doping on the thin film and device were investigated and discussed. To construct a planar heterojunction perovskite solar cell, a two-step process was employed to fabricate the $MAPbI_3$ film. PbI_2 was first deposited on PEDOT:PSS film, then the MAI was spin-coated on the PbI_2 layer, followed by thermal annealing. Water and potassium halides were added into the PbI_2 precursor solution to elucidate their influence on the performance of perovskite solar cells. As a result, the PCE of devices made from these additive-enhanced perovskites increased from 8.8% (based on pristine perovskite) to 13.9%.

2. Materials and Methods

2.1. Chemicals

Lead (II) iodide (PbI_2, 99.9985%), potassium iodide (KI, 99.995%), and potassium chloride (KCl, 99.997%) were purchased from Alfa Aesar. Anhydrous N,N-dimethylformide (DMF, 99.8%), 2-propanol (IPA, 99.5%), and chlorobenzene (CB, 99.8%) were purchased from Sigma-Aldrich (Saint Louis, MO, USA). Potassium bromide (KBr, IR spectroscopic) was purchased from Honeywell (Morristown, NJ, USA). Ultra-pure wa ter (Resistivity > 18.2 MΩ·cm at 25 °C, Baker Analyzed LC/MS Reagent) was purchased from J.T. Baker (Radnor, PA, USA). Poly(3,4-ethylenedioxythiophene) polystyrene sulfonate (PEDOT:PSS), methylammonium iodide (MAI, >98%), bathocuproine (BCP, >99.5%), and phenyl-C61-butyric acid methyl ester ($PC_{61}BM$, >99.5%) were purchased from Uni-onward (New Taipei City, Taiwan). All materials were used as received.

2.2. Device Fabrication

The fabrication process of the referenced standard perovskite solar cell (Ref) was as follows: patterned ITO-coated (indium tin oxide) glass substrates were cleaned ultrasonically in acetone and isopropanol for 10 min, and then dried in an oven for 15 min at 110 °C. After a UV-ozone treatment of 30 min, the substrates were transferred into a N_2-filled glovebox. Filtered PEDOT:PSS was spin-coated at 2000 rpm onto the ITO substrate and baked on a hot plate at 110 °C for 10 min. The PbI_2 precursor was prepared by dissolving PbI_2 in the DMF solvent (370 mg/mL) and heated to 70 °C on a hot plate. Following this, the hot PbI_2-DMF precursor was spin-coated on top of the PEDOT:PSS layer and baked on a hot plate at 90 °C for 10 min. The MAI (45 mg/mL in IPA) precursor was spin-coated on the crystalline PbI_2 layer and annealed at 110 °C for 1 h to form the $MAPbI_3$ perovskite. After annealing for 20 min, the pure IPA solution was spun on the $MAPbI_3$ layer to wash out redundant MAI, and then the annealing process continued for 40 min. The $PC_{61}BM$ (20 mg/mL in CB) was spin-coated on the perovskite. BCP of 10 nm and Al of 150 nm were sequentially evaporated on the top with a deposition rate of 0.4 Å/s and 3–5 Å/s, respectively. For the co-doping samples, H_2O with a 1, 2, or 3 volume percent, and potassium halide of various concentrations were added into the PbI_2- DMF solution. The water added into the precursor was ultra-pure water.

2.3. Characterizations

The crystallinity of the $MAPbI_3$ film was examined by X-ray Diffraction (XRD) with ultraX 18 (Tokyo, Japan). The microstructure of the films was characterized by high resolution scanning electron microscope (HR-SEM) SU8000 (HITACHI, Tokyo, Japan). The electrical analysis was performed by E5270B Precision IV Analyzer (Keysight, Santa Rosa, CA, USA) under AM 1.5 G illumination. The absorption analysis was performed by UV-vis-NIR spectrophotometer U-4100 (HITACHI, Tokyo, Japan). The photoluminescence (PL) was performed by Jobin Yvon LabRAM HR micro-Raman system (HORIBA, Kyoto, Japan) and the excitation laser wavelength was set to 532 nm. The Fourier transform infrared spectroscopies (FTIRs) were performed by Nicolet FTIR instrument (Thermo Fisher Scientific, Waltham, MA, USA). The Raman spectroscopy was performed by Raman microscope (Renishaw, Wotton-under-Edge, UK). The depth profiles were performed by Auger Electron Spectroscopy MICROLAB 350 (Thermo Fisher Scientific, Waltham, MA, USA).

3. Results and Discussion

The photovoltaic properties of the devices (structure: glass/ITO/PEDOT:PSS/$MAPbI_3$/$PC_{61}BM$/BCP/Al) were investigated. Figure 1 displays the photocurrent density–voltage (J–V) curves of the perovskite solar cells prepared with and without the incorporation of water and potassium halide additives in the PbI_2-DMF precursor. The concentration of each cell was at the optimal conditions for cell properties. Table 1 lists the corresponding cell parameters. The cell performance was measured using an aperture metal mask of 2.2 mm × 3.2 mm to a device active area of 2 mm × 3 mm. The PCE of device that was prepared using the 2 vol% water-doped PbI_2 precursor increased from 8.8% (without water) to 12.0% with a short-circuit current density (J_{SC}) of 22.5 mA·cm^{-2}, an open-circuit voltage (V_{OC}) of 0.9 V, and a fill factor (FF) of 59.2%. The photovoltaic performance, as a function of the water content, is illustrated in Figure S1, and the corresponding photovoltaic parameters are listed in Table S1. It was found that the PCE of devices that were prepared using a water content below 3 vol% were improved, and that the best performance was achieved by adding 2 vol% water in PbI_2. The V_{OC} and FF increased slightly, and the J_{SC} increased markedly by increasing the water content to 2 vol%, resulting in the best PCE. Further increasing the water content to 3 vol% degraded the cell property. It has been reported that a planar heterojunction solar cell based on a high-quality perovskite film has a power conversion efficiency of 18% with a remarkably high FF value of 0.85 [17]. The study indicated that adding small amounts of water into the PbI_2-DMF precursor made the solution more uniform, forming a smooth PbI_2 film on top of the PEDOT:PSS, with high crystallinity and large crystalline domains. The perovskite film fabricated from the high-quality PbI_2 film was highly pure and dense, without any pinhole. Our research showed similar observations. Water addition improved the morphology and crystallinity of the PbI_2 films, as revealed in the SEM images (Figure S2). Moreover, the thickness of the PbI_2 films increased with the content of water. The thickness was 156, 192, 189, and 194 nm for 0, 1, 2, and 3 vol% water incorporated PbI_2 films, respectively. Figure S3 shows the $MAPbI_3$ films fabricated from the water-doped PbI_2 films. The grain size of all of the three water-incorporated films, Figure S3b–d, was larger than the film without water (Figure S3a). It has been observed that a rapid crystallization of organo-halide perovskites into the expected tetragonal cell for $MAPbI_3$ occurred on exposure to small amounts of moisture [20]. You et al. have proposed a growth mode via thermal annealing of the perovskite precursor film in a humid environment (e.g., ambient air) to greatly improve the film quality, grain size, carrier mobility, and lifetime [21]. They indicated that, due to the strong hydroscopic nature of MAI, exposing the perovskite precursor to moisture during film formation could result in the accumulation of moisture within grain boundaries, inducing grain boundary creep and, subsequently, merging adjacent grains together. In addition, moisture could also provide an aqueous environment to enhance the diffusion length of the precursor ions, further promoting perovskite grain growth. The PbI_2 films fabricated by the water-containing precursor possibly retained residual water or a water-related function group, which helped perovskite grain growth, until the deposition of MAI.

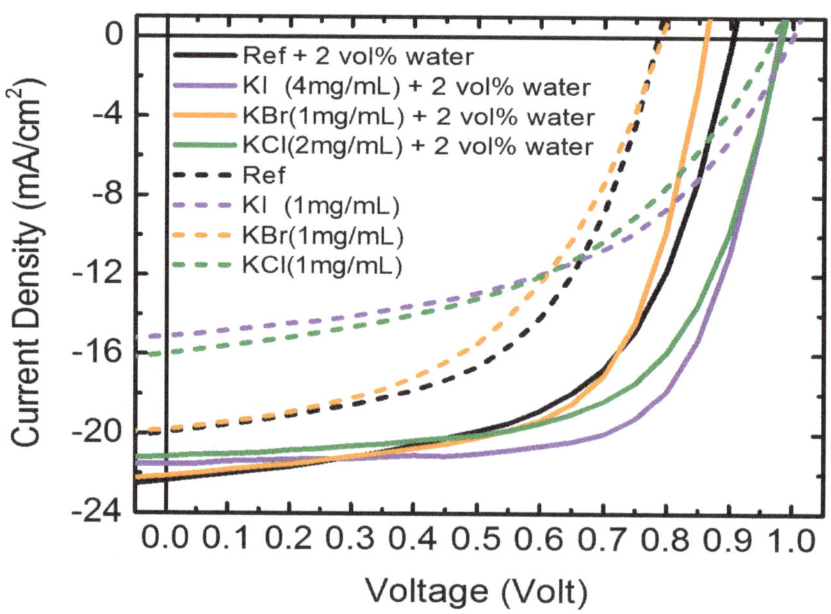

Figure 1. Photocurrent density–voltage (J–V) curves of perovskite solar cells prepared without and with water and/or potassium halide additives under their optimized concentrations.

Table 1. The corresponding photovoltaic parameters of devices in Figure 1.

Samples	V_{OC} (V)	J_{SC} (mA/cm^2)	FF	PCE (%)	R_{SH} (Ω·cm^2)	R_S (Ω·cm^2)
Ref	0.79 ± 0.03	19.7 ± 1.4	56.3 ± 1.6	8.8 ± 1.1	340 ± 99	7.2 ± 0.7
KI (1 mg/mL)	0.99 ± 0.02	15.1 ± 0.8	50.1 ± 1.5	7.5 ± 0.6	262 ± 40	8.8 ± 1.6
KBr (1 mg/mL)	0.79 ± 0.05	19.8 ± 3.9	52.2 ± 4.8	8.2 ± 2.5	220 ± 70	11.7 ± 3.3
KCl (1 mg/mL)	0.96 ± 0.03	16.0 ± 1.0	52.4 ± 2.1	8.0 ± 0.6	337 ± 137	12.5 ± 1.8
Ref + water	0.90 ± 0.01	22.5 ± 1.3	59.2 ± 0.6	12.0 ± 0.6	168 ± 40	3.6 ± 0.1
KI (4 mg/mL) + water	0.97 ± 0.01	21.3 ± 1.2	67.0 ± 2.2	13.9 ± 0.7	555 ± 39	5.5 ± 0.7
KBr (1 mg/mL) + water	0.85 ± 0.04	22.1 ± 1.0	63.4 ± 1.9	11.9 ± 1.2	264 ± 46	3.8 ± 0.3
KCl (2 mg/mL) + water	0.97 ± 0.01	21.0 ± 1.6	62.8 ± 0.8	12.8 ± 1.2	1217 ± 106	4.7 ± 0.3

Conversely, the PCE of photovoltaic devices fabricated by the PbI_2-DMF precursor with potassium halides (KI, KBr, and KCl) as the only additives became worse. As displayed in Figure 1, the PCE deteriorated from 8.8% for pristine $MAPbI_3$ to 7.5%, 8.2%, and 8.0%, by adding 1 mg/mL KI, KBr, and KCl, respectively. For KCl- and KI-doped samples, the degradation of PCE was caused by the decrease in the J_{SC} and the fill factor, while for KBr-doped samples, only the fill factor decreased. Figure S4 illustrates the changes in photovoltaic performance of the potassium halide-doped devices as a function of the amount of potassium halide added. Notably, the devices were further improved by incorporating potassium halide into the water-PbI_2-DMF precursor. By co-doping with 2 vol% water and potassium halides, the PCE was improved from 8.8% (pristine perovskite) to 13.9% (4 mg/mL KI), 11.9% (1 mg/mL KBr) and 12.8% (2 mg/mL KCl). The changes in the photovoltaic performance of the water and potassium halide co-doped devices as a function of the potassium halide additives is also illustrated in Figure 2. The performance of devices was enhanced by adding KI or KCl into the 2 vol% water-doped PbI_2-DMF precursor, while KBr-doped devices had inferior properties than those without KBr. Using 4 mg/mL KI as the additive provided the best performance of the device. Figure S4 further plots the cell parameters as functions of the content of the potassium halides for devices prepared by PbI_2 without water; all of the cell parameters degraded and Table S2 lists the corresponding photovoltaic parameters of the devices.

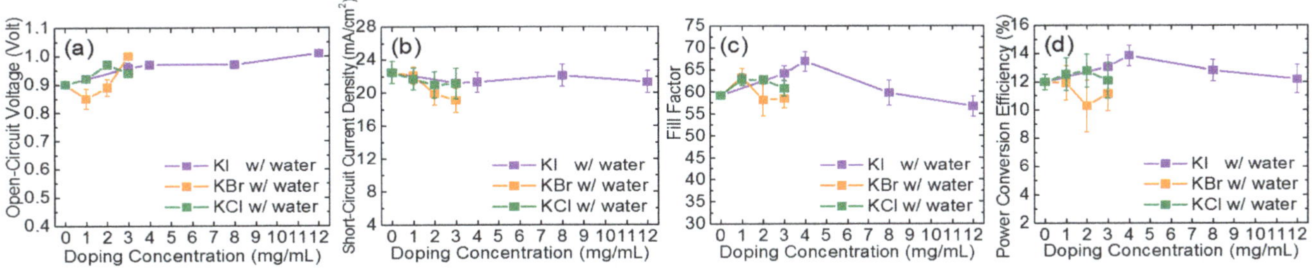

Figure 2. The changes in **(a)** V_{OC}; **(b)** J_{SC}; **(c)** FF (fill factor); **(d)** power conversion efficiency (PCE) of the potassium halide-doped devices with 2 vol% water co-doping as a function of doping amounts of KI, KBr, and KCl.

Figure 3 shows comprehensive XRD analyses of the potassium halides-doped PbI_2 and $MAPbI_3$ films with and without water incorporated with PbI_2. The PbI_2 and $MAPbI_3$ films were deposited on ITO/PEDOT:PSS. Without water, the XRD intensity increased markedly for all the KCl-, KBr-, and KI-doped PbI_2 films (Figure 3a). However, the XRD of the $MAPbI_3$ films prepared on the potassium-doped PbI_2 were similar to that without doping (Figure 3c). Water incorporation increased the XRD intensity of PbI_2, but did not affect the XRD intensity of the perovskite film (Figure 3a,c). Co-doping of water and potassium halide increased the XRD intensity of both the PbI_2 and $MAPbI_3$ significantly (Figure 3b,d). The increase in the XRD intensity was also accompanied by an increase in the grain size.

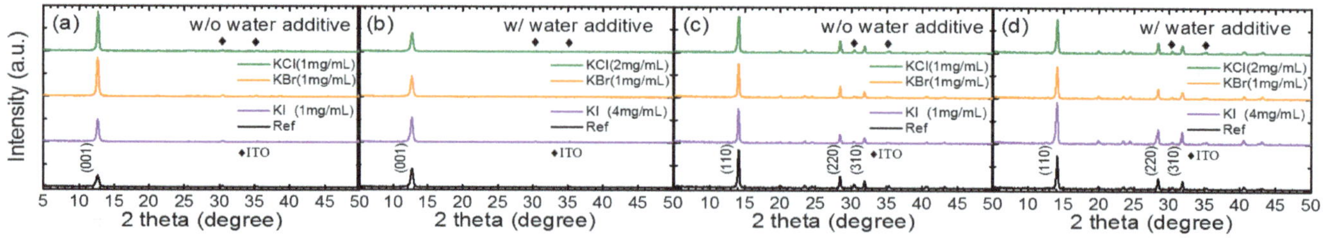

Figure 3. X-ray Diffraction (XRD) of PbI_2 prepared by incorporating (**a**) KCl, KBr, and KI (without water) and (**b**) KCl, KBr, and KI with water additives. XRD of $MAPbI_3$ on PbI_2 prepared using the same conditions as (**a**) and (**b**) are shown in (**c**) and (**d**), respectively.

Figure 4 shows the top-view and cross-section SEM images of the PbI_2 films that were co-doped with 2 vol% water and potassium halides (KI: 4 mg/mL, KBr: 1 mg/mL, and KCl: 2 mg/mL). The film thickness was unchanged by the potassium halides, but the surface coverage was improved and the grain size was enlarged, which agreed with the XRD observation. Figure 5 displays the SEM images of the co-doped films. Adding potassium halides (especially KI and KBr) to the 2 vol% water-incorporated PbI_2 precursor enhanced $MAPbI_3$ grain growth, which was denser with higher continuity, allowing for effective charge generation and dissociation in perovskite films. Such a highly continuous and large grain structure was also beneficial for carriers to transport through the film.

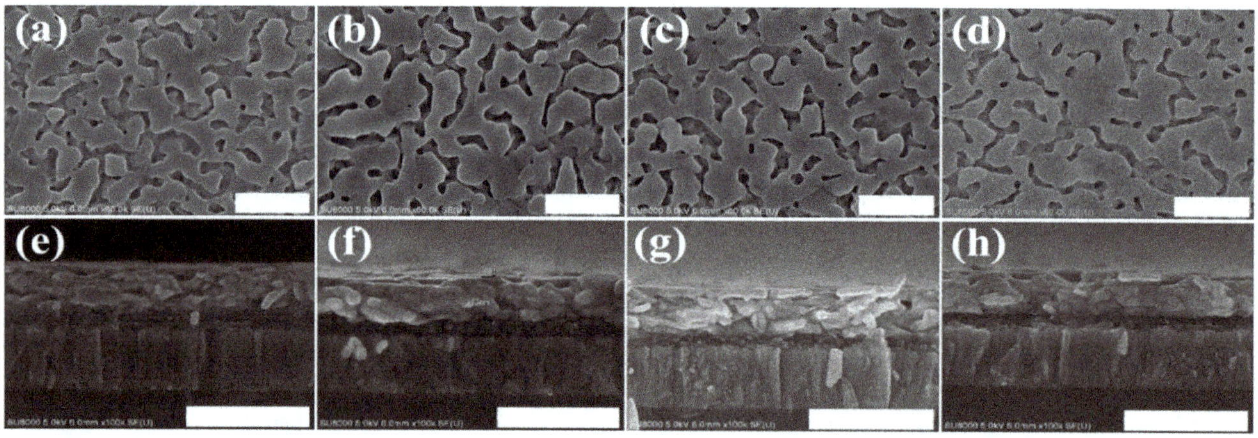

Figure 4. Plane and cross-section scanning electron microscope (SEM) images, respectively, of PbI_2 prepared with (**a**,**e**) 2% water, (**b**,**f**) 2% water + KI (4 mg/mL) (**c**,**g**) 2% water + KBr (1 mg/mL), and (**d**,**h**) 2% water + KCl (2 mg/mL). (Scale bar = 500 nm).

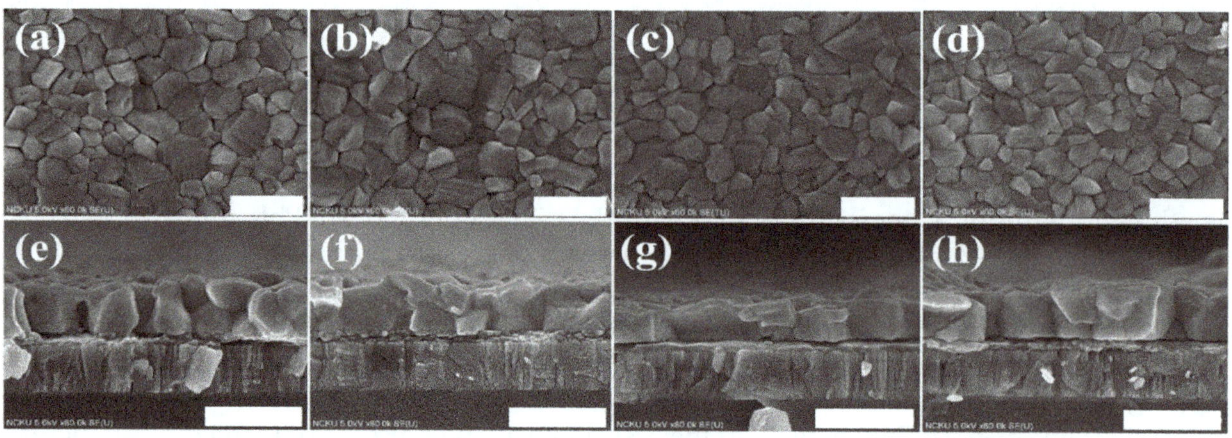

Figure 5. Plane and cross-section SEM images, respectively, of $MAPbI_3$ on PbI_2 with (**a**,**e**) 2% water, (**b**,**f**) 2% water + KI (4 mg/mL) (**c**,**g**) 2% water + KBr (1 mg/mL), and (**d**,**h**) 2% water + KCl (2 mg/mL). (Scale bar = 500 nm).

To exploit the effect of co-doping water and potassium halides in the PbI_2 layer, the Fourier transform infrared spectroscopy and Raman spectroscopy were adopted, as shown in Figure 6. The PbI_2 films were prepared on Si/PEDOT:PSS because glass absorbs the IR light. In Figure 6a, characteristic peaks related to the C=O stretching were found around 1715 cm^{-1} [22]. According to the literature, the C=O vibration shifts to lower frequency around 1650 cm^{-1} for DMF, which forms Lewis adducts due to the reaction of the PbI_2 layer [23,24]. In our sample, the C=O stretching was found to be blue-shifted, which might be caused by the formation of a PbI_2-DMF-H_2O adduct by incorporating water through a Lewis acid–base reaction. The other possibility was that the PbI_2-DMF precursor reacted with the PEDOT:PSS under-layer because of the incorporation of water by partially dissolving the PEDOT:PSS near the PbI_2/PEDOT:PSS interface. A similar observation has been reported by Winther et al. [25]. Figure 6b shows the Raman spectra of the PbI_2 layer with or without potassium halides and water. Si-related signals were found at 521 cm^{-1} (Si–Si LO), 940, and 987 cm^{-1} (Si–OH) [26]. Notably, the O–H and C–H stretching of PEDOT split into two peaks at 2865 and 2945 cm^{-1}, owing to the incorporation of metal halides and water, particularly for KBr- and KCl-doping [27]. Evidence from FTIR and Raman spectroscopies indicated that the doping of water and potassium halides could form some new adducts and change the interfacial chemistry near the PbI_2/PEDOT:PSS.

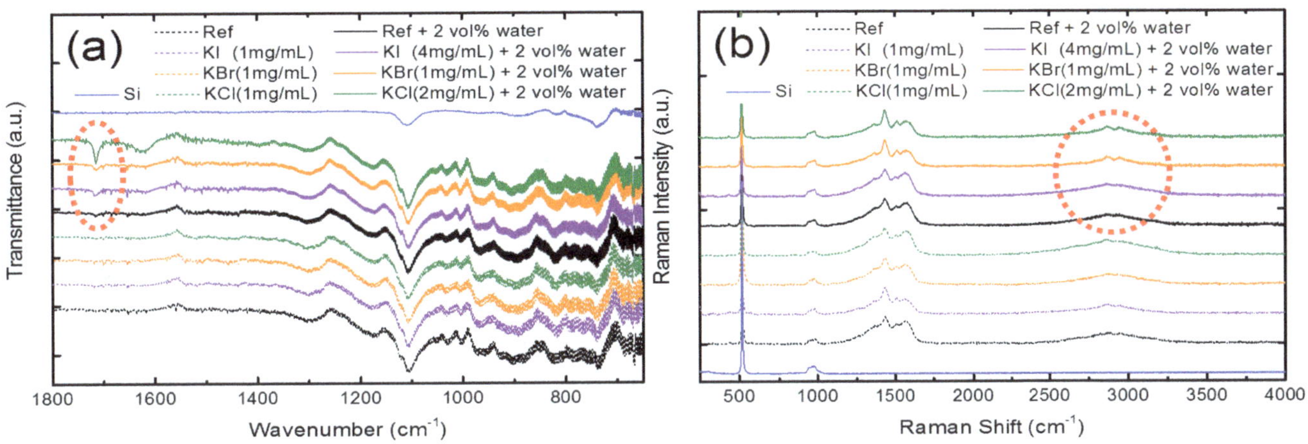

Figure 6. (**a**) Fourier transform infrared (FTIR) and (**b**) Raman spectroscopies of PbI_2 films doped using potassium halides with and without water.

Figure 7 displays the absorption spectra (Figure 7a) and PL spectra (Figure 7b,c) of the perovskite-glass structures prepared on the PbI_2 layers that were doped with water, potassium halides, and potassium halides and water co-dopants. In the visible region (470–800 nm), the absorbance of the $MAPbI_3$ on PbI_2, prepared with the co-doped potassium halides and water, increased, indicating improved crystallinity. With potassium-only additives, the absorbance of $MAPbI_3$ decreased. The PL-quenching effect of KBr-doping was observed (Figure 7b), while PL enhancement was found in the KI- and KCl-doped films. The steady-state photoluminescence (PL) was an effective way to detect the trap states within the perovskite layer. The higher PL intensity indicated fewer traps or defects within the films and improved crystallinity. Figure 7c shows the PL spectra of the co-doped water and potassium halides for $MAPbI_3$ films. The PL intensity of the perovskite films obviously increased by co-doping, except for the KBr-doped film. In particular, the PL intensity of the co-doped KI and water increased by five times. Therefore, it was concluded that the co-doping of water and potassium halides increased the grain size of $MAPbI_3$ and eliminated radiative defects that contributed to the strong PL response.

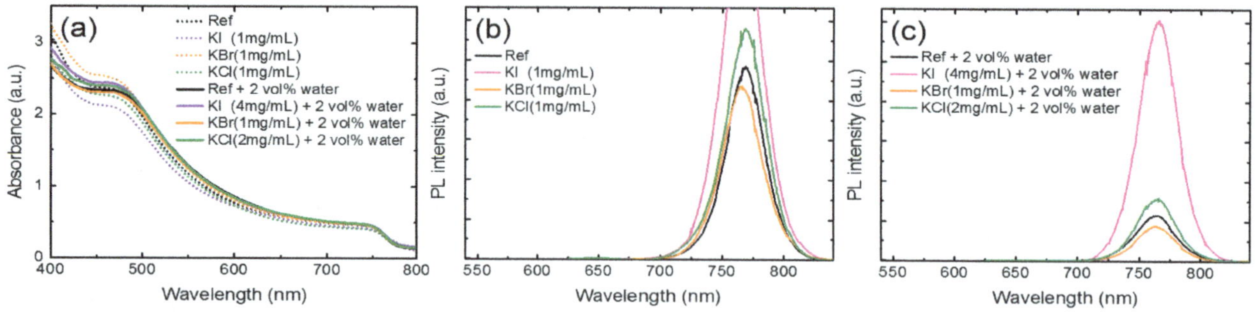

Figure 7. (**a**) Absorption spectra and (**b**,**c**) photoluminescence (PL) spectra of perovskite films prepared using PbI_2 with different potassium halide and water additives. The samples were measured with this structure: glass/ITO/PEDOT:PSS/$MAPbI_3$/$PC_{61}BM$.

Auger electron spectroscopy (AES) was used to investigate the elementary depth distribution of the full device structure using KI and 2 vol% water co-doping, as shown in Figure 8. The potassium signal was found to be uniformly distributed across the perovskite and penetrated into the PEDOT:PSS. It has been reported that KI can provide an extra I-ion source that affects the coordination with Pb^{2+} and compensates for the I vacancy [15]. Therefore, in addition to the positive effect due to water- and potassium halides-induced grain growth, the enhancement of PL intensity can be associated with the passivation of the K^+ cation and halide anions on the grain boundaries of perovskite film [15,16] and the perovskite interface, thus, reducing the trapped states. In other words, the addition of potassium halides and water improved the quality of the perovskite by reducing the traps and interfacial radiative recombination centers, allowing for effective charge generation and collection. Such a highly continuous and large grain structure was also beneficial for carriers to transport through the film. This improvement reduced series resistance (R_S) and increased shunt resistance (R_{SH}), improving the J_{SC}, FF, and V_{OC} as observed in Table 1. Some studies partially attributed good photovoltaic performance to a significant absorption improvement because the use of additives led to a denser perovskite film with less pinholes [14,28]. However, our research results showed water and/or potassium halide additives only slightly enhanced the absorption of perovskite. The pristine $MAPbI_3$, prepared by our process, exhibited a compact microstructure with pure tetragonal structure phases, and the absorption effect was considered to be minor. It is worth noting that the KBr additive was harmful to the device, where even its physical properties, shown in Figures 3–5, seemed as good as KI. We attributed this to the possible formation of a small amount of $MAPbBr_3$, through the incorporation of Br even though no secondary phase was found in the XRD spectra. The existence of $MAPbBr_3$ within the $MAPbI_3$ film may cause an energy barrier and inhibit charge transport from the perovskite layer to the ITO [12].

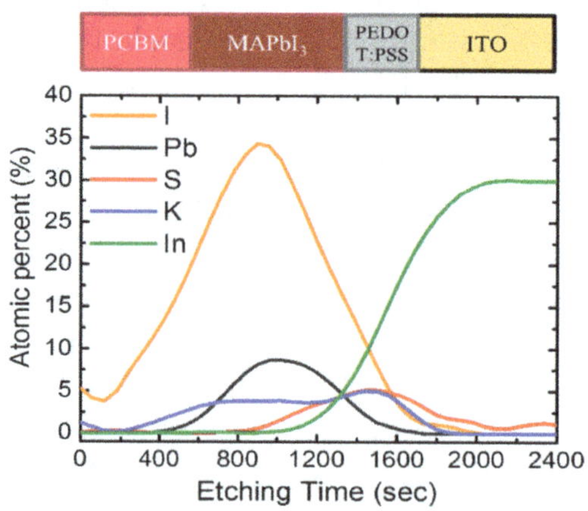

Figure 8. Auger electron spectroscopy (AES) depth profile of $MAPbI_3$ prepared on PEDOT:PSS coated ITO (indium tin oxide) glass.

However, incorporation of potassium halides had little influence on the microstructure of the perovskite without any water additive. As shown in Figure S5, the surface morphology and the cross-section images of the films were similar, regardless of the types of potassium halides. The grain size was unchanged by the potassium halide additives, which is in agreement with the unchanged XRD FWHM (full width at half maximum) observation shown in Figure 3. The additive effect of each potassium halide can be clearly observed in Figures S6–S8. To conclude, the potassium halides had positive influences on the film quality and the device, only if they were co-doped with the water additive. Without water, the potassium halide additive made the solution more inhomogeneous, worsening the film properties. Because the solubility of potassium halides in PbI_2-DMF was low, they were unevenly dispersed within the solution without water, forming an inhomogeneous film. The different optimal concentration of each potassium halide for the photovoltaic properties was influenced by their solubility.

4. Conclusions

In conclusion, a way to improve the perovskite solar cells (structure: glass/ITO/PEDOT:PSS/ $MAPbI_3/PC_{61}BM$/BCP/Al) was proposed by co-doping water and potassium halides in the PbI_2 layer, which was coated on the PEDOT:PSS layer based in a two-step sequential process. When only potassium halides were added to PbI_2, the PCE of the devices became worse, while the PCE of the devices prepared using the 2 vol% water-doped PbI_2 precursor increased from 8.8% (without water) to 12.0%. By co-doping with the 2 vol% water and a potassium halide, the PCE was improved from 8.8% (pristine perovskite) to 13.9% (4 mg/mL KI), 11.9% (1 mg/mL KBr), and 12.8% (2 mg/mL KCl). XRD results showed that the incorporation of water and a potassium halide improved the crystallinity and enlarged the grain size. SEM images showed that the grain of PbI_2 became coarse and continuous upon co-doping. FTIR spectra showed characteristic peaks, related to C=O stretching, around 1715 cm^{-1}, which was probably caused by the formation of a PbI_2–DMF–H_2O adduct or an interfacial reaction near the PbI_2/PEDOT:PSS interface. Raman spectra revealed that O–H and C–H stretching of PEDOT split into two peaks at 2865 and 2945 cm^{-1}, owing to metal halides and water incorporation. Obvious PL enhancement was caused by the co-doping of water and KI, reducing the defect density. Together with the AES observations, it was found that KI distributed uniformly within the perovskite layer and penetrated into the PEDOT:PSS layer, which suggested that the elimination of the defects in the film and interface upon KI doping was one of the reasons to improve the KI–water co-doped solar cell.

Supplementary Materials

Figure S1: The tendency of photovoltaic performance of the water-doped devices as a function of water amount; Table S1: The corresponding photovoltaic parameters of devices in Figure S1; Figure S2: SEM morphology and cross-section of PbI_2 films with (a,e) 0%, (b,f) 1%, (c,g) 2%, and (d,h) 3% water; Figure S3: SEM morphology of perovskite films prepared by PbI_2 with (a) 0%, (b) 1%, (c) 2%, and (d) 3% water; Figure S4: The (a) V_{OC}; (b) J_{SC}; (c) FF; (d) PCE of the potassium halide-doped devices as a function of doping amounts of KI, KBr and KCl; Table S2: The corresponding photovoltaic parameters of devices in Figure S4 and Figure 2; Figure S5: The SEM cross-sectional images of the potassium halide-doped $MAPbI_3$ films (without water additive), (a) ref, (b) 1 mg/mL KI, (c) 1 mg/mL KBr, and (d) 1 mg/mL KCl; Figure S6: SEM surface morphology of the perovskite films with different KI and water additives in the PbI_2 layer; Figure S7: SEM surface morphology of the perovskite films with different KBr and water additives in the PbI_2 layer; and Figure S8: SEM surface morphology of the perovskite films with different KCl and water additives in the PbI_2 layer.

Author Contributions: Conceptualization, Y.-T.C. and C.-F.S.; Data curation, H.-T.W. and C.-C.L.; Formal analysis, H.-T.W. and Y.-T.C.; Investigation, H.-T.W. and S.-H.W.; Methodology, Y.-T.C.; Project administration, H.-T.W. and C.-F.S.; Resources, C.-C.L., S.-H.W. and C.-F.S.; Supervision, C.-C.L. and C.-F.S.; Validation, H.-T.W. and Y.-T.C.; Writing—original draft, H.-T.W. and C.-C.L.; Writing—review & editing, H.-T.W., S.-H.W. and C.-F.S.

Acknowledgments: The authors are grateful for the support of the Ministry of Science and Technology of the Republic of China under Contract No. MOST 106-2221-E-006-225- and MOST 107-2221-E-006-160-, Taiwan.

This work was also financially supported by the Hierarchical Green-Energy Materials (Hi-GEM) Research Center, from The Featured Areas Research Center Program within the framework of the Higher Education Sprout Project by the Ministry of Education (MOE) and the Ministry of Science and Technology (MOST 107-3017-F-006-003) in Tainan. The authors are also grateful for the support of the Department of Industrial Technology, Ministry of Economic Affairs, Taiwan.

References

1. Kojima, A.; Teshima, K.; Shirai, Y.; Miyasaka, T. Organometal Halide Perovskites as Visible-Light Sensitizers for Photovoltaic Cells. *J. Am. Chem. Soc.* **2009**, *131*, 6050–6051. [CrossRef]
2. Shin, S.S.; Yeom, E.J.; Yang, W.S.; Hur, S.; Kim, M.G.; Im, J.; Seo, J.; Noh, J.H.; Seok, S.I. Colloidally prepared La-doped $BaSnO_3$ electrodes for efficient, photostable perovskite solar cells. *Science* **2017**, *356*, 167–171. [CrossRef]
3. Saliba, M.; Matsui, T.; Seo, J.-Y.; Domanski, K.; Correa-Baena, J.-P.; Nazeeruddin, M.K.; Zakeeruddin, S.M.; Tress, W.; Abate, A.; Hagfeldt, A.; et al. Cesium-containing triple cation perovskite solar cells: Improved stability, reproducibility and high efficiency. *Energy Environ. Sci.* **2016**, *9*, 1989–1997. [CrossRef] [PubMed]
4. Chiang, C.-H.; Nazeeruddin, M.K.; Grätzel, M.; Wu, C.-G. The synergistic effect of H_2O and DMF towards stable and 20% efficiency inverted perovskite solar cells. *Energy Environ. Sci.* **2017**, *10*, 808–817. [CrossRef]
5. Zuo, C.; Vak, D.; Angmo, D.; Ding, L.; Gao, M. One-step roll-to-roll air processed high efficiency perovskite solar cells. *Nano Energy* **2018**, *46*, 185–192. [CrossRef]
6. Bi, C.; Wang, Q.; Shao, Y.; Yuan, Y.; Xiao, Z.; Huang, J. Non-wetting surface-driven high-aspect-ratio crystalline grain growth for efficient hybrid perovskite solar cells. *Nat. Commun.* **2015**, *6*, 7747. [CrossRef] [PubMed]
7. Moore, D.T.; Sai, H.; Tan, K.W.; Smilgies, D.-M.; Zhang, W.; Snaith, H.J.; Wiesner, U.; Estroff, L.A. Crystallization Kinetics of Organic–Inorganic Trihalide Perovskites and the Role of the Lead Anion in Crystal Growth. *J. Am. Chem. Soc.* **2015**, *137*, 2350–2358. [CrossRef]
8. Fei, C.; Guo, L.; Li, B.; Zhang, R.; Fu, H.; Tian, J.; Cao, G. Controlled growth of textured perovskite films towards high performance solar cells. *Nano Energy* **2016**, *27*, 17–26. [CrossRef]
9. Song, X.; Wang, W.; Sun, P.; Ma, W.; Chen, Z.-K. Additive to regulate the perovskite crystal film growth in planar heterojunction solar cells. *Appl. Phys. Lett.* **2015**, *106*, 033901. [CrossRef]
10. Wu, Y.; Xie, F.; Chen, H.; Yang, X.; Su, H.; Cai, M.; Zhou, Z.; Noda, T.; Han, L. Thermally Stable $MAPbI_3$ Perovskite Solar Cells with Efficiency of 19.19% and Area over 1 cm^2 achieved by Additive Engineering. *Adv. Mater.* **2017**, *29*, 1701073. [CrossRef]
11. Ke, W.; Xiao, C.; Wang, C.; Saparov, B.; Duan, H.-S.; Zhao, D.; Xiao, Z.; Schulz, P.; Harvey, S.P.; Liao, W.; et al. Employing Lead Thiocyanate Additive to Reduce the Hysteresis and Boost the Fill Factor of Planar Perovskite Solar Cells. *Adv. Mater.* **2016**, *28*, 5214–5221. [CrossRef] [PubMed]
12. Wang, P.; Wang, J.; Zhang, X.; Wang, H.; Cui, X.; Yuan, S.; Lu, H.; Tu, L.; Zhan, Y.; Zheng, L. Boosting the performance of perovskite solar cells through a novel active passivation method. *J. Mater. Chem. A* **2018**, *6*, 15853–15858. [CrossRef]
13. Tang, Z.; Bessho, T.; Awai, F.; Kinoshita, T.; Maitani, M.M.; Jono, R.; Murakami, T.N.; Wang, H.; Kubo, T.; Uchida, S.; et al. Hysteresis-free perovskite solar cells made of potassium-doped organometal halide perovskite. *Sci. Rep.* **2017**, *7*, 12183. [CrossRef] [PubMed]
14. Nam, J.K.; Chai, S.U.; Cha, W.; Choi, Y.J.; Kim, W.; Jung, M.S.; Kwon, J.; Kim, D.; Park, J.H. Potassium Incorporation for Enhanced Performance and Stability of Fully Inorganic Cesium Lead Halide Perovskite Solar Cells. *Nano Lett.* **2017**, *17*, 2028–2033. [CrossRef]
15. Abdi-Jalebi, M.; Andaji-Garmaroudi, Z.; Cacovich, S.; Stavrakas, C.; Philippe, B.; Richter, J.M.; Alsari, M.; Booker, E.P.; Hutter, E.M.; Pearson, A.J.; et al. Maximizing and stabilizing luminescence from halide perovskites with potassium passivation. *Nature* **2018**, *555*, 497. [CrossRef]
16. Son, D.-Y.; Kim, S.-G.; Seo, J.-Y.; Lee, S.-H.; Shin, H.; Lee, D.; Park, N.-G. Universal Approach toward Hysteresis-Free Perovskite Solar Cell via Defect Engineering. *J. Am. Chem. Soc.* **2018**, *140*, 1358–1364. [CrossRef]
17. Wu, C.-G.; Chiang, C.-H.; Tseng, Z.-L.; Nazeeruddin, M.K.; Hagfeldt, A.; Gratzel, M. High efficiency stable inverted perovskite solar cells without current hysteresis. *Energy Environ. Sci.* **2015**, *8*, 2725–2733. [CrossRef]

18. Xiu, G.; Meng, L.; Xiao-Bo, S.; Heng, M.; Zhao-Kui, W.; Liang-Sheng, L. Controllable Perovskite Crystallization by Water Additive for High-Performance Solar Cells. *Adv. Funct. Mater.* **2015**, *25*, 6671–6678. [CrossRef]
19. Boopathi, K.M.; Mohan, R.; Huang, T.-Y.; Budiawan, W.; Lin, M.-Y.; Lee, C.-H.; Ho, K.-C.; Chu, C.-W. Synergistic improvements in stability and performance of lead iodide perovskite solar cells incorporating salt additives. *J. Mater. Chem. A* **2016**, *4*, 1591–1597. [CrossRef]
20. Bass, K.K.; McAnally, R.E.; Zhou, S.; Djurovich, P.I.; Thompson, M.E.; Melot, B.C. Influence of moisture on the preparation, crystal structure, and photophysical properties of organohalide perovskites. *Chem. Commun.* **2014**, *50*, 15819–15822. [CrossRef] [PubMed]
21. You, J.; Yang, Y.; Hong, Z.; Song, T.-B.; Meng, L.; Liu, Y.; Jiang, C.; Zhou, H.; Chang, W.-H.; Li, G.; et al. Moisture assisted perovskite film growth for high performance solar cells. *Appl. Phys. Lett.* **2014**, *105*, 183902. [CrossRef]
22. Sison, E.S.; Rahman, M.D.; Durham, D.L.; Hermanowski, J.; Ross, M.F.; Jennison, M.J. Dielectric and chemical characteristics of electron-beam-cured photoresist. In Proceedings of the SPIE's 1995 Symposium on Microlithography, Santa Clara, CA, USA, 9 June 1995; p. 14.
23. Cao, X.; Li, C.; Li, Y.; Fang, F.; Cui, X.; Yao, Y.; Wei, J. Enhanced performance of perovskite solar cells by modulating the Lewis acid–base reaction. *Nanoscale* **2016**, *8*, 19804–19810. [CrossRef] [PubMed]
24. Cao, X.B.; Li, Y.H.; Fang, F.; Cui, X.; Yao, Y.W.; Wei, J.Q. High quality perovskite films fabricated from Lewis acid–base adduct through molecular exchange. *RSC Adv.* **2016**, *6*, 70925–70931. [CrossRef]
25. Winther-Jensen, B.; West, K. Vapor-Phase Polymerization of 3,4-Ethylenedioxythiophene: A Route to Highly Conducting Polymer Surface Layers. *Macromolecules* **2004**, *37*, 4538–4543. [CrossRef]
26. Xu, J.; Wang, C.; Wang, T.; Wang, Y.; Kang, Q.; Liu, Y.; Tian, Y. Mechanisms for low-temperature direct bonding of Si/Si and quartz/quartz via VUV/O_3 activation. *RSC Adv.* **2018**, *8*, 11528–11535. [CrossRef]
27. Wilamowska, M.; Kujawa, M.; Michalska, M.; Lipińska, L.; Lisowska-Oleksiak, A. Electroactive polymer/graphene oxide nanostructured composites; evidence for direct chemical interactions between PEDOT and GOx. *Synth. Met.* **2016**, *220*, 334–346. [CrossRef]
28. Chang, J.; Lin, Z.; Zhu, H.; Isikgor, F.H.; Xu, Q.-H.; Zhang, C.; Hao, Y.; Ouyang, J. Enhancing the photovoltaic performance of planar heterojunction perovskite solar cells by doping the perovskite layer with alkali metal ions. *J. Mater. Chem. A* **2016**, *4*, 16546–16552. [CrossRef]

11

The Way to Pursue Truly High-Performance Perovskite Solar Cells

Jia-Ren Wu [1,†], Diksha Thakur [1,†], Shou-En Chiang [1], Anjali Chandel [1], Jyh-Shyang Wang [1,2], Kuan-Cheng Chiu [1,2] and Sheng Hsiung Chang [1,2,*]

1 Department of Physics, Chung Yuan Christian University, Taoyuan 32023, Taiwan
2 Center for Nano Technology, Chung Yuan Christian University, Taoyuan 32023, Taiwan
* Correspondence: shchang@cycu.edu.tw
† These authors contributed equally to this work.

Abstract: The power conversion efficiency (PCE) of single-junction solar cells was theoretically predicted to be limited by the Shockley–Queisser limit due to the intrinsic potential loss of the photo-excited electrons in the light absorbing materials. Up to now, the optimized GaAs solar cell has the highest PCE of 29.1%, which is close to the theoretical limit of ~33%. To pursue the perfect photovoltaic performance, it is necessary to extend the lifetimes of the photo-excited carriers (hot electrons and hot holes) and to collect the hot carriers without potential loss. Thanks to the long-lived hot carriers in perovskite crystal materials, it is possible to completely convert the photon energy to electrical power when the hot electrons and hot holes can freely transport in the quantized energy levels of the electron transport layer and hole transport layer, respectively. In order to achieve the ideal PCE, the interactions between photo-excited carriers and phonons in perovskite solar cells has to be completely understood.

Keywords: perovskite solar cells; hot-carrier characteristics; quantized electron transport layer; quantized hole transport layer

---

1. Introduction

The bounded electrons of inorganic and organic semiconductors can be efficiently and instantaneously excited from the ground state to the excited state when the photon energy of the incident lightwaves is higher than the absorption bandgap. However, the photo-excited electrons (hot electrons) in the light-absorbing materials (LAMs) have to relax to the meta-stable state (conduction band minimum (CBM) or lowest unoccupied molecular orbital (LUMO)) due to the ultrafast thermallization process [1–4], which results in the intrinsic potential loss and thereby limits the power conversion efficiency (PCE) of single-junction solar cells to be a moderate value of 33.7% [5]. The physical concept of the Shockley–Queisser (S–Q) limit can be understood as the following descriptions. When then LAM has a large bandgap, the broadband sun light cannot be efficiently absorbed by the wide-bandgap material. Therefore, the photocurrent density of solar cells can be increased with a decrease in the absorption bandgap of the active layer. For example, the photocurrent density of single-crystalline Si solar cells (~42 mA/cm^2) is always higher than that of single-crystalline GaAs solar cells (~29 mA/cm^2) because the absorption bandgap of crystalline Si (1.1 eV) is lower than that of crystalline GaAs (1.43 eV). When the active layer is a wide-bandgap material, the photo-excited electrons have to relax to the meta-stable state, which indicates that the highest potential difference between the cathode electrode and the anode electrode is equal to E_g/e. $E_g$and e are the absorption bandgap of the active layer and the electric charge, respectively. Usually, the potential difference between the cathode electrode and the anode electrode equals to the open-circuit voltage (V_{OC}) which is defined by the current density–voltage (J–V) curve of solar cells. As we know that the V_{OC} of a solar cell is proportional to the absorption bandgap of the active layer. For example, the V_{OC} of single-crystalline

Si solar cells (~0.738 V) is lower than that of single-crystalline GaAs solar cells (~1.127 V). According to the S–Q limit, it is impossible to simultaneously obtain the high V_{OC} and the high photocurrent density (short-circuit current density, J_{SC}), which results in an optimal absorption bandgap of 1.34 eV for the highest PCE of 33.7%.

In the past several decades, physical and chemical scientists were trying to achieve the highest PCE of 33.7% by using the different types of solar cells. When the exciton binding energy of LAMs is lower than the thermal energy, the planar thin-film structure can be used to construct the high-performance solar cells, such as the crystalline Si [6,7], crystalline GaAs [8,9] and crystalline InP [10,11] solar cells. When the exciton binding energy of the LAMs is higher than the thermal energy, the P:N nanocomposite thin-film structures have to be used to increase the photovoltaic performance, such as the organic bulk-heterojunction solar cells [12–14] and dye-sensitized solar cells (DSSCs) [15–17]. Although the PCE of organic photovoltaics (OPVs) and DSSCs is significantly lower than that of the planar thin-film inorganic semiconductor-based solar cells, the cost-effective OPVs and DSSCs still received a lot of attentions in the past two decades. Thanks to the fundamental investigations on the OPVs and DSSCs, the PCE of perovskite solar cells has dramatically increased from 3.8% [18] to 25.2% [19] by using the solution-processed methods.

It is amazing that the high-efficiency perovskite solar cells can be realized by using the low-temperature solution-processed methods because the presence of high-density defects in the active layer [20–22] usually can simultaneously reduce the V_{OC}, J_{SC} and fill factor (FF) of solar cells. It is well known that high-efficiency perovskite solar cells can be explained mainly due to the large absorption coefficient [23,24], moderate refractive index [25,26], low exciton binding energy [27,28], long exciton (carrier) lifetime [29,30] and long exciton (carrier) diffusion length [31,32]. In addition, the high PCE of perovskite solar cells also relieson the efficient energy transfer at the perovskite/electron transport layer (ETL) and perovskite/hole transport layer (HTL) interfaces. The highest PCE of perovskite solar cells is theoretically predicted to be an attractive value of 31% [33], which is also limited by the prediction from the S–Q limit.

To pursue truly high-performance solar cells, it is necessary to reduce the intrinsic potential loss via increasing the hot-carrier lifetimes of LAMs. The hot-electron lifetimes of GaAs, Si and InP crystals are 1.5 ps [34], 0.18 ps [35] and 3.4 ns [36], respectively. In general, the lower phonon energy corresponds to the longer hot-electron lifetime [37,38]. The ultrashort hot-electron lifetimes mean that the photo-excited electrons must relax to the meta-stable state to form excitons. In recent reports, the lifetime for the hot electrons in perovskite crystals has been related to the +1 cation [39]. Furthermore, the hot-electron lifetime (diffusion length) of $MAPbI_3$ thin films was determined to be longer than 20 ps (600 nm) by using transient absorbance spectroscopy [40]. The long-lived hot-carrier mediated light emission was also observed in formamidinium tin triiodide perovskites [41]. The existence of long-lived hot electrons means that it is possible to realize truly high-performance solar cells when crystalline perovskite thin films are used as the LAM.

In this review, we discuss the hot-carrier characteristics and the ways for hot-carrier extractions in the energy–space diagrams. A theoretical point of view is proposed in order to understand how the single-junction hot-carrier solar cells can be realized. Finally, the practical issues are discussed in order to assess the possibility for the realization of perovskite-based hot-carrier solar cells.

2. Light-Materials' Interactions: Excited Bounded Electrons

Photon energy can be efficiently converted to electrical power by using p-type materials due to the high absorption coefficient. Figure 1 shows the carrier dynamics of photo-excited electrons in an energy–space diagram. When the incident photons are absorbed by a p-type material, the electrons in the ground state can transit to the excited state to form hot carriers. The hot carriers can be viewed as the oscillating charged particles, which can coherently and incoherently collide with the lattice vibrations (photons). The coherent collisions between the hot carriers and phonons can result in Raman scattering emissions. The incoherent collisions between the hot carriers and photons can result in the photoluminescence (PL) emissions. In addition, the hot-carrier mediated PL emissions can be observed in the perovskite thin film due to the slow thermalization process [41].

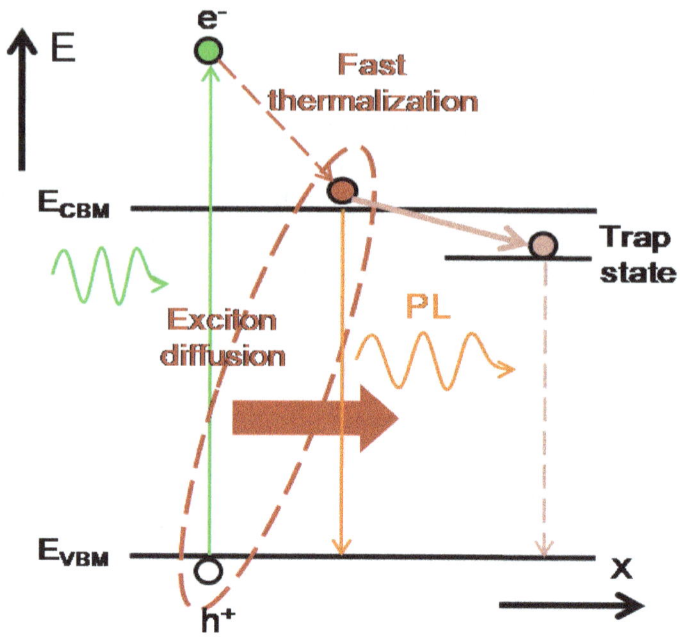

Figure 1. Photo-excited carrier dynamics in an energy–space diagram.

In general, the hot-carrier lifetime of organic LAMs is shorter than 100 fs. Therefore, it is not easy to observe the hot-carrier dynamics of organic materials by using the femtosecond time-resolved photoluminescence (FTR-PL) technique due to the limited instrument response function (IRF ~150 fs) [42,43]. The ultrashort hot-carrier lifetime is mainly due to the large optical phonon energy of organic materials [44], which results in an extremely short hot-carrier diffusion length. It means that the hot carriers rapidly decay to the meta-stable state in organic materials via the thermalization (downhill relaxation) process [2,45]. Then, the electrons in the lowest unoccupied molecular orbital (LUMO) and the holes in the highest occupied molecular orbital (HOMO) are mutually attracted to form excitons. In conjugated small organic molecules, the exciton binding energy can be reduced by increasing the conjugation due to the delocalization effect [46]. In general, the exciton-binding energy of organic LAMs can be a wide range from 0.3 eV to 1.0 eV [47,48], which depends on the degree of delocalization of electron-hole pairs. This means that the larger exciton binding energy corresponds to the shorter (smaller) exciton radius (volume). From the concept of allowable excitation density, the smaller exciton volume results in the higher exciton generation rate (absorption coefficient). Due to the high density of excitons, the exciton diffusion length and exciton lifetime of organic materials can be lower than 1 nm and 1 ns, respectively.

It can be predicted that the hot-carrier lifetime of inorganic materials [34–36] is longer than that of organic materials because the optical phonon energy of inorganic materials [49,50] is lower. The optical phonon energy and hot-carrier lifetime of various materials are listed in Table 1 [34–38,49–53]. During the hot-carrier thermalization process in a polar semiconductor as shown in Figure 2, the energy of the hot carriers has to be firstly transferred to the longitudinal optical (LO) phonons. Then, the transition from the LO phonons to acoustic phonons results in the lattice heating. With the propagation of acoustic phonons, the thermal energy can transfer to the surroundings. This means that there are three ways that can be used to slow down the hot-carrier thermalization process. The energy transfer rate from hot carriers to LO phonons is intrinsic fast in non-polar materials [54,55], such as Si and Ge. In a polar $CH_3NH_3PbI_3$ ($MAPbI_3$) crystal thin film, the energy transfer rate from hot carriers to LO phonons can be delayed due to the formation of hot polarons [55], which results in a long hot-carrier cooling time. As we know that hot polarons are quasiparticles, which describes the interaction between the hot carriers and polar lattices.

Therefore, the formation of hot polarons can delay the transfer rate from hot carriers to LO phonons. This was firstly explained due to a phonon bottleneck effect [56]. In addition, the propagation of acoustic phonons in perovskite crystals is theoretically predicted to be slow due to the use the insulating organic cation [57], which also can generate the up-conversion of acoustic phonons to re-heat the hot carriers and thereby increases the hot-carrier lifetime [53]. Up to now, the methods to delay the energy transition from LO phonons to acoustic phonons have not yet been proposed for increasing the hot-carrier lifetime in polar perovskite crystals.

Table 1. Optical phonon energies (E_{Phonon}) and hot-carrier lifetimes (τ_{hc}) of various inorganic materials and organic materials. TPA-TTAR-A: triphenylamine-tetrathienoacene-acceptor.

Materials	GaAs	Si	InP	P3HT	TPA-TTAR-A	$CH_3NH_3PbI_3$	$HC(NH_2)_2PbI_3$
E_{Phonon} (meV)	40	60	42	None	None	25	11.5
τ_{hc} (ps)	1.5	0.18	3400	<0.1	1.01	20	124
Ref.	[34]	[35,36]	[37,38]	[49]	[50]	[40,51]	[52,53]

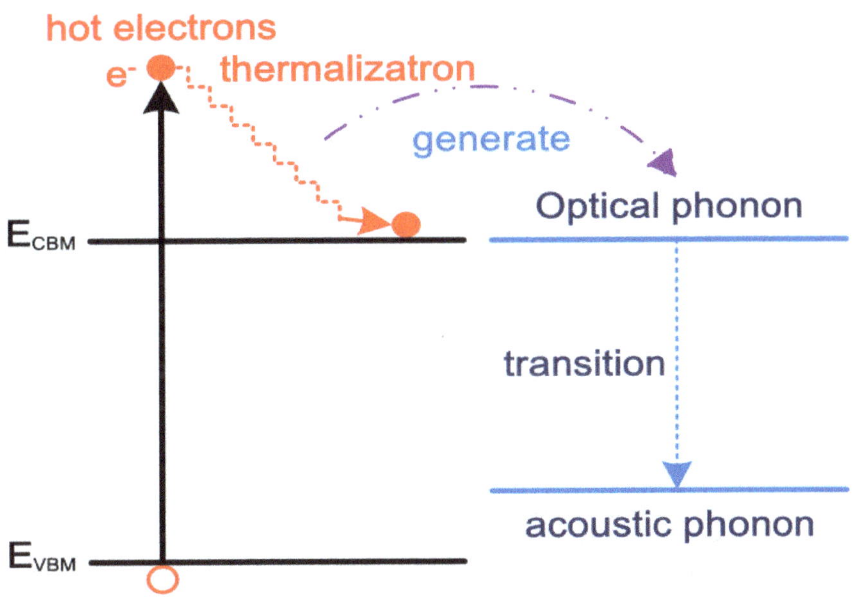

Figure 2. Hot carrier-optical phonon energy transfer and thermalization process.

3. Hot-Carrier Extraction at a Light-Absorbing Material/Electron Transport Layer (LAM/ETL) Interface

The hot-electron injection from organic fused thiophene-based dyes to TiO_2 nanoparticles (NPs) was investigated for the first time by using a FTR-PL technique [58], which was used to explain the abnormal high V_{OC} of 0.93 V in DSSCs with an iodide/triliodide based electrolyte [50]. The dyes can be adsorbed on the surface of TiO_2 NPs due to the electrical attraction between the anchoring group of dyes and the oxygen defect of TiO_2 NPs. Therefore, it can be understood that the high V_{OC} is due to the hot-electron injection from the dyes to the higher quantized energy levels of the TiO_2 NPs, as shown in Figure 3. In this study, the average diameter of the TiO_2 NPs is about 10 nm [59], which is about 3 times of the exciton radius. Therefore, it can be predicted that the quantized energy levels can be created in the TiO_2 quantum dots (QDs) [60].

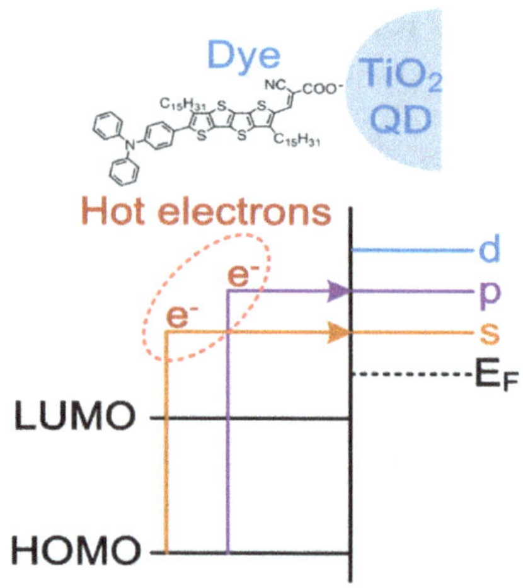

Figure 3. Hot-electron injection from dyes to the quantized energy levels of TiO_2 quantum dots (QDs).

The efficient hot-electron extraction in OPVs has not yet been reported in literature, which is probably due to the fact that the diffusion process is needed for the hot carriers to reach the region of charge transfer radius [61,62]. For example, the hot electrons in poly(3-hexylthiophene-2,5,-diyl) (P3HT) polymers is rapidly decayed from the excited state to the LUMO energy level due to the ultrafast self-localization process (~100 fs) [2,43], which suppresses the hot-electron diffusion. This means that it is possible to realize organic hot-carrier photovoltaics when the ultrafast self-localization process can be reduced by increasing the delocalization quantum-assisted transport within long polymer chains [46].

The efficient hot-electron extraction in perovskite solar cells has not yet been discussed in literature. However, the abnormal high V_{OC} (=1.61 V) of inverted-type $MAPbBr_3$ based solar cells is probably due to the efficient hot-electron extraction from the $MAPbBr_3$ nano-crystals to the indene-C_{60}-bisadduct (ICBA) ETL [63]. In this study, the poly(3,4-ethylenedioxythiophene):polystyrene sulfonate (PEDOT:PSS (1:6 wt%)) thin film and ICBA thin film are used as the HTL and ETL, respectively. The Fermi level of PEDOT:PSS thin film and the LUMO energy level of ICBA thin film are −5.1 eV [64,65] and −3.9 eV [66,67], respectively, which results in a S.-Q limited V_{OC} of 1.2 eV. The high V_{OC} of $MAPbBr_3$ based solar cells means that the hot electrons have to be extracted by the LUMO+1 and/or LUMO+2 of the ICBA thin film because the PEDOT:PSS thin film is a metal-like conductive polymer [68]. The photovoltaic performances of high-V_{OC} perovskite based solar cells are listed in Table 2 [63,69,70]. The three types of perovskite solar cells both contain bromide elements in the active layer, which suggests that the bromide-based perovskite thin films probably have longer hot-carrier lifetimes. In addition, the long hot-carrier lifetime and diffusion length were observed in a $MAPbI_3$ perovskite thin film by using transient absorbance spectroscopy [40], which means that the realization of high-performance hot-carrier perovskite solar cells is possible.

Table 2. Photovoltaic performance of bromide-based perovskite solar cells.

Active Layer	$MAPbBr_3$	$CsPbI_2Br$	$CsPb_{0.97}Tb_{0.03}Br_3$
ETL/LUMO	ICBA/−3.9 eV	TiO_2/−4.1 eV	TiO_2/−4.1 eV
HTL/E_F or HOMO	PEDOT:PSS/−5.1 eV	Spiro-OMe TAD/−5.2 eV	NiO_x/−5.1 eV
V_{OC} (V)	1.61	~1.3	1.57
J_{SC} (mA/cm^2)	6.04	~12	8.21
FF (%)	77.0	~74	79.6
Ref.	[63]	[69]	[70]

4. Hot-Carrier Extraction at a LAM/Hole Transport Layer (HTL) Interface

As we know that the bounded electrons are not excited from the valence band maximum (E_{VBM}) when the photon energy of the incident lightwaves is higher than the absorption bandgap of materials.

Therefore, the hot-hole relaxation process has to be considered in order to realize the hot-carrier solar cells. Figure 4 shows the hot-hole and hot-electron dynamics in the energy diagram. For example, the hot-hole and hot-electron relaxation times of a $MAPbI_3$ thin film are 100–500 fs and 1–5 ps [71], respectively, which means that it is more difficult to collect the hot holes in perovskite thin films by using a hot-hole selective layer due to the sub-picosecond relaxation time. Fortunately, there is experimental evidence of transient absorbance spectra to show that the hot holes in a $CsPbI_3$ thin film can be efficiently extracted by the capping layer of P3HT thin film within a few 100 fs [72]. However, the hot-hole extraction process from the perovskite thin film to the P3HT thin film is not yet completely understood. Further experiments are needed to demonstrate that the hot-hole extraction can increase the V_{OC} of perovskite solar cells.

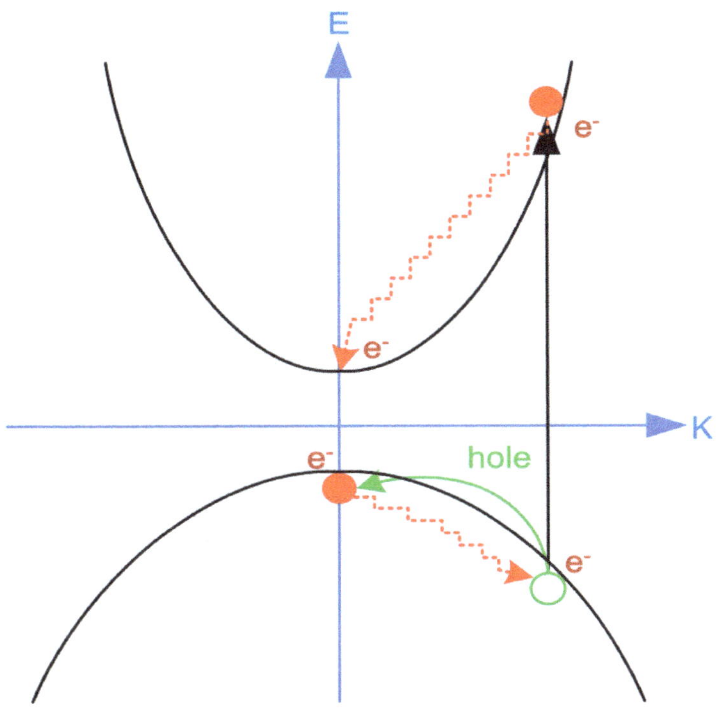

Figure 4. Energy diagram of hot-hole and hot-electron relaxations.

The extraction efficiency of hot holes is also related to the carrier diffusion coefficient of perovskite thin films because the hot holes have to diffuse into the region of the charge transfer radius at the perovskite/HTL interface. According to the Einstein relation ($D = \mu k_B T/e$), the carrier diffusion coefficient (D) is proportional to the carrier mobility (μ), where $k_B T$ is the thermal energy and e is the electric charge. The carrier drift equation ($\mu = e\tau/m^*$) shows that the carrier mobility is proportional (inversely proportional) to the carrier relaxation time (effective mass), where $\tau$and $m^*$are the carrier relaxation time and carrier effective mass, respectively. Therefore, the smaller hot-hole effective mass corresponds to the longer hot-hole diffusion length and thereby results in the higher extraction efficiency of hot holes. Interestingly, the hole mobility can be higher than the electron mobility when the LAM is $CsPbBr_3$ or $CsPbCl_3$, which is due to the lower hole effective mass [73]. However, the propagation characteristics of hot holes are not yet completely understood.

5. Theoretical Point of View

The highest PCE of single-junction hot-carrier solar cells was theoretically predicted to be 66% under one sun illumination [74]. Figure 5a shows the energy diagram of a single-junction hot-carrier solar cell. In the single-junction solar cell, the high-energy and low-energy incident photons are absorbed by the electrons in the deeper levels and in the shallower levels of the valence band, respectively. When the electrons in the valence band are excited to the conduction band, the hot electrons in the higher energy levels and in the lower energy levels both have to be directly extracted in order to avoid ultrafast potential loss. As to the hot holes, they also have to be directly extracted in order to keep the original potential.

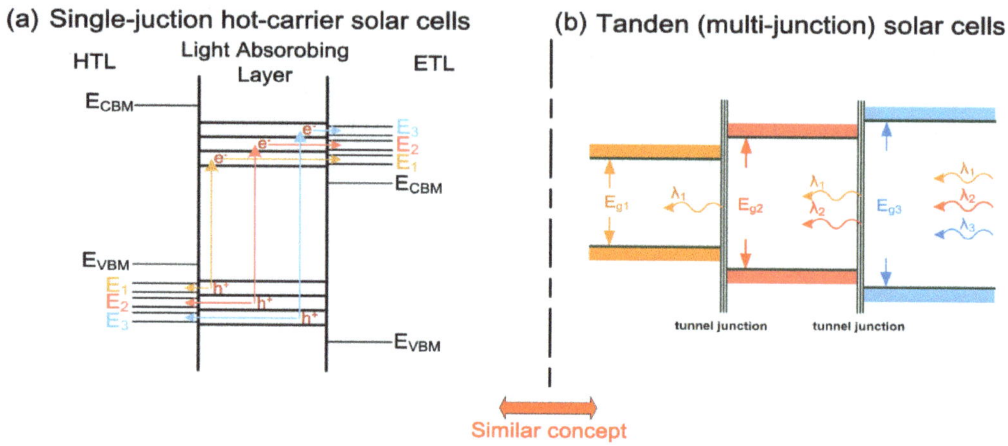

Figure 5. Energy diagrams. (**a**) Single-junction hot-carrier solar cell. (**b**) Tandem solar cell.

The concept of single-junction hot-carrier solar cells is similar to the tandem (multi-junction) solar cells which has a highly theoretical PCE of 68% under one sun illumination [75]. Up to now, the highest PCE of tandem solar cells is 39.2% under one sun illumination, which is significantly higher than the highest PCE (29.1%) of single-junction GaAs solar cells [19]. When the different absorption ranges in a single LAM are viewed as the individual materials with the different absorption bandgaps, the energy diagram of a tandem solar cell can be plotted in Figure 5b. The PCE of tandem solar cells is strongly related to the performance of the tunnel junctions [76] with an ultrafast carrier dynamic [77]. In single-junction hot-carrier solar cells, the double-barrier resonant tunneling structure was proposed as the energy selective contact of hot carrier solar cells [78] due to the sub-picosecond carrier extraction ability [79]. Therefore, it can be believed that the highly efficient single-junction hot-carrier solar cells can be realized when the ultrafast carrier lifetimes of hot carriers in the LAM can be increased from the sub-picosecond time scale to sub-nanosecond time scale.

In order to efficiently collect the hot carriers at the different energy levels, a multi-band ETL and a multi-band HTL have to be used to extract the dispersive hot carriers under a broadband excitation. T 5ashows the energy diagram of an ideal hot-carrier solar cell. The E_{CBM} of HTL (E_{VBM} of ETL) has to be higher (lower) than the E_3 in the conduction band of ETL (E_3 in the valence band of HTL) in order to block the hot electrons (hot holes), which can help the collection of hot carriers. If the high-energy hot carriers can be efficiently collected by the E_3, the hot electrons (hot holes) have to freely transport within the E_3 of ETL (HTL) without the energy transitions from the E_3 to the E_2 and/or E_1. Fortunately, the photo-excited carriers can be collected and transport in the quantized energy levels of Si-doped quantum dots, which increases the V_{OC} from 0.78 V to 0.91 V [80]. Therefore, it can be expected that the hot electrons and hot holes can freely transport in the quantized energy levels of ETL and HTL, respectively. In addition, n-type graphene quantum dots [81–83] and p-type graphene quantum dots [84–86] might have the potential as the multi-band ETL and multi-band HTL of hot-carrier solar cells, respectively.

In addition, a theoretical approach is used to calculate the J–V curves of single-junction solar cells when the selective contacts are used to collect the hot carriers [87]. Their simulation results show that the $e{\times}V_{OC}$ of hot-carrier solar cells can larger than the absorption bandgap of LAMs. For example, $e{\times}V_{OC}$ of the GaAs solar cell is 1.85 eV which is larger than the absorption bandgap of crystalline GaAs.

6. Experimental Challenges and Opportunities

Up to now, the LAMs of highly efficient perovskite solar cells were fabricated by using solution-processed methods [26,28,88–91], which means that the defect density of LAMs remains high. Although the shallow defects of perovskite thin films do not significantly influence the photovoltaic performance [92,93], the formation of defects can increase the exciton binding energy and optical phonon energy of perovskite thin films [51,94]. Therefore, the defect-mediated phononic properties of perovskite thin films have to be investigated in order to understand how to realize hot-carrier solar cells.

The grain size of solution-processed perovskite thin films can be increased from several hundred nanometers to several micrometers by adding the small molecules [95] or with the solvent annealing process [96]. The average crystal domain size of perovskite thin films is smaller than averaged grain size, which indicates that the perovskite thin films are composed of multi-crystalline grains [97]. It means that the residual stress in a multi-crystalline perovskite thin film [98] can also influence the phononic properties, which might dominate the hot-carrier characteristics. In other words, the defect-mediated phononic properties [99] and/or crystal distortion-mediated phononic properties [100] have to be considered when the perovskite thin films are fabricated on top of amorphous substrates or poly-crystalline substrates by using solution processes or thermal evaporation methods. However, the preferred oriented perovskite thin films can be fabricated on top of the single-crystalline substrates by using the spin-coating method [101], which was observed by measuring the two-dimensional X-ray diffraction patterns.

Conceptually, the existences of defects and lattice distortions should decrease the hot-carrier lifetimes in the LAMs, which are predicted to impede the development of hot-carrier solar cells. Therefore, the development of single-crystalline perovskite bulks plays an important step for the realization of high-performance optoelectronic devices. Two years ago, the single-crystalline perovskites were grown on top of various substrates, such as indium tin oxide (ITO), quartz and silicon wafer, which were used as the X-ray detector [102]. The strategy is to modify the surface of the substrates with a NH_3-Br-teminated self-assembling molecules monolayer, which provides a seeding layer to grow the single-crystalline $MAPbBr_3$. The long carrier lifetime of the single-crystalline $MAPbBr_3$ is 692 ns, which indicates the low defect density. Therefore, the hot-carrier lifetimes of single-crystalline perovskites can be expected to be longer than that of poly-crystalline perovskite thin films.

When the substrate (Au/p^+-type wafer) is the anode side, the single-crystalline perovskite has to be grown on top of a quantized HTL. Then, a quantized ETL has to be fabricated on top of the single-crystalline perovskite. A transparent conductive oxide has to be deposited on top of the device as the cathode electrode. Therefore, it can be imagined that the device architecture is Au/p^+-type wafer/quantized HTL/single-crystalline perovskite/quantized ETL/transparent conductive cathode. The p^+-type wafer has to be a large-bandgap material in order to block the hot electrons from the single-crystalline perovskite. The quantized HTL and the quantized ETL can be a p-type quantum well (QW) structure [103] and a double-barrier resonant-tunneling structure [78], respectively. In addition, the transparent conductive cathode is used to block the hot holes from the single-crystalline perovskite. The potential candidates as the p^+-type substrate, HTL, ETL and transparent conductive anode are listed in Table 3. The Au coated p^+-type GaN, AlN or SiC wafer can be used as the anode electrode. The epitaxial growth process of GaN/AlN QW [104] or AlGaN QDs [105] on top of the GaN or AlN wafer is a mature technique by using metal organic chemical-vapor deposition (MOCVD) or molecular beam epitaxy (MBE). However, the barrier high and physical size of the QW or QDs has to be investigated and designed in order to be used as the quantized HTL for the collection of hot holes. P-type graphene QDs can be produced by using microwave-assisted heating method [106] or pulsed laser ablation method [107]. And, the p-type graphene QDs thin film can be deposited on top of the p^+-type wafer by using the spin-coating method. Although, the single-crystalline $MAPbBr_3$ has been demonstrated that can be grown on various substrates with a surface modification method [102]. The contact at the $MAPbI_3$/HTL interface, which should strongly influence the collection efficiency of hot holes, has not yet investigated. It is predicted that the [6,6]-phenyl-C_{61}-butyric acid methyl ester (PCBM)/bathocuproine (BCP) QW, ZnO QDs [108] or TiO_2 QDs [109] can be used to collect the hot electrons from the single-crystalline perovskite. PCBM molecules can be dissolved in low-polarity solvents, such as chlorobenzene and toluene, which can be directly spin-coated on top of $MAPbI_3$ thin films. BCP molecules, ZnO QDs and TiO_2 QDs are usually dissolved in isopropanol (IPA) which is a polar solvent. Therefore, the BCP, ZnO QDs or TiO_2 QDs thin film cannot be directly spin-coated on top of $MAPbI_3$ thin films. The contact at the hydrophobic ETL/hydrophilic $MAPbI_3$ interface can be improved by slightly roughening the surface of $MAPbI_3$ thin film [91]. Then, the Al-doped ZnO, Ga-doped ZnO or Al-Ga co-doped ZnO thin film can be used as the transparent conductive cathode because of the metal-like electrical conductivity which can directly collect the hot electrons without the additional potential loss. However, the high-quality transparent conductive cathodes are usually deposited by using the radio-frequency magnetron sputtering method, which can damage the $MAPbI_3$ thin film due to the excessive energy bombardment during the deposition process [110]. To resist the excessive energy bombardment, an inorganic thin film has to be used as the buffer layer in between the $MAPbI_3$ thin film and transparent conductive cathode [110].

Table 3. Potential candidates as the p^+-type substrate, hole transport layer (HTL), electron transport layer (ETL) and transparent conductive anode.

p^+-Type Substrate	HTL	ETL	Transparent Conductive Cathode
GaN	GaN/AlGaN QW	PCBM/BCP QW	Al-doped ZnO
AlN	AlGaN QDs	ZnO QDs	Ga-doped ZnO
SiC	p-type graphene QDs	TiO_2 QDs	Al-Ga co-doped ZnO

The realization of hot-carrier solar cells has to rely on the long-lived hot carriers in the active layer and the efficient collections of hot carriers by the HTL and ETL. The hot-carrier lifetime and hot-carrier diffusion length of crystalline perovskite thin films can be longer than 100 ps and 600 nm, respectively, which indicates that the hot carriers are possibly collected without potential loss. However, the hot carriers have to be rapidly collected before the ultrafast thermalization process, which means that the collection times of hot carriers have to be faster than the hot-carrier lifetimes. For example, the hot-carrier lifetime of the LAM and the hot-carrier collection time of a double-barrier resonant tunneling structure can be 100 ps and 0.4ps, respectively. Furthermore, the hot-carrier collection efficiency can be calculated by $h = 1/t_1/(1/t_1 + 1/t_2)$ [58], where t_1 and t_2 are hot-carrier collection time and hot-carrier lifetime, respectively. Therefore, the calculated hot-carrier collection efficiency equals to 99.6%, which means that it is worthwhile to develop a double-barrier resonant tunneling structure as the selective contact of hot-carrier perovskite solar cells. In this review, we suggest that the quantized HTL and quantized ETL can be used to collect the dispersive hot holes and hot electrons, respectively. The collection times of hot carriers will dominate the collection efficiency. Therefore, the barrier height and physical size of the quantized HTL and quantized ETL has to be varied in order to decrease the collection times of hot carriers. Conceptually, the collection times of hot carriers are also influenced by the crystallinity (carrier mobility) of the HTL and ETL. Therefore, the formation of high-quality quantized HTL, quantized ETL and perovskite thin film is necessary in order to realize hot-carrier solar cells.

7. Conclusions

Recent progress in the understanding of the light-perovskite interactions shows that the realization of highly efficient hot-carrier solar cells is possible because the long-lived hot polarons can be formed and thereby delays the thermalization process. Conceptually, the extraction of hot electrons and hot holes can increase the open-circuit voltage (V_{OC}). The hot-electron injection was firstly observed in the dye-sensitized solar cells, which significantly increased the V_{OC} from 0.75 V to 0.93 V.

The long-lived hot electrons in the organic fused thiophene-based dyes were observed by using the femtosecond time-resolved photoluminescence technique.

The highest V_{OC} of $CH_3NH_3PbBr_3$ solar cells is 1.61 V, which is far larger than the potential difference (1.2 V) between the LUMO energy level of the electron transport layer (ICBA thin film) and the Fermi level of the hole transport layer (PEDOT:PSS thin film). The abnormal high V_{OC} of perovskite solar cells can be explained due to the efficient collection of hot electrons by the LUMO+1 and/or LUMO+2 of the ICBA thin film. In addition, the P3HT polymer thin film was used as the hot-hole selective contact layer, which was observed by using the femtosecond transient absorbance spectra.

To realize truly high-performance hot-carrier perovskite solar cells, it is necessary to simultaneously collect the hot electrons and hot holes without the additional potential loss. We have proposed a device architecture which might be used to achieve the desired power conversion efficiency. The device architecture is Au/p^+-type GaN wafer/quantized hole transport layer (QHTL)/crystalline perovskite/quantized electron transport layer (QETL)/transparent conductive cathode. Ideally, the hot holes and hot electrons in the crystalline perovskite have to be collected by the QHTL and QETL, respectively. Therefore, it is necessaryto investigate the ultrafast energy transfer dynamics of hot carriers from the crystalline perovskite to the QHTL and QETL by varying the barrier high and physical size of the quantum wells or quantum dots.

Author Contributions: Conceptualization was made by S.H.C., J.-S.W. and K.-C.C. References were collected by J.-R.W., D.T., S.-E.C. and A.C. Draft preparation was made by J.-R.W. and D.T. Writing—Review and Editing was made by S.H.C.

References

1. Katayama, K.; Inagaki, Y.; Sawada, T. Ultrafast two-step thermalization processes of photoexcited electrons at a gold surface: Application of a wavelength-selective transient reflecting grating method. *Phys. Rev. B* **2000**, *61*, 7332–7335. [CrossRef]
2. Xie, Y.; Li, Y.; Xiao, L.; Qiao, Q.; Dhakal, R.; Zhang, Z.; Gong, Q.; Galipeau, D.; Yan, X. Femtosecond Time-Resolved Fluorescence Study of P3HT/PCBM Blend Films. *J. Phys. Chem. C* **2010**, *114*, 14590–14600. [CrossRef]
3. Bräm, O.; Cannizzo, A.; Chergui, M. Ultrafast fluorescence studies of dye sensitized solar cells. *Phys. Chem. Chem. Phys.* **2012**, *14*, 7934–7937. [CrossRef] [PubMed]
4. Kirton, P.; Keeling, J. Thermalization and breakdown of thermalization in photon condensates. *Phys. Rev. A* **2015**, *91*, 033826. [CrossRef]
5. Shockley, W.; Queisser, H.J. Detailed Balance Limit of Efficiency of p-n Junction Solar Cells. *J. Appl. Phys.* **1961**, *32*, 510–519. [CrossRef]
6. Blakers, A.; Zin, N.; McIntosh, K.R.; Fong, K. High Efficiency Silicon Solar Cells. *Energy Procedia* **2013**, *33*, 1–10. [CrossRef]
7. De Wolf, S.; Cuevas, A.; Battaglia, C. High-efficiency crystalline silicon solar cells: Status and perspectives. *Energy Environ. Sci.* **2016**, *9*, 1552–1576.
8. Moon, S.; Kim, K.; Kim, Y.; Heo, J.; Lee, J. Highly efficient single-junction GaAs thin-film solar cell on flexible substrate. *Sci. Rep.* **2016**, *6*, 30107. [CrossRef]
9. Liang, D.; Kang, Y.; Huo, Y.; Chen, Y.; Cui, Y.; Harris, J.S. High-Efficiency Nanostructured Window GaAs Solar Cells. *Nano Lett.* **2013**, *13*, 4850–4856. [CrossRef] [PubMed]
10. Yamamoto, A.; Yamaguchi, M.; Uemura, C. High efficiency homojunction InP solar cells. *Appl. Phys. Lett.* **1985**, *47*, 975–977. [CrossRef]
11. Yin, X.; Battaglia, C.; Lin, Y.; Chen, K.; Hettick, M.; Zheng, M.; Chen, C.-Y.; Kiriya, D.; Javey, A. 19.2% Efficient InP Heterojunction Solar Cell with Electron-Selective TiO_2 Contact. *ACS Photon.* **2014**, *1*, 1245–1250. [CrossRef] [PubMed]
12. He, Y.; Chen, H.-Y.; Hou, J.; Li, Y. Indene–C60Bisadduct: A New Acceptor for High-Performance Polymer Solar Cells. *J. Am. Chem. Soc.* **2010**, *132*, 1377–1382. [CrossRef] [PubMed]

13. Guo, X.; Cui, C.; Zhang, M.; Huo, L.; Huang, Y.; Hou, J.; Li, Y. High efficiency polymer solar cells based on poly(3-hexylthiophene)/indene-C70 bisadduct with solvent additive. *Energy Environ. Sci.* **2012**, *5*, 7943–7949. [CrossRef]
14. Zhang, H.; Yao, H.; Hou, J.; Zhu, J.; Zhang, J.; Li, W.; Yu, R.; Gao, B.; Zhang, S.; Hou, J. Over 14% Efficiency in Organic Solar Cells Enabled by Chlorinated Nonfullerene Small-Molecule Acceptors. *Adv. Mater.* **2018**, *30*, 1800613. [CrossRef] [PubMed]
15. Roy, P.; Kim, D.; Lee, K.; Spiecker, F.; Schmuki, P. TiO_2 nanotubes and their application in dye-senstized solar cells. *Nanoscale* **2010**, *2*, 45–49. [CrossRef] [PubMed]
16. Liao, J.-Y.; He, J.-W.; Xu, H.; Kuang, D.-B.; Su, C.-Y. Effect of TiO_2 morphology on photovoltaic performance of dye-sensitized solar cells: Nanoparticles, nanofibers, hierarchical spheres and ellipsoid spheres. *J. Mater. Chem.* **2012**, *22*, 7910–7918. [CrossRef]
17. Fan, Y.-H.; Ho, C.-Y.; Chang, Y.-J. Enhancement of Dye-Sensitized Solar Cells Efficiency Using Mixed-Phase TiO_2 Nanoparticles as Photoanode. *Scanning* **2017**, *2017*, 9152973. [CrossRef]
18. Kojima, A.; Teshima, K.; Shirai, Y.; Miyasaka, T. Organometal Halide Perovskites as Visible-Light Sensitizers for Photovoltaic Cells. *J. Am. Chem. Soc.* **2009**, *131*, 6050–6051. [CrossRef] [PubMed]
19. NREL Efficiency Chart. This Plot is Courtesy of the National Renewable Energy Laboratory, Golden. Available online: https://www.nrel.gov/pv/assets/pdfs/best-reserch-cell-efficiencies.20190802.pdf (accessed on 5 August 2019).
20. Bi, C.; Shao, Y.; Yuan, Y.; Xiao, Z.; Wang, C.; Gao, Y.; Huang, J. Understanding the formation and evolution of interdiffusion grown organolead halide perovskite thin films by thermal annealing. *J. Mater. Chem. A* **2014**, *2*, 18508–18514. [CrossRef]
21. Barker, A.J.; Sadhanala, A.; Deschler, F.; Gandini, M.; Senanayak, S.P.; Pearce, P.M.; Mosconi, E.; Pearson, A.J.; Wu, Y.; Kandada, A.R.S.; et al. Defect-Assisted Photoinduced Halide Segregation in Mixed-Halide Perovskite Thin Films. *ACS Energy Lett.* **2017**, *2*, 1416–1424. [CrossRef]
22. Ran, C.; Xu, J.; Gao, W.; Huang, C.; Dou, S. Defects in metal triiodide perovskite materials towards high-performance solar cells: Origin, impact, characterization, and engineering. *Chem. Soc. Rev.* **2018**, *47*, 4581–4610. [CrossRef] [PubMed]
23. Chang, S.H.; Lin, K.-F.; Chiang, C.-H.; Chen, S.-H.; Wu, C.-G. Plasmonic Structure Enhanced Exciton Generation at the Interface between the Perovskite Absorber and Copper Nanoparticles. *Sci. World J.* **2014**, *2014*, 128414. [CrossRef] [PubMed]
24. De Wolf, S.; Holovsky, J.; Moon, S.-J.; Löper, P.; Niesen, B.; Ledinsky, M.; Haug, F.-J.; Yum, J.-H.; Ballif, C. Organometallic Halide Perovskites: Sharp Optical Absorption Edge and Its Relation to Photovoltaic Performance. *J. Phys. Chem. Lett.* **2014**, *5*, 1035–1039. [CrossRef] [PubMed]
25. Xie, Z.; Liu, S.; Qin, L.; Pang, S.; Wang, W.; Yan, Y.; Yao, L.; Chen, Z.; Wang, S.; Du, H.; et al. Refractive index and extinction coefficient of CH3NH3PbI3 studied by spectroscopic ellipsometry. *Opt. Mater. Express* **2015**, *5*, 29–43.
26. Chang, S.H.; Huang, W.-C.; Chen, C.-C.; Chen, S.-H.; Wu, C.-G. Effects of anti-solvent (iodobenzene) volume on the formation of $CH_3NH_3PbI_3$ thin films and their application in photovoltaic cells. *Appl. Surf. Sci.* **2018**, *445*, 24–29. [CrossRef]
27. Galkowski, K.; Mitioglu, A.; Miyata, A.; Plochocka, P.; Portugall, O.; Eperon, G.E.; Wang, J.T.-W.; Stergiopoulos, T.; Stranks, S.D.; Snaith, H.J.; et al. Determination of the exciton binding energy and effective masses for methylammonium and formamidinium lead tri-halide perovskite semicroductors. *Energy Environ. Sci.* **2016**, *9*, 962–970. [CrossRef]
28. Chen, C.-C.; Chang, S.H.; Chen, L.-C.; Tsai, C.-L.; Cheng, H.-M.; Huang, W.-C.; Chen, W.-N.; Lu, Y.-C.; Tseng, Z.-L.; Chiu, K.Y.; et al. Interplay between nucleation and crystal growth during the formation of $CH_3NH_3PbI_3$ thin films and their application in solar cells. *Sol. Energy Mater. Sol. Cells* **2017**, *159*, 583–589. [CrossRef]
29. Yang, Y.; Yang, M.; Li, Z.; Crisp, R.; Zhu, K.; Beard, M.C. Comparison of recombination dynamics in $CH_3NH_3PbBr_3$ and $CH_3NH_3PbI_3$ perovskite thin films: Influence of exciton binding energy. *J. Phys. Chem. Lett.* **2015**, *6*, 4688–4692. [CrossRef] [PubMed]
30. Chen, C.-C.; Chang, S.H.; Chen, L.-C.; Kao, F.-S.; Cheng, H.-M.; Yeh, S.-C.; Chen, C.-T.; Wu, W.-T.; Tseng, Z.-L.; Chuang, C.L.; et al. Improving the efficiency of inverted mixed-organic-cation perovskite absorber based

photovoltaics by tailing the surface roughness of PEDOT: PSS thin film. *Sol. Energy* **2016**, *134*, 445–451. [CrossRef]

31. Stranks, S.D.; Eperon, G.E.; Grancini, G.; Menelaou, C.; Alcocer, M.J.P.; Leijtens, T.; Herz, L.M.; Petrozza, A.; Snaith, H.J. Electron-Hole Diffusion Lengths Exceeding 1 Micrometer in an Organometal Trihalide Perovskite Absorber. *Science* **2013**, *342*, 341–344. [CrossRef]
32. Guo, Z.; Manser, J.S.; Wan, Y.; Kamat, P.V.; Huang, L. Spatial and temporal imaging of long-range charge transport in perovskite thin films by ultrafast microscopy. *Nat. Commun.* **2015**, *6*, 7471. [CrossRef] [PubMed]
33. Sha, W.E.I.; Ren, X.; Chen, L.; Choy, W.C.H. The efficiency limit of $CH_3NH_3PbI_3$ perovskite solar cells. *Appl. Phys. Lett.* **2015**, *106*, 221104. [CrossRef]
34. Bernardi, M.; Vigil-Fowler, D.; Ong, C.S.; Neaton, J.B.; Louie, S.G. Ab initio study of hot electrons in GaAs. *Proc. Natl. Acad. Sci. USA* **2015**, *112*, 5291–5296. [CrossRef] [PubMed]
35. Cushing, S.K.; Zürch, M.; Kraus, P.M.; Carneiro, L.M.; Lee, A.; Chang, H.-T.; Kaplan, C.J.; Leone, S.R. Hot phonon and carrier relaxation in Si(100) determined by transient extreme ultraviolet spectroscopy. *Struct. Dyn.* **2018**, *5*, 054302. [CrossRef] [PubMed]
36. Lock, D.; Rusimova, K.R.; Pan, T.L.; Palmer, R.E.; Sloan, P.A. Atomically resolved real-space imaging of hot electron dynamics. *Nat. Commun.* **2015**, *6*, 8365. [CrossRef] [PubMed]
37. Gadret, E.G.; de Lima, M.M., Jr.; Madureira, J.R.; Chiaramonte, T.; Cotta, M.A.; Iikawa, F.; Cantarero, A. Optical phonon modes of wurtzite InP. *Appl. Phys. Lett.* **2013**, *102*, 122101. [CrossRef]
38. Wang, Y.; Jackson, H.E.; Smith, L.M.; Burgess, T.; Paiman, S.; Gao, Q.; Tan, H.H.; Jagadish, C. Carrier Thermalization Dynamics in Single Zincblende and Wurtzite InP Nanowires. *Nano Lett.* **2014**, *14*, 7153–7160. [CrossRef] [PubMed]
39. Madjet, M.E.-A.; Berdiyorov, G.R.; El-Mellouhi, F.; Alharbi, F.H.; Akimov, A.V.; Kais, S. Cation Effect on Hot Carrier Cooling in Halide Perovskite Materials. *J. Phys. Chem. Lett.* **2017**, *8*, 4439–4445. [CrossRef] [PubMed]
40. Guo, Z.; Wan, Y.; Yang, M.; Snaider, J.; Zhu, K.; Huang, L. Long-range hot-carrier transport in hybrid perovskites visualized by ultrafast microscopy. *Science* **2017**, *356*, 59–62. [CrossRef]
41. Fang, H.-H.; Adjokatse, S.; Shao, S.; Even, J.; Loi, M.A. Long-lived hot-carrier light emission and large blue shift in formamidinium tin triiodide perovskites. *Nat. Commun.* **2018**, *9*, 243. [CrossRef]
42. Murakami, H. Femtosecond time-resolved fluorescence up-conversion spectrometer corrected for wavelength-dependent conversion efficiency using continuous white light. *Rev. Sci. Instrum.* **2006**, *77*, 113105. [CrossRef]
43. Xu, J.; Knutson, J.R. Ultrafast fluorescence spectroscopy via upconversion: Applications to biophyiscs. *Methods Enzymol.* **2008**, *450*, 159–183. [PubMed]
44. Maultzsch, J.; Reich, S.; Thomsen, C.; Requardt, H.; Ordejon, P. Phonon dispersion in graphite. *Phys. Rev. Lett.* **2004**, *92*, 075501. [CrossRef] [PubMed]
45. Banerji, N.; Cowan, S.; Vauthey, E.; Heeger, A.J. Ultrafast Relaxation of the Poly(3-hexylthiophene) Emission Spectrum. *J. Phys. Chem. C* **2011**, *115*, 9726–9739. [CrossRef]
46. Kaufmann, C.; Kim, W.; Nowak-Król, A.; Hong, Y.; Kim, D.; Würthner, F. Ultrafast Exciton Delocalization, Localization, and Excimer Formation Dynamics in a Highly Defined Perylene Bisimide Quadruple π-Stack. *J. Am. Chem. Soc.* **2018**, *140*, 4253–4258. [CrossRef]
47. Knupfer, M. Exciton binding energies in organic semiconductors. *Appl. Phys. A* **2003**, *77*, 623–626. [CrossRef]
48. Nayak, P.K. Exicton binding energy in small organic conjugated molecule. *Synth. Met.* **2013**, *174*, 42–45. [CrossRef]
49. Chen, K.; Barker, A.J.; Reish, M.E.; Gordon, K.C.; Hodgkiss, J.M. Broadband Ultrafast Photoluminescence Spectroscopy Resolves Charge Photogeneration via Delocalized Hot Excitons in Polymer:Fullerene Photovoltaic Blends. *J. Am. Chem. Soc.* **2013**, *135*, 18502–18512. [CrossRef]
50. Zhou, N.; Prabakaran, K.; Lee, B.; Chang, S.H.; Harutyunyan, B.; Guo, P.; Butler, M.R.; Timalsina, A.; Bedzyk, M.J.; Ratner, M.A.; et al. Metal-Free Tetrathienoacene Sensitizers for High-Performance Dye-Sensitized Solar Cells. *J. Am. Chem. Soc.* **2015**, *137*, 4414–4423. [CrossRef] [PubMed]
51. Wu, K.; Bera, A.; Ma, C.; Du, Y.; Yang, Y.; Li, L.; Wu, T. Temperature-dependent excitonic photoluminescence of hybrid organometal halide perovskite films. *Phys. Chem. Chem. Phys.* **2014**, *16*, 22476–22481. [CrossRef] [PubMed]
52. Wright, A.D.; Verdi, C.; Milot, R.L.; Eperon, G.E.; Pérez-Osorio, M.A.; Snaith, H.J.; Giustino, F.; Johnston, M.B.;

Herz, L.M. Electron–phonon coupling in hybrid lead halide perovskites. *Nat. Commun.* **2016**, *7*, 11755. [CrossRef] [PubMed]
53. Yang, J.; Wen, X.; Xia, H.; Sheng, R.; Ma, Q.; Kim, J.; Tapping, P.; Harada, T.; Kee, T.W.; Huang, F.; et al. Acoustic-optical phonon up-conversion and hot-phonon bottleneck in lead-halide perovskites. *Nat. Commun.* **2017**, *8*, 14120. [CrossRef] [PubMed]
54. Ernst, T. Controlling the Polarity of Silicon Nanowire Transistors. *Science* **2013**, *340*, 1414–1415. [CrossRef] [PubMed]
55. Masselink, W.T.; Fischer, R.; Klem, J.; Henderson, T.; Pearah, P.; Morkoc, H. Polar semiconductor quantum wells on onopolar substrates: (Al,Ga) As/GaAs on (100)Ge. *Appl. Phys. Lett.* **1984**, *45*, 457. [CrossRef]
56. Yang, Y.; Ostrowski, D.P.; France, R.M.; Zhu, K.; van de Lagemaat, J.; Luther, J.M.; Beard, M.C. Observation of a hot-phonon bottleneck in lead-iodide perovskites. *Nat. Photonics* **2016**, *10*, 53–59. [CrossRef]
57. Frost, J.M.; Whalley, L.D.; Walsh, A. Slow Cooling of Hot Polarons in Halide Perovskite Solar Cells. *ACS Energy Lett.* **2017**, *2*, 2647–2652. [CrossRef] [PubMed]
58. Chang, S.H.; Chiang, C.-H.; Tseng, Z.-L.; Chiu, K.-Y.; Tai, C.-Y.; Wu, C.-G. Unraveling simultaneously enhanced open-circuit voltage and short-circuit current density in P3HT:ICBA:2,3-pyridinediol blended film based photovoltaics. *J. Phys. D Appl. Phys.* **2015**, *48*, 195104. [CrossRef]
59. Lee, B.; Stoumpos, C.C.; Zhou, N.; Hao, F.; Malliakas, C.; Yeh, C.-Y.; Marks, T.J.; Kanatzidis, M.G.; Chang, R.P.H. Air-stable molecular semiconductor Iodosalts for solar cell applications: Cs_2SnI_6 as a hole conductor. *J. Am. Chem. Soc.* **2014**, *136*, 15379–15385. [CrossRef] [PubMed]
60. Zou, B.; Zhang, Y.; Xiao, L.; Li, T. Self-trapped state and phonon localization in TiO_2 quantum dot with a dipole layer. *J. Appl. Phys.* **1993**, *73*, 4689–4690. [CrossRef]
61. Chang, S.H.; Chiang, C.-H.; Cheng, H.-M.; Tai, C.-Y.; Wu, C.-G. Broadband charge transfer dynamics in P3HT:PCBM blended film. *Opt. Lett.* **2013**, *38*, 5342–5345. [CrossRef] [PubMed]
62. Wang, H.-Y.; Gao, B.-R.; Wang, L.; Yang, Z.-Y.; Du, X.-B.; Chen, Q.-D.; Song, J.-F.; Sun, H.-B. Exciton diffusion and charge transfer dynamics in nano phase-separated P3HT/PCBM blend films. *Nanoscale* **2011**, *3*, 2280–2285. [CrossRef]
63. Wu, C.-G.; Chiang, C.-H.; Chang, S.H. A perovskite cell with a record-high-VOC of 1.61 V based on solvent annealed $CH_3NH_3PbBr_3$/ICBA active layer. *Nanoscale* **2016**, *8*, 4077–4085. [CrossRef] [PubMed]
64. Chang, S.H.; Lin, K.-F.; Chiu, K.Y.; Tsai, C.-L.; Cheng, H.-M.; Yeh, S.-C.; Wu, W.-T.; Chen, W.-N.; Chen, C.-T.; Chen, S.-H.; et al. Improving the efficiency of $CH_3NH_3PbI_3$ based photovoltaics by tuning the work function of the PEDOT:PSS hole transport layer. *Sol. Energy* **2015**, *122*, 892–899. [CrossRef]
65. Chang, S.H.; Chen, W.-N.; Chen, C.-C.; Yeh, S.-C.; Cheng, H.-M.; Tseng, Z.-L.; Chen, L.-C.; Chiu, K.Y.; Wu, W.-T.; Chen, C.-T.; et al. Manipulating the molecular structure of PEDOT chains through controlling the viscosity of PEDOT:PSS solutions to improve the photovoltaic performance of $CH_3NH_3PbI_3$ solar cells. *Sol. Energy Mater. Sol. Cells* **2017**, *161*, 7–13. [CrossRef]
66. Yip, H.-L.; Jen, A.K.-Y. Recent advances in solution-processed interfacial materials for efficient and stable polymer solar cells. *Energy Environ. Sci.* **2012**, *5*, 5994–6011. [CrossRef]
67. Fang, J.; Deng, D.; Zhang, J.; Zhang, Y.; Lu, K.; Wei, Z. High open-circuit voltage ternary organic solar cells based on ICBA as acceptor and absorption-complementary donors. *Mater. Chem. Front.* **2017**, *1*, 1223–1228. [CrossRef]
68. Xia, Y.; Sun, K.; Ouyang, J. Solution-Processed Metallic Conducting Polymer Films as Transparent Electrode of Optoelectronic Devices. *Adv. Mater.* **2012**, *24*, 2436–2440. [CrossRef]
69. Zhang, L.; Li, B.; Yuan, J.; Wang, M.; Shen, T.; Huang, F.; Wen, W.; Cao, G.; Tian, J. High-Voltage-Efficiency Inorganic Perovskite Solar Cells in a Wide Solution-Processing Window. *J. Phys. Chem. Lett.* **2018**, *9*, 3646–3653. [CrossRef]
70. Yuan, H.; Zhao, Y.; Duan, J.; Wang, Y.; Yang, X.; Tang, Q. All-inorganic $CsPbBr_3$ perovskite solar cell with 10.26% efficiency by spectra engineering. *J. Mater. Chem. A* **2018**, *6*, 24324–24329. [CrossRef]
71. Hedley, G.J.; Quati, C.; Harwell, J.; Predhdo, O.V.; Beljonne, D.; Samuel, I.D.W. Hot-hole cooling controls the intial ultrafast relaxation in methylammonium lead iodide perovskite. *Sci. Rep.* **2018**, *8*, 8115. [CrossRef]
72. Shen, Q.; Ripolles, T.S.; Even, J.; Zhang, Y.; Ding, C.; Liu, F.; Izuishi, T.; Nakazawa, N.; Toyoda, T.; Ogomi, Y.; et al. Ultrafast selective extraction of hot holes from cesium lead iodide perovskite films. *J. Energy Chem.* **2018**, *27*, 1170–1174. [CrossRef]

73. Kang, Y.; Han, S. Intrinsic Carrier Mobility of Cesium Lead Halide Perovskites. *Phys. Rev. Appl.* **2018**, *10*, 044013. [CrossRef]
74. Kahmann, S.; Loi, M.A. Hot carrier solar cells and the potential of perovskites for breaking the Shockley–Queisser limit. *J. Mater. Chem. C* **2019**, *7*, 2471–2486. [CrossRef]
75. De Vos, A. Detailed balance limit of the efficiency of tandem solar cells. *J. Phys. D Appl. Phys.* **1980**, *13*, 839–846. [CrossRef]
76. Wheeldon, J.F.; Valdivia, C.E.; Walker, A.W.; Kolhatar, G.; Jaouad, A.; Turala, A.; Riel, B.; Masson, D.; Puetz, N.; Fafard, S.; et al. Performance comparison of AlGaAs, GaAs and InGaP tunnel junctions for concentrated multijunction solar cells. *Prog. Photovolt.* **2011**, *19*, 442–452. [CrossRef]
77. Nagae, M. Response Time of Metal-Insulator-Metal Tunnel Junctions. *Jpn. J. Appl. Phys.* **1972**, *11*, 1611–1621. [CrossRef]
78. Bannai, R.; Kikuchi, A.; Kishino, K.; Lee, C.-M.; Chyi, J.-I. AlN/GaN double-barrier resonant tunneling diodes grown by rf-plasma-assisted molecular-beam epitaxy. *Appl. Phys. Lett.* **2002**, *81*, 1729–1731.
79. Sollner, T.C.L.G. Resonant tunneling through quantum wells at frequencies up to 2.5 THz. *Appl. Phys. Lett.* **1983**, *43*, 588. [CrossRef]
80. Lam, P.; Hatch, S.; Wu, J.; Tang, M.; Dorogan, V.G.; Mazur, Y.I.; Salamo, G.J.; Ramiro, Í.; Seeds, A.; Liu, H. Voltage recovery in charged InAs/GaAs quantum dot solar cells. *Nano Energy* **2014**, *6*, 159–166. [CrossRef]
81. Shi, X.; Dong, G.; Fang, M.; Wang, F.; Lin, H.; Yen, W.-C.; Chan, K.S.; Chueh, Y.-L.; Ho, J.C. Selective n-type doping in graphene via the aluminium nanoparticle decoration approach. *J. Mater. Chem. C* **2014**, *2*, 5417–5421. [CrossRef]
82. Rani, P.; Jindal, V.K. Designing band gap of graphene by B and N dopant atoms. *RSC Adv.* **2013**, *3*, 802–812. [CrossRef]
83. Tian, P.; Tang, L.; Teng, K.; Lau, S. Graphene quantum dots from chemistry to applications. *Mater. Today Chem.* **2018**, *10*, 221–258. [CrossRef]
84. Qian, F.; Li, X.; Tang, L.; Lai, S.K.; Lu, C.; Lau, S.P. Postassium doping: Tuning the optical properties of graphene quantum dots. *AIP Adv.* **2016**, *6*, 075116. [CrossRef]
85. Qian, J.; Shen, C.; Yan, J.; Xi, F.; Dong, X.; Liu, J. Tailoring the electronic properties of graphene quantum dots by p doping and their enhanced performance in metal-free composite photocatalyst. *J. Phys. Chem. C* **2018**, *122*, 349–358. [CrossRef]
86. Lee, S.I.; Song, W.; Kim, Y.; Song, I.; Jung, D.S.; Jung, M.W.; Cha, M.-J.; Park, S.E.; An, K.-S.; Park, C.-Y. P-Type Doping of Graphene Films by Hybridization with Nickel Nanoparticles. *Jpn. J. Appl. Phys.* **2013**, *52*, 75101. [CrossRef]
87. Tsai, C.-Y. Carrier heating and its effects on the current-voltage relations of conventional and hot-carrier solar cells: A physical model incorporating energy transfer between carriers, photons, and phonons. *Sol. Energy* **2019**, *188*, 450–463. [CrossRef]
88. Liang, P.-W.; Liao, C.-Y.; Chueh, C.-C.; Zuo, F.; Williams, S.T.; Xin, X.-K.; Lin, J.-J.; Jen, A.K.-Y. Additive Enhanced Crystallization of Solution-Processed Perovskite for Highly Efficient Planar-Heterojunction Solar Cells. *Adv. Mater.* **2014**, *26*, 3748–3754. [CrossRef] [PubMed]
89. Xiao, M.; Huang, F.; Huang, W.; Dkhissi, Y.; Zhu, Y.; Etheridge, J.; Bach, U.; Cheng, Y.-B.; Spiccia, L.; Gray-Weale, A.; et al. A Fast Deposition-Crystallization Procedure for Highly Efficient Lead Iodide Perovskite Thin-Film Solar Cells. *Angew. Chem.* **2014**, *126*, 10056–10061. [CrossRef]
90. Jeon, N.J.; Noh, J.H.; Kim, Y.C.; Yang, W.S.; Ryu, S.; Seok, S.I. Solvent engineering for high-performance inorganic–organic hybrid perovskite solar cells. *Nat. Mater.* **2014**, *13*, 897–903. [CrossRef]
91. Chang, S.H.; Wong, S.-D.; Huang, H.-Y.; Yuan, C.-T.; Wu, J.-R.; Chiang, S.-E.; Tseng, Z.-L.; Chen, S.-H. Effects of the washing-enhanced nucleation process on the material properties and performance of perovskite solar cells. *J. Alloys Compd.* **2019**, *808*, 151723. [CrossRef]
92. Yin, W.-J.; Shi, T.; Yan, Y. Unique Properties of Halide Perovskites as Possible Origins of the Superior Solar Cell Performance. *Adv. Mater.* **2014**, *26*, 4653–4658. [CrossRef] [PubMed]
93. Shao, Y.; Xiao, Z.; Bi, C.; Yuan, Y.; Huang, J. Origin and elimination of photocurrent hysteresis by fullerene passivation in $CH_3NH_3PbI_3$ planar heterojunction solar cells. *Nat. Commun.* **2014**, *5*, 5784. [CrossRef] [PubMed]

94. March, S.A.; Clegg, C.; Riley, D.B.; Webber, D.; Hill, I.G.; Hall, K.C. Simultaneous observation of free and defect-bound excitons in $CH_3NH_3PbI_3$ using four-wave mixing spectroscopy. *Sci. Rep.* **2016**, *6*, 39139. [CrossRef] [PubMed]
95. Hsu, H.-L.; Hsiao, H.-T.; Juang, T.-Y.; Jiang, B.-H.; Chen, S.-C.; Jeng, R.-J.; Chen, C.-P. Carbon Nanodot Additives Realize High-Performance Air-Stable p-i-n Perovskite Solar Cells Providing Efficiencies of up to 20.2%. *Adv. Energy Mater.* **2018**, *8*, 1802323. [CrossRef]
96. Chiang, C.-H.; Wu, C.-G. Film Grain-Size Related Long-Term Stability of Inverted Perovskite Solar Cells. *ChemSusChem* **2016**, *9*, 2666–2672. [CrossRef] [PubMed]
97. Chen, C.-C.; Chang, S.H.; Chen, L.-C.; Cheng, H.-M.; Tseng, Z.-L.; Wu, C.-G. Manipulating multicrystalline grain size in $CH_3NH_3PbI_3$ thin films for application in photovoltaics. *Sol. Energy* **2016**, *139*, 518–523. [CrossRef]
98. Chang, S.H.; Chen, C.-C.; Chen, L.-C.; Tien, C.-L.; Cheng, H.-M.; Huang, W.-C.; Lin, H.-Y.; Chen, S.-H.; Wu, C.-G. Unraveling the multifunctional capabilities of PCBM thin films in inverted-type $CH_3NH_3PbI_3$ based photovoltaics. *Sol. Energy Mater. Sol. Cells* **2017**, *169*, 40–46. [CrossRef]
99. Murphy, K.F.; Piccione, B.; Zanjani, M.B.; Lukes, J.R.; Gianola, D.S. Strain- and Defect-Mediated Thermal Conductivity in Silicon Nanowires. *Nano Lett.* **2014**, *14*, 3785–3792. [CrossRef]
100. Cohn, J.L.; Neumeier, J.J.; Popoviciu, C.P.; McClellan, K.J.; Leventouri, T. Local lattic distorsions and thermal transport in perovskite maganites. *Phys. Rev. B* **1997**, *56*, R8495–R9498. [CrossRef]
101. Chen, L.-C.; Chen, C.-C.; Chang, S.H.; Lee, K.-L.; Tseng, Z.-L.; Chen, S.-H.; Kuo, H.-C. Formation and characterization of preferred oriented perovskite thin films on single-crystalline substrates. *Mater. Res. Express* **2018**, *5*, 066403. [CrossRef]
102. Wei, W.; Zhang, Y.; Xu, Q.; Wei, H.; Fang, Y.; Wang, Q.; Deng, Y.; Li, T.; Gruverman, A.; Cao, L.; et al. Monolithic integration of hybrid perovskite single crystals with heterogenous substrate for highly sensitive X-ray imaging. *Nat. Photon.* **2017**, *11*, 315–321. [CrossRef]
103. Park, S.-H.; Ahn, D.; Park, C.-Y. Intersubband absorption of p-type wurtzite GaN/AlN quantum well for fiber-optics telecommunication. *J. Appl. Phys.* **2017**, *122*, 184303. [CrossRef]
104. Fan, Z.Y.; Li, J.; Nakarmi, M.L.; Lin, J.Y.; Jiang, H.X. AlGaN/GaN/AlN quantum-well field-effect transistors with highly resistive AlN epilayers. *Appl. Phys. Lett.* **2006**, *88*, 73513. [CrossRef]
105. Hirayama, H.; Tanaka, S.; Aoyagia, Y. Fabrication of self-assembling InGaN and AlGaN quantum dots on AlGaN surfaces using anti-surfactant. *Microelectron. Eng.* **1999**, *49*, 287–290. [CrossRef]
106. Tsai, M.-K.; Hu, S.-Y.; Lee, J.-W.; Lee, Y.-C.; Lee, M.-H.; Shen, J.-L. Morphology control and characteristics of ZnO/ZnS nanorod arrays synthesised by microwave-assistd heating. *Micro Nano Lett.* **2016**, *11*, 192–195. [CrossRef]
107. Lin, T.N.; Santiago, S.R.M.; Yuan, C.T.; Huang, H.Y.; Shen, J.L. Origin of tunable photoluminescence from graphene quantum dots synthesized via pulsed laser ablation. *Phys. Chem. Chem. Phys.* **2016**, *18*, 22599–22605.
108. Patra, M.; Manoth, M.; Singh, V.; Gowd, G.S.; Choudhry, V.; Vadera, S.; Kumar, N. Synthesis of stable dispersion of ZnO quantum dots in aqueous medium showing visible emission from bluish green to yellow. *J. Lumin.* **2009**, *129*, 320–324. [CrossRef]
109. Wang, X.; Jiang, S.; Huo, X.; Xia, R.; Muhire, E.; Gao, M. Facile preparation of a TiO_2 quantum dot/graphitic carbon nitride heterojunction with highly efficient photocatalytic activity. *Nanotechnology* **2018**, *29*, 205702. [CrossRef] [PubMed]
110. Lee, K.-M.; Chen, K.-S.; Wu, J.-R.; Lin, Y.-D.; Yu, S.-M.; Chang, S.H. Hihgly efficient and stable semi-transparent perovskite solar modules with a trilayer anode electrode. *Nanoscale* **2018**, *10*, 17699–17704. [CrossRef]

12

An Analysis of Fill Factor Loss Depending on the Temperature for the Industrial Silicon Solar Cells

Kwan Hong Min [1,2], Taejun Kim [3], Min Gu Kang [1], Hee-eun Song [1], Yoonmook Kang [4], Hae-Seok Lee [4], Donghwan Kim [2], Sungeun Park [1,*] and Sang Hee Lee [1,*]

1. Photovoltaics Laboratory, Korea Institute of Energy Research, Daejeon 34129, Korea; steel1217@nate.com (K.H.M.); mgkang@kier.re.kr (M.G.K.); hsong@kier.re.kr (H.-e.S.)
2. Department of Materials Science and Engineering, Korea University, Seoul 02841, Korea; solar@korea.ac.kr
3. Hyundai Energy Solutions, PV R&D Center, Eumseong, Cheongju 27711, Korea; taejunkim@hyundai-es.co.kr
4. Department of Energy Environment Policy and Technology, Green School, Graduate School of Korea Energy and Environment, Korea University, Seoul 02841, Korea; ddang@korea.ac.kr (Y.K.); lhseok@korea.ac.kr (H.-S.L.)

* Correspondence: separk@kier.re.kr (S.P.); aktkd88@naver.com (S.H.L.)

Abstract: Since the temperature of a photovoltaic (PV) module is not consistent as it was estimated at a standard test condition, the thermal stability of the solar cell parameters determines the temperature dependence of the PV module. Fill factor loss analysis of crystalline silicon solar cell is one of the most efficient methods to diagnose the dominant problem, accurately. In this study, the fill factor analysis method and the double-diode model of a solar cell was applied to analyze the effect of J_{01}, J_{02}, R_s, and R_{sh} on the fill factor in details. The temperature dependence of the parameters was compared through the passivated emitter rear cell (PERC) of the industrial scale solar cells. As a result of analysis, PERC cells showed different temperature dependence for the fill factor loss of the J_{01} and J_{02} as temperatures rose. In addition, we confirmed that fill factor loss from the J_{01} and J_{02} at elevated temperature depends on the initial state of the solar cells. The verification of the fill factor loss analysis was conducted by comparing to the fitting results of the injection dependent-carrier lifetime.

Keywords: fill factor loss analysis; double-diode model; PERC; temperature dependence; recombination current density; parasitic resistance

---

1. Introduction

The fill factor (FF), open circuit voltage (V_{oc}) and short circuit current (J_{sc}) of a solar cell are important parameters because they determine the maximum power that a solar cell can generate. It is well known that the fill factor of silicon solar cells is influenced by the recombination current and parasitic resistance. In order to clarify the effect of these factors on a fill factor, the double-diode model is generally used as a theoretical concept. In the double-diode model, the photo-generated current density reduces due to the saturation current densities (J_{01}, J_{02}), and parasitic resistance (R_s, R_{sh}) as the equation shows below.

$$J = J_{ph} - J_{01}\left[\exp\left\{\frac{q(V+JR_s)}{n_1kT}\right\}-1\right] - J_{02}\left[\exp\left\{\frac{q(V+JR_s)}{n_2kT}\right\}-1\right] - \frac{(V+JR_s)}{R_{sh}}, \quad (1)$$

where J is the diode circuit current density, J_{ph} is the photo-generated current density, J_{01} is the recombination current density in quasi-neutral region, V is the diode voltage, q is the elementary charge, k is Boltzmann's constant, n_1 is the ideality factor for the diode 1, n_2 is the ideality factor for the diode 2, and T is the absolute temperature. The J_{01} in the equation is the information of the recombination current from the base material (quasi-neutral region) and both side surface of the solar cell. J_{02} generally describes the Shockley-Read-Hall (SRH) recombination at the p-n junction (space charge region), but there are also other recombination sources such as edge recombination [1] and localized high defect region [2,3].

Analyzing the recombination and resistive loss of a solar cell from the perspective of fill factor is a very efficient method because it directly shows the gain of the conversion efficiency from the loss factor. Hence, there were some effort to extract fill factor loss at the space charge region (J_{02}) by determining fill factor loss from J_{01} (FF_{J01}) and subtracting the pseudo fill factor (pFF) that is not influenced by the series resistance [4,5]. Afterward, A. Khanna proposed a method to quantify the fill factor loss of silicon solar cells due to parasitic resistance as well as J_{02}, because there was a limitation that the shunt resistance must be extremely high to extract an exact fill factor loss due to J_{02} in this method. In addition, since this approach does not need to assume n_2 value during the fitting procedure, it showed that a more accurate analysis is possible. The theoretical derivation is well described in [6].

This fill factor loss analysis was used in this experiment to investigate the source of fill factor loss depending on the elevated temperature, because the module efficiency assessed at the standard test condition (STC, 25 °C) reduces at the actual operating temperature [7]. Figure 1 shows the measured PV module temperature change according to the seasonal temperature in Daejeon and Seoul of South Korea. The PV module temperature rises to 60 °C which will highly degrade the module power when it is operating. The V_{oc} reduction is well known for the main source of efficiency loss. V_{oc} is a function of the saturation current (I_0) which is strongly dependent on the temperature as it can be seen in Equations (2) and (3) [8]. Where q is the elementary charge, A is the area, D is the diffusion constant, n_i is the intrinsic carrier concentration, L is the diffusion length, N_D is the doping concentration, and T is kelvin temperature. By extracting the fill factor loss of each parameter (J_{01}, J_{02}, R_s, and R_{sh}), it will be able to provide an accurate diagnosis and appropriate solution to reduce the temperature dependence of the PV module.

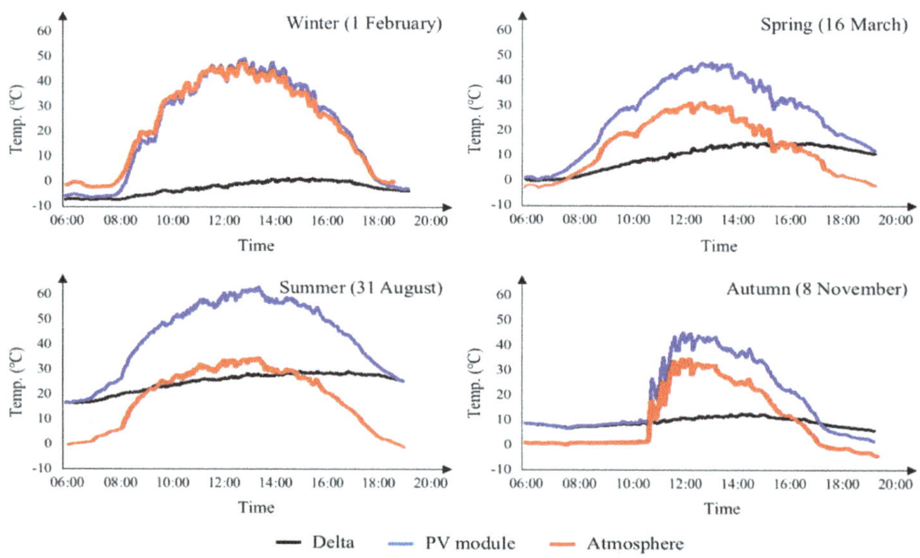

Figure 1. Photovoltaic (PV) module temperature change according to seasonal temperature.

For the analysis of fill factor loss, industrial passivated emitter and rear cell (PERC), which is leading product in PV industries, was used [9]. Many researches for optimization of both side surface passivation [10–16] and laser ablation on rear aluminum oxide (AlO_x) layer [17,18] for the local back surface field formation (LBSF) [19] have improved efficiency of the PERC and reached 22.8% [20]. In this paper, temperature dependence of two industrial type PERC samples were compared to inspect dominant factor of fill factor loss.

$$I_0 = qA\frac{Dn_i^2}{LN_D}, \tag{2}$$

$$n_i(T) = 9.15 \times 10^9 \left(\frac{T + 273.15}{300}\right)^2 \exp\left(\frac{-6880}{T + 273.15}\right). \tag{3}$$

2. Experimental

2.1. Analysis Process of Fill Factor Loss

Figure 2 briefly summarizes the steps of the fill factor loss analysis. The analysis of the fill factor loss starts with the calculation of FF_{J01}. The FF_{J01} can be determined when the effect of J_{02}, R_s and R_{sh} is absent. With this assumption, Equation (1) changes to Equation (4).

$$J = J_{ph} - J_{01}\left[\exp\left\{\frac{qV}{nkT}\right\} - 1\right]. \tag{4}$$

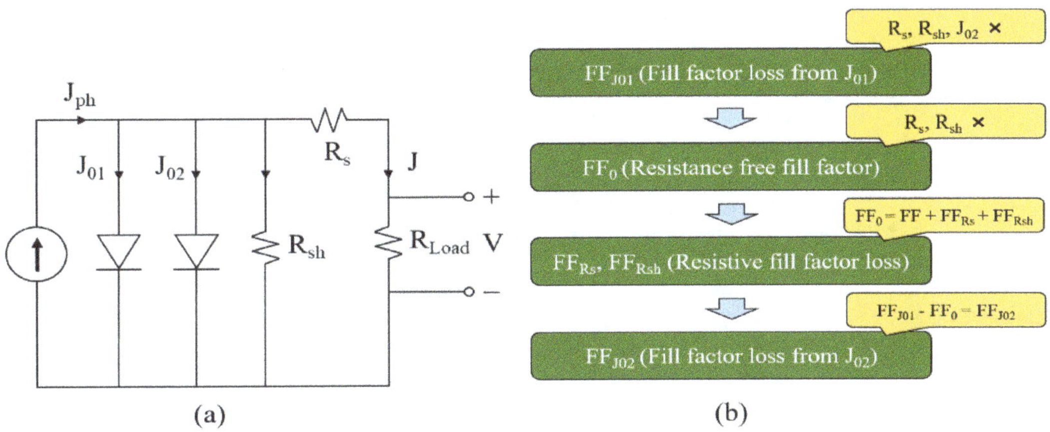

Figure 2. (**a**) Equivalent circuit of solar cells with double-diode model, and (**b**) process of fill factor loss analysis.

By imposing the boundary condition $J = 0$ at $V = V_{oc}$ and $J = J_{sc}$ at $V = 0$, relationship between J and V can be expressed in Equation (5), which FF_{J01} is allowed to be obtained.

$$J = J_{sc} - \frac{J_{sc}}{\exp\left(\frac{qV_{oc}}{nkT}\right) - 1}\left[\exp\left\{\frac{qV}{nkT}\right\} - 1\right]. \tag{5}$$

Afterwards, resistance free fill factor (FF_0), which does not have loss from the parasitic resistance, is derived as in Equation (6) by considering the double-diode model at maximum power point (MPP). Where, J_{mpp} and V_{mpp} is the current density and voltage at power maximum point, respectively. Equation (6) is composed of three terms: real fill factor, fill factor loss from R_s (FF_{Rs}), and fill factor loss from R_{sh} (FF_{Rsh}), respectively from the left term.

$$FF_0 = FF + \frac{J_{mpp}^2 R_s}{V_{oc} J_{sc}} + \frac{\left(V_{mpp} + J_{mpp} R_s\right)^2}{R_{sh} V_{oc} J_{sc}}. \tag{6}$$

As a last step, the fill factor loss from J_{02} can be readily determined from the difference of FF_{J01} and FF_0 Equation (7), because the difference comes from the absence of the J_{02} effect.

$$FF_{J02} = FF_{J01} - FF_0. \tag{7}$$

2.2. Sample Measurements with Elevated Temperature

Two industrial solar cells with PERC structure (Figure 3) were prepared as samples (PERC A and B). The PERC A and B were fabricated on different production lines, but the front contact pattern was unified to exclude optical loss due to the front contact. In case of the PERC A, silver paste for the front contact contains higher glass frit contents compared to the PERC B. The characteristics of illuminated *J-V* were measured by using an in-house solar simulator, which can control the temperature of substrate. Even though this fill factor analysis method assumed that J_{01} follows an n factor of 1, there were changes in the n factor depending on the temperature. Therefore, n factor at 1-sun, extracted from the Suns-V_{oc} measurement, was used for analysis. The temperature of cells during the Suns-V_{oc} measurement was controlled by using a box that has an ambient heater. The conditions of the experimental temperature were set to 25 °C, 45 °C, and 65 °C.

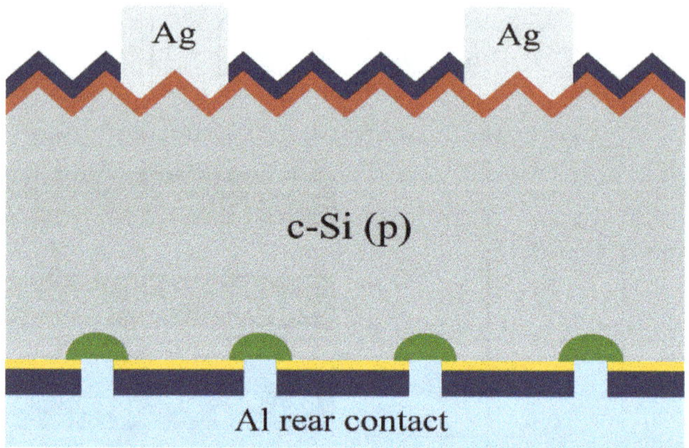

Figure 3. Schematic of the industrial PERC solar cell structure.

3. Results and Discussion

3.1. Temperature Dependence

The two PERC solar cell samples were named by the PERC A and B. As a result of the illuminated *J-V* measurements (Table 1), V_{oc} and fill factor of PERC B were 17 mV and 1.93 % higher at 25 °C, respectively. J_{sc} difference was not significant (~1.3 mA/cm^2) and it comes from the different fraction of front metal contact. Graphs in Figure 4 show the drop of V_{oc} and fill factor (depending on temperature), and also presents the average temperature coefficient as well. The average drop of V_{oc} and fill factor with the PERC A cell were slightly higher than the PERC B cell. Especially, the larger fill factor drops occurred at 45 °C for the PERC A and 65 °C for the PERC B. Illuminated *J-V* measurements alone do not have enough evidence to accurately diagnose what the main problem is.

Table 1. Results of illuminated *J-V* measurement and n factor (from Suns-V_{oc}) of PERC A and B according to temperature dependence.

Parameters	PERC A			PERC B		
Temperature [°C]	25	45	65	25	45	65
V_{oc} [mV]	654	616	578	671	635	596
J_{sc} [mA/cm^2]	38.49	38.74	39.05	39.79	40.08	40.21
V_{mpp} [mV]	564	517	478	580	526	494
J_{mpp} [mA/cm^2]	35.84	35.92	36.18	37.85	38.07	38.20
FF [%]	80.28	77.70	76.60	82.21	80.53	78.62
R_s [Ω·cm^2]	1.58	1.77	1.631	1.49	1.42	1.60
R_{sh} [Ω·cm^2]	10252	2796	2982	9553	26562	5126
η [%]	20.21	18.57	17.28	21.96	20.49	18.87
n factor (Suns-V_{oc})	1.07	1.14	1.17	1.05	1.12	1.19

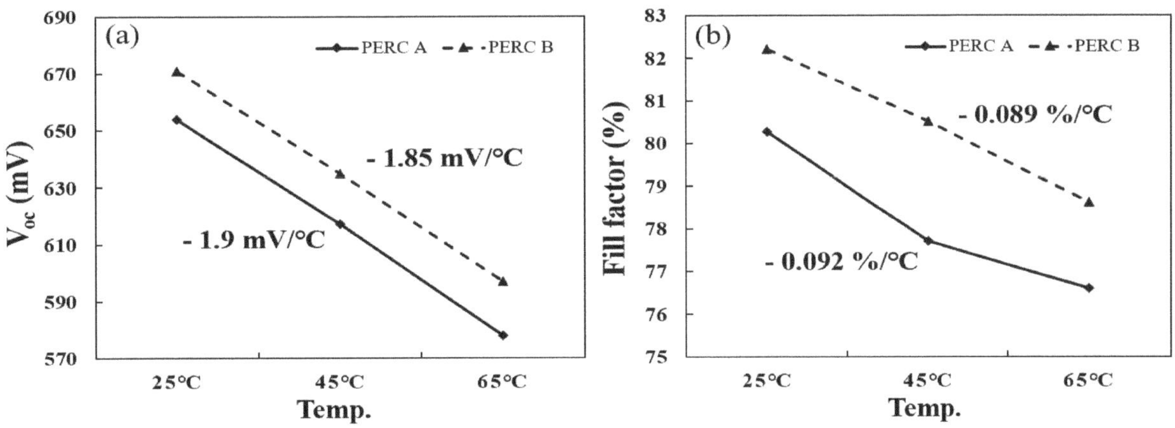

Figure 4. Temperature coefficient of (**a**) V_{oc} and (**b**) fill factor.

3.2. Results of Fill Factor Loss Analysis

Table 2 organizes the values of each fill factor loss according to the parameters from double-diode model. From this data, Figures 5 and 6 shows the effect of each parameter on the total fill factor loss. At initial temperature, percentage of FF_{J01} was higher than FF_{J02} for the PERC B cell. On the other hand, FF_{J02} had a higher percentage of total fill factor loss than FF_{J01} for the PERC A cell. As the temperature rose, FF_{J01} ratio of both PERC A and B increased while FF_{J02} ratio decreased. This can be explained by the temperature dependence of J_{01} and J_{02} in Equations (8) and (9).

$$J_{01}(T) = J_{01}(25\ ^\circ\mathrm{C}) \times \left(\frac{n_i(T)}{n_i(25\ ^\circ\mathrm{C})}\right)^2, \tag{8}$$

$$J_{02}(T) = J_{02}(25\ ^\circ\mathrm{C}) \times \left(\frac{n_i(T)}{n_i(25\ ^\circ\mathrm{C})}\right). \tag{9}$$

J_{01} is proportional to square of $n_i(T)$, which increases sensitively according to the temperature, while J_{02} is directly proportional to $n_i(T)$. Due to the difference of the temperature dependence, the effect of J_{01} becomes dominate as temperature goes up. However, the FF_{J02} ratio and the absolute value are still higher than FF_{J01} at 65 °C for the PERC A cell. A possible explanation is that the initial value of FF_{J02} (1.86%) was twice higher than that of FF_{J01} (0.82%) and helped increase at a high temperature.

Table 2. Results of fill factor loss analysis of PERC A and PERC B according to temperature (the values are absolute).

***FF* Loss Analysis**	**PERC A**			**PERC B**		
Temperature [°C]	25	45	65	25	45	65
Ideal *FF* [%]	83.78	82.89	82.36	84.08	83.43	82.67
Total *FF* loss [%]	3.50	5.19	5.78	1.88	2.91	4.06
FF_{J01} [%]	0.82	1.70	2.10	0.58	1.42	2.30
FF_{J02} [%]	1.86	2.47	2.69	0.48	0.70	0.75
FF_{Rs} [%]	0.81	0.97	0.94	0.80	0.78	0.98
FF_{Rsh} [%]	0.01	0.05	0.05	0.02	0.01	0.03
FF_0 [%]	81.00	78.72	77.57	83.02	81.31	79.62
Real *FF* (LIV) [%]	80.28	77.70	76.60	82.21	80.53	78.62

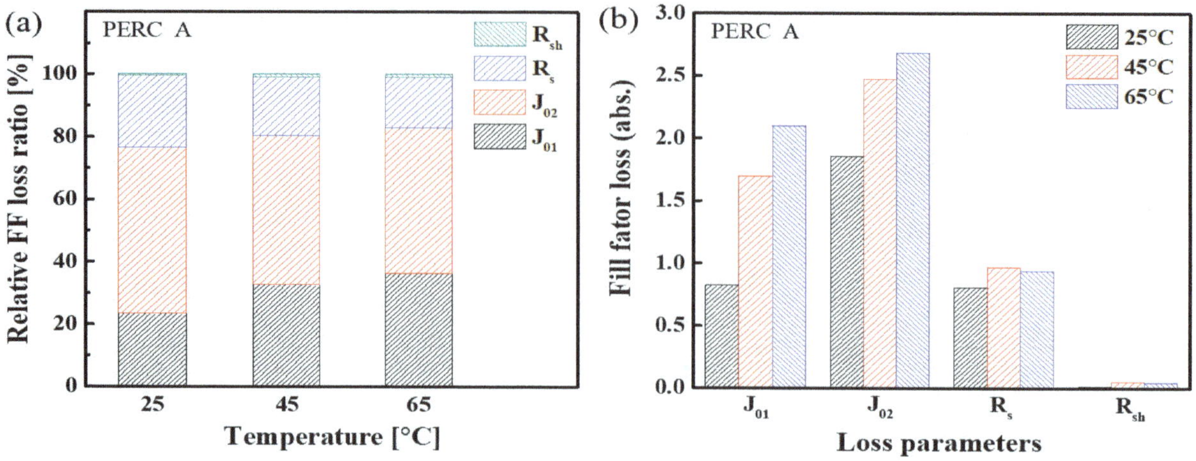

Figure 5. Fill factor loss analysis results of PERC A: (**a**) Normalized fill factor ratio; (**b**) Absolute fill factor loss.

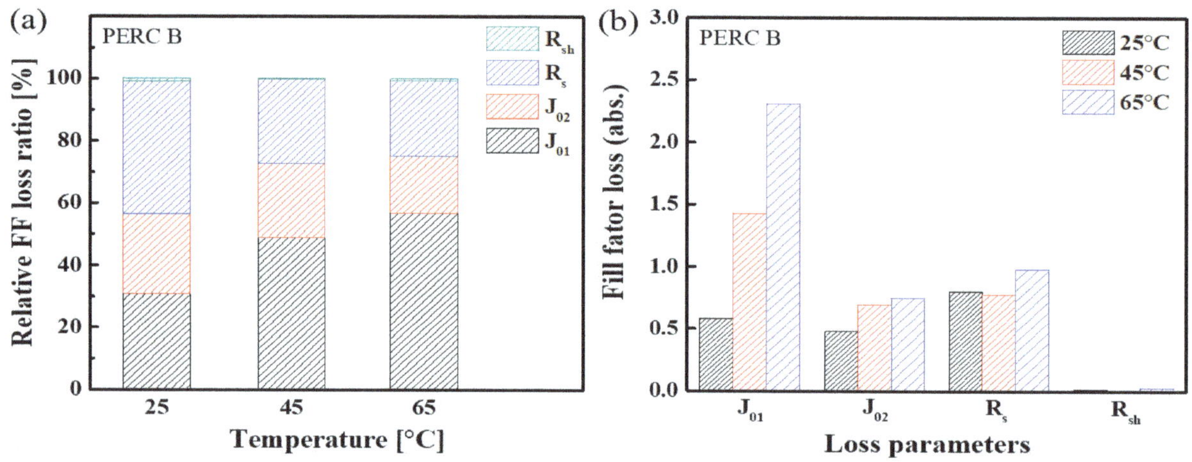

Figure 6. Fill factor loss analysis results of PERC B: (**a**) Normalized fill factor ratio; (**b**) Absolute fill factor loss.

In addition, the temperature dependence of FF_{J02} for the PERC B cell (Figure 6b) reveals that the FF_{J02} will not increase significantly at elevated temperature if the initial ratio to the total fill factor loss is relatively small. On the other hand, the difference of FF_{J01} between 25 °C and 65 °C increased up to 20%, even though the initial difference was only 7%. This means that the effect of J_{01} on fill factor will be accelerated if the initial loss is relatively high.

In case of the PERC B cell, one can diagnose that the recombination of emitter, quasi-neutral region, or surface passivation have to be improved. Reduction of FF_{Rs} ratio for the PERC B cell will further reduce the temperature dependence, even though there were small variation at the elevated temperature. The ratio and temperature dependence of FF_{Rsh} were negligible for both PERC A and B cells.

3.3. Verification Through Injection Dependent-Carrier Lifetime

In order to verify the results of the fill factor loss analysis, the lifetime was measured through Sinton instrument's Suns-V_{oc} equipment at 25 °C. The recombination current density of the samples was evaluated by using a lifetime fitting method. Through the specific resistance and thickness of the wafer, the Auger recombination and the radiative recombination are represented as lifetime according to the excess carrier concentration as graphs in Figure 7 show.

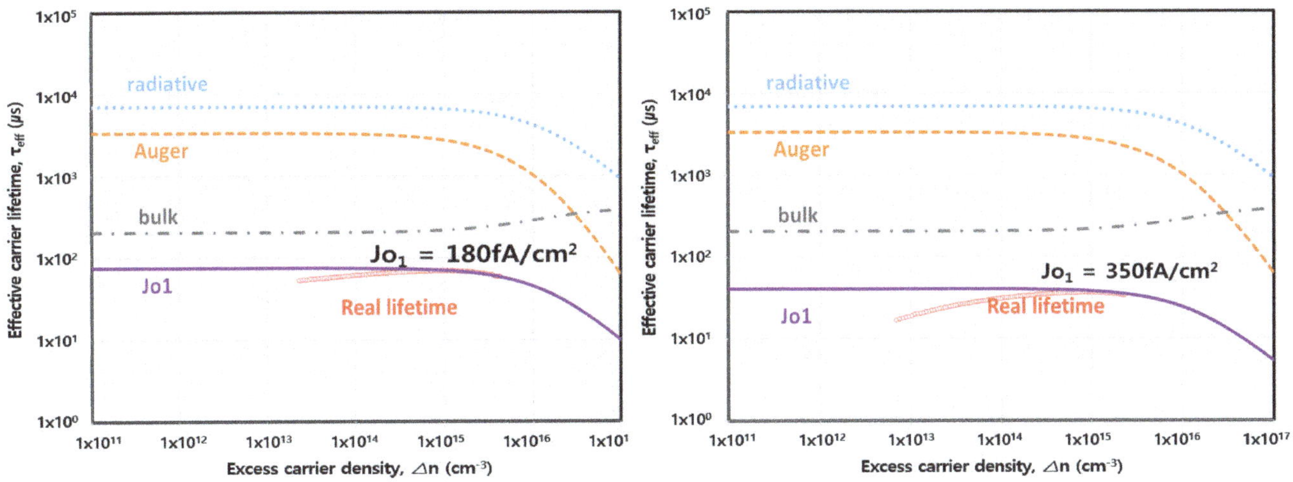

Figure 7. Fitting results of injection dependent-carrier lifetime for the PERC A (**Left**) and B (**Right**).

The expressions for the intrinsic carrier lifetime and bulk lifetime can be expressed as follows.

$$\tau = \frac{\Delta n}{R} \tag{10}$$

$$\frac{1}{\tau_{intrinsic+bulk}} = \frac{1}{\tau_{rad}} + \frac{1}{\tau_{Auger}} + \frac{1}{\tau_{SRH}} \tag{11}$$

Here, intrinsic recombination can be expressed as an Equation (12) [21].

$$R_{Intrinsic} = R_{Auger} + R_{Radiative} = np\left(1.8 \times 10^{-24} n_0{}^{0.65} + 6 \times 10^{-25} p_0{}^{0.65} + 3 \times 10^{-27} \Delta n^{0.8} + 9.5 \times 10^{-15}\right) \tag{12}$$

Where n and p are the free electron and hole concentrations, n_0 and p_0 are the electron and hole concentration in thermal equilibrium, and Δn is the excess carrier densities. The bulk lifetime of a wafer can be measured with quasi-steady-state photoconductance (QSSPC) equipment. To measure the bulk lifetime of the wafer, the doped layer was removed from the wafer using KOH solution after $POCl_3$ diffusion, and then the Al_2O_3/SiN_x stack passivation layer was deposited on both sides. Bulk lifetime can be calculated by the following two formulas [22,23].

$$J_o = \frac{S_{eff}}{qn_i^2} \times (N_A + \Delta n), \tag{13}$$

$$\frac{1}{\tau_{bulk}} = \frac{1}{\tau_{eff}} - \frac{2S_{eff}}{W}, \tag{14}$$

where W is the sample thickness, S_{eff} is the surface recombination velocity. The S_{eff} values were calculated as 9.8 cm/s at the 10^{15} cm^{-3} injection level which is commonly used [22]. As a result of our calculation, the bulk lifetime was 343 μs for our wafer. The graphs in Figure 7 shows these three lifetimes of radiative, Auger and bulk recombination. If the parameter of J_{01} is added to this formula, lifetime graph at high injection can be fit by using the formula below:

$$\frac{1}{\tau_{eff}} - \frac{1}{\tau_{intrinsic}} = \frac{1}{\tau_{bulk}} + \frac{J_{01}(N_A + \Delta n)}{qn_i{}^2 w}. \tag{15}$$

By fitting the lifetime graph through the J_{01} value, the J_{01} value was calculated as 350 fA/cm^2 for the PERC A cell and 180 fA/cm^2 for the PERC B cell. Similar to the results of the fill factor loss analysis, the J_{01} value was significantly higher for the PERC A cell.

Furthermore, we could draw the theoretical graph of FF_0 as function of J_{02} (25 °C) by applying the J_{01} value from Figure 7 to the J-V equation of double diode model without resistance Equation (16). The J_{02} (25 °C) of the PERC A cell was extracted by finding the point that the FF_0 from the fill factor loss analysis equal to the graph (black line in Figure 8). In this way, the J_{02} (25 °C) of the PERC A and B cells were assessed as 27 nA/cm^2 and 7.35 nA/cm^2, respectively. The difference of the J_{02} value was also expected from the graph in Figure 7. The graphs generally showed poor fitting results at low-level injection region. This means that the PERC A cell has poor properties of J_{02} or shunt resistance because these are well known as main reasons for low lifetime of solar cells in the low-level injection region [24].

$$J = J_{ph} - J_{01}\left[\exp\left\{\frac{qV}{n_1kT}\right\} - 1\right] - J_{02}\left[\exp\left\{\frac{qV}{n_2kT}\right\} - 1\right]. \quad (16)$$

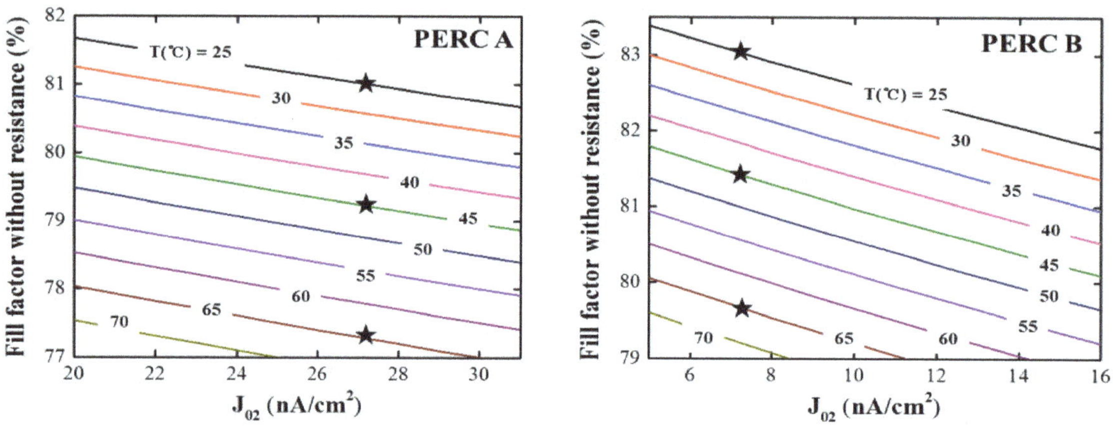

Figure 8. Theoretical calculation of FF_0 as function of the J_{02} (25 °C). The star marks are the FF_0 value extracted from fill factor loss analysis (Table 2).

In case of the temperature above 25 °C, J_{01} and J_{02} values were calculated by using Equations (8) and (9). Through these J_{01} and J_{02} values, temperature conditions of the J_{02} (25 °C)-FF_0 graph were extended from 30 to 70 °C. We confirmed that the FF_0 from Table 2 matched quite well with the theoretical graph. Consequently, we confirmed that the fill factor loss analysis was conducted precisely by checking the theoretically extracted J_{01} and J_{02} values.

4. Conclusions

We conducted fill factor loss analysis to investigate the temperature dependence of industrial PERC cells based on the double diode model. While analyzing the loss of fill factor for the industrial PERC A and B cells, it was confirmed that both samples showed temperature dependence of fill factor (~0.092%/ °C). However, the main parameter that dominantly contribute to temperature dependence were different depending on the initial state of the samples.

For PERC A, J_{02} was the main parameter of fill factor loss in initial state (25 °C). However, the contribution of the J_{01} to the fill factor loss increased as the temperature rose. It could be estimated that the PERC A cell has potential to improve fill factor by optimizing J_{02} or shunt resistance. On the other hand, PERC B has a higher FF_{Rs} ratio at initial state (25 °C) than other parameters, but the ratio of FF_{Rs} reduced due to the rapid increase in the FF_{J01}. Therefore, we believe that the J_{01} has to be reduced for the PERC B cell by optimizing the doping process, passivation quality, and metal contacts.

In addition, we verified the accuracy of the fill factor loss analysis by comparing it to the fitting results of the injection dependent-carrier lifetime. The J_{01} and J_{02} values at 25 °C were extracted, and it was confirmed that the values of FF_0 obtained from the fill factor loss analysis were in good agreement with the theoretically calculated values. In conclusion, it seems very important to reduce initial fill factor loss from the J_{01}, because it has a high temperature dependence. In case of the J_{02} and R_s, they seem less detrimental for the reduction of fill factor at an elevated temperature. Also, we confirmed that the fill factor loss analysis used in this experiment is very efficient to diagnose the temperature dependence of the solar cells, as well as their initial state.

Author Contributions: K.H.M. designed experiments, performed the experiments and wrote manuscripts; Formal analysis, T.K.; data curation, M.G.K.; investigation, H.-e.S.; review and editing, Y.K. and H.-S.L.; supervision, D.K.; project administration and supervision, S.P.; writing-review and editing and project administration, S.H.L. All authors have read and agreed to the published version of the manuscript.

Acknowledgments: This work was conducted under the New and Renewable Energy Technology Development Program of the Korea Institute of Energy Technology Evaluation and Planning (KETEP) through a grant funded by the Ministry of Knowledge Economy, Korea (Project No. 20183030019460), and by the Technology Development Program to Solve Climate Changes of the National Research Foundation (NRF) funded by the Ministry of Science, ICT & Future Planning (Project No. 2017M1A2A2086911).

References

1. McIntosh, K. Lumps, humps and bumps: Three detrimental effects in the current–voltage curve of silicon solar cells, in: Centre for Photovoltaic Engineering. *Univ. N. S. W. Aust.* **2001**, 171.
2. Breitenstein, O.; Rakotoniaina, J.; Al Rifai, M.H.; Werner, M. Shunt types in crystalline silicon solar cells. *Prog. Photovolt. Res. Appl.* **2004**, *12*, 529–538. [CrossRef]
3. Steingrube, S.; Breitenstein, O.; Ramspeck, K.; Glunz, S.; Schenk, A.; Altermatt, P.P. Explanation of commonly observed shunt currents in c-Si solar cells by means of recombination statistics beyond the Shockley-Read-Hall approximation. *J. Appl. Phys.* **2011**, *110*, 014515. [CrossRef]
4. Greulich, J.; Glatthaar, M.; Rein, S. Fill factor analysis of solar cells' current–voltage curves. *Prog. Photovolt. Res. Appl.* **2010**, *18*, 511–515. [CrossRef]
5. Hoenig, R.; Glatthaar, M.; Clement, F.; Greulich, J.; Wilde, J.; Biro, D. New measurement method for the investigation of space charge region recombination losses induced by the metallization of silicon solar cells. *Energy Procedia* **2011**, *8*, 694–699. [CrossRef]
6. Khanna, A.; Mueller, T.; Stangl, R.A.; Hoex, B.; Basu, P.K.; Aberle, A.G. A fill factor loss analysis method for silicon wafer solar cells. *IEEE J. Photovolt.* **2013**, *3*, 1170–1177. [CrossRef]
7. Correa-Betanzo, C.; Calleja, H.; De León-Aldaco, S. Module temperature models assessment of photovoltaic seasonal energy yield. *Sustain. Energy Technol. Assess.* **2018**, *27*, 9–16. [CrossRef]
8. Sproul, A.; Green, M. Improved value for the silicon intrinsic carrier concentration from 275 to 375 K. *J. Appl. Phys.* **1991**, *70*, 846–854. [CrossRef]
9. The International Technology Roadmap for Photovoltaic (ITRPV). Available online: https://itrpv.vdma.org (accessed on 8 May 2020).
10. Glunz, S.W. High-efficiency crystalline silicon solar cells. *Adv. Optoelectron.* **2007**, *2007*, 1–15. [CrossRef] [PubMed]
11. Cuevas, A.; Kerr, M.; Schmidt, J. Overview of PECVD Silicon Nitride for Solar Cells. In Proceedings of the ANZSES 2003—Destination Renewables from Research to Market, Melbourne, Austrilia, 26–29 November 2003.
12. Schmidt, J.; Merkle, A.; Bock, R.; Altermatt, P.P.; Cuevas, A.; Harder, N.-P.; Hoex, B.; Van De Sanden, R.; Kessels, E.; Brendel, R. Progress in the surface passivation of silicon solar cells. In Proceedings of the 23rd European Photovoltaic Solar Energy Conference, Valencia, Spain, 1–5 September 2008; pp. 1–5.
13. Dingemans, G.; Kessels, W. Status and prospects of Al2O3-based surface passivation schemes for silicon solar cells. *J. Vac. Sci. Technol. A* **2012**, *30*, 040802. [CrossRef]
14. Hoex, B.; Schmidt, J.; Pohl, P.; Van de Sanden, M.; Kessels, W. Silicon surface passivation by atomic layer deposited Al_2O_3. *J. Appl. Phys.* **2008**, *104*, 044903. [CrossRef]
15. Hoex, B.; Gielis, J.; Van de Sanden, M.; Kessels, W. On the c-Si surface passivation mechanism by the negative-charge-dielectric Al_2O_3. *J. Appl. Phys.* **2008**, *104*, 113703. [CrossRef]
16. Huang, H.; Lv, J.; Bao, Y.; Xuan, R.; Sun, S.; Sneck, S.; Li, S.; Modanese, C.; Savin, H.; Wang, A. 20.8% industrial PERC solar cell: ALD Al_2O_3 rear surface passivation, efficiency loss mechanisms analysis and roadmap to 24%. *Sol. Energy Mater. Sol. Cells* **2017**, *161*, 14–30. [CrossRef]
17. Engelhart, P. Laser processing for high-efficiency silicon solar cells. In Proceedings of the Laser-based Micro- and Nanopackaging and Assembly III, San Jose, CA, USA, 28–29 January 2009; p. 72020S. [CrossRef]

18. Grohe, A.; Knorz, A.; Nekarda, J.; Jäger, U.; Mingirulli, N.; Preu, R. Novel laser technologies for crystalline silicon solar cell production. In Proceedings of the Laser-based Micro- and Nanopackaging and Assembly III, San Jose, CA, USA, 28–29 January 2009; p. 72020P. [CrossRef]
19. Riegel, S.; Mutter, F.; Lauermann, T.; Terheiden, B.; Hahn, G. Review on screen printed metallization on p-type silicon. *Energy Procedia* **2012**, *21*, 14–23. [CrossRef]
20. Canadian Solar Sets a 22.80% Conversion Efficiency World Record for P-Type Large Area Multi-Crystalline Silicon Solar Cell. Available online: http://investors.canadiansolar.com/news-releases?items_per_page=10&page=3 (accessed on 1 June 2020).
21. Kerr, M.J.; Cuevas, A. General parameterization of Auger recombination in crystalline silicon. *J. Appl. Phys.* **2002**, *91*, 2473–2480. [CrossRef]
22. Al-Amin, M.; Grant, N.E.; Pointon, A.I.; Murphy, J.D. Iodine—Ethanol Surface Passivation for Measurement of Millisecond Carrier Lifetimes in Silicon Wafers with Different Crystallographic Orientations. *Phys. Status Solidi A* **2019**, *216*, 1900257. [CrossRef]
23. Aberle, A.G. Surface passivation of crystalline silicon solar cells: A review. *Prog. Photovolt. Res. Appl.* **2000**, *8*, 473–487. [CrossRef]
24. Lee, J.H.; Min, K.H.; Kang, M.G.; Jeong, K.T.; Lee, J.I.; Song, H.-E.; Park, S.; Park, J.-S. Efficiency characteristics of a silicon oxide passivation layer on p-type crystalline silicon solar cell at low illumination. *Curr. Appl. Phys.* **2019**, *19*, 683–689. [CrossRef]

13

Enhanced Silver Nanowire Composite Window Electrode Protected by Large Size Graphene Oxide Sheets for Perovskite Solar Cells

Hongye Chen [1], Min Li [1], Xiaoyan Wen [1], Yingping Yang [1], Daping He [1], Wallace C. H. Choy [2] and Haifei Lu [1,*]

1 School of Science, Wuhan University of Technology, Wuhan 430070, China; hongyechen@whut.edu.cn (H.C.); minli@whut.edu.cn (M.L.); wenxy@whut.edu.cn (X.W.); ypyang@whut.edu.cn (Y.Y.); hedaping@whut.edu.cn (D.H.)

2 Department of Electrical and Electronic Engineering, The University of Hong Kong, Pok Fu Lam Road, Hong Kong SAR, China; chchoy@eee.hku.hk

* Correspondence: haifeilv@whut.edu.cn

Abstract: Despite the outstanding features of high transmittance and low sheet resistance from silver nanowire (Ag NW) based transparent electrodes, their applications in perovskite solar cells (PVSCs) as window electrodes encounter significant obstacles due to the stability issue brought by the corrosion of halogen species from perovskite layer. In this study, we used large size graphene oxide (LGO) sheets as the protective barrier for bottom Ag NW nano-network. Contributed by the LGO with average size of 60 μm, less GO sheet was necessary for forming the fully covered protective barrier with fewer cracks, which consequently improved the optical transparency and anticorrosive ability of the composite electrode compared to the one from relatively small size GO. Our experiments demonstrated the composite electrode of Ag NW/LGO. The glass substrate exhibited transmittance of 83.8% and 81.8% at 550 nm before and after partial reduction, which maintained 98.4% and 95.1% average transmittance (AVT) of the pristine Ag NW electrode. Meanwhile, we utilized the steady hot airflow to assist the fast solvent evaporation and the uniform GO film formation on Ag NW electrode. Before the application of composite electrode in organic-inorganic hybrid perovskite solar cells, the operational stability of composite electrodes from different sizes of GO with perovskite film fabricated on top were characterized under continuing external bias and light irradiation. Experimental results indicate that the Ag NW electrode protected by LGO could maintain original resistance for more than 45 h. Finally, the PVSC fabricated on Ag NW/LGO based composite electrode yielded a power conversion efficiency (PCE) of 9.62%, i.e., nearly 85% of that of the reference device fabricated on the commercial indium-tin oxide (ITO) glass. Our proposed low temperature and solution processed bottom electrode with improved optical transparency and operational stability can serve as the very beginning layer of optoelectronic devices, to promote the development of low cost and large area fabrication perovskite solar cells.

Keywords: Ag nanowires; graphene oxide; perovskite solar cell; transparent bottom electrode

---

1. Introduction

Organic-inorganic hybrid perovskite has been demonstrated as a promising candidate of light absorber because of its excellent properties for photovoltaic devices. The power conversion efficiency (PCE) has achieved over 23% since it was first introduced in 2009 with PCE of 3.8% [1–7]. With the increase in efficiency and potential commercial value, researchers have paid more attention to controlling the cost of preparation, by avoiding the use of high energy consuming instruments and adopting low temperature and solution processed methods for each layer. For example, the low temperature and solution processed interfacial layers, including electron [8–14] and hole [15–19] transporting layers, have been intensively studied to improve the efficiency of PVSCs. Carbon based materials, such as carbon paste [20,21], carbon nanotubes [22–25] and graphene [26], have also been used as top electrodes in PVSCs to replace conventional top metal electrodes formed by thermal evaporation, to reduce the cost of device fabrication and improve the stability of device simultaneously. Besides, metal nanowire electrode through spray-coating or spin-coating method is another favored candidate of top electrode for semi-transparent PVSCs [27,28].

For the bottom electrode of PVSCs, the most commonly used transparent conductive electrodes at present are indium-tin oxide (ITO) and fluorine-doped tin oxide (FTO) films fabricated by evaporation or sputtering in high vacuum environment, which however have the following problems. First, due to the large amount of electrical energy consumption and machine loss during the preparation process, the fabrication cost of ITO and FTO is relatively high. Second, the indium element required for ITO is a rare element. The rising of its price could be predicted in the long-term. Third, the metal oxide electrode has natural brittleness. The optoelectronic device based on this type of electrode will be greatly limited in flexibility and folding characteristics. As a result, several candidate materials have been investigated as bottom electrode to replace ITO or FTO, such as carbon nanotubes [29], graphene [30,31], conductive polymers [32] and metal nanowires [33–36]. Among these materials, silver nanowire (Ag NW) network is regarded as the most promising candidate because of the low sheet resistance and high transmittance. Moreover, silver nanowires have excellent flexibility, making them suitable for flexible optoelectronic device. However, the application of silver nanowire as bottom electrode to perovskite solar cells has encountered various obstacles, especially the chemical stability upon the perovskite materials because the halide species released from the perovskite layer will easily penetrate through the conventional interfacial layer, corrode the bottom silver nanowires, and thus lead to degradation of the electrical conduction [37–39].

To address the chemical stability issue, various attempts have been made to protect the bottom silver nanowire electrode. For instance, Han et al. [38] proposed a dense fluorine-doped ZnO (FZO) layer deposited on the surface of Ag NWs by pulsed laser deposition (PLD) to enhance the corrosion resistance of the electrode and ensure the conductivity of the electrode, achieving a PCE of 3.29% from perovskite device. Kim et al. [39] demonstrated a sandwich protection structure of ITO/Ag NW/ITO by successive-spin-coating method and annealing at 250 °C to enhance the conductivity of the composite electrode, resulting in a PCE of 8.44% from PVSC. Based on the sandwich protection structure, Lee et al. [37] proposed using amorphous Al-doped zinc oxide (AZO) instead of ITO as a protective layer for the purpose of reducing the use of indium and achieved a PCE of 13.93% from PVSC. In addition, our previous research [40] reported an anti-corrosive film composed of self-assembled graphene oxide (GO) flakes on a silver nano-network under ambient conditions, which can effectively prevent halide corrosion against the Ag NWs, and finally revealed a PCE of 9.23% by optimizing the amount of graphene oxide and reducing agent.

We demonstrated the formation of fully covered film composed of large size graphene oxide sheets (LGO) as the protective barrier for bottom Ag NW nano-network through utilizing a time-saving strategy of steady hot airflow, which is beneficial for the film formation with good uniformity. We found that the anti-corrosive abilities of the composite Ag NW electrodes from large size GO sheets were significantly enhanced compared to those composited from small size GO sheets. They were experimentally demonstrated by characterizing the resistance of composite electrode under continuing external bias and light irradiation. Importantly, the composite electrode of Ag NW/LGO after partially reduction showed an excellent optical transmittance, reaching 81.8% at 550 nm and maintaining 95.1% average transmittance (AVT) of the pristine Ag NW electrode. Finally, the PVSC fabricated with Ag NW/LGO based composite electrode demonstrated a power conversion efficiency (PCE) of 9.62%, corresponding to nearly 85% of the efficiency from reference device on commercial ITO glass.

2. Materials and Methods

2.1. Materials

Graphene oxide powder was purchased from Anhui Lianruan Education Technology Co., Ltd. (Hefei, China). ITO glasses were purchased from YINGKOU OPV TECH NEW ENERGY Co., Ltd. (Yingkou, China). Poly(3,4-ethylenedioxythiophene):poly(styrene sulfonate) PEDOT:PSS (Heraeus Clevios PVP AI 4083), Lead (II) iodide(PbI_2, >99.99%) and 2,9-dimethyl-4,7-diphenyl-1,10-Phenanthroline (BCP, >99%) were purchased from Xi'an Polymer Light Technology Corp. (Xian, China). Methylammonium iodide (MAI, 99%) was purchased from Greatcell Solar PTY Ltd. (Queanbeyan, Australia). [6,6]-Phenyl-C61-butyric acid methyl ester (PCBM, >99.5%) was purchased from Luminescence Technology Corp. (Shanghai, China). Dimethyl sulfoxide (DMSO, anhydrous, $\geq$99.9%), chlorobenzene (anhydrous, 99.8%) and 2-Propanol (anhydrous, 99.5%) were purchased from Sigma-Aldrich Co. LLC. (Shanghai, China). γ-Butyrolactone (GBL, anhydrous, 99.9%) was purchased from Shanghai Aladdin Biochemical Technology Co., Ltd. (Shanghai, China). Ethanol, acetone and sodium borohydride were purchased from Sinopharm Chemical Reagent Co., Ltd. (Shanghai, China). All chemicals were used without further purification.

2.2. Preparation of Large Size GO

Large size GO sheets were prepared using a centrifugal classification method according to previous report [41]. Briefly, appropriate amount of GO powder was taken for preparing an aqueous dispersion with a concentration of 2 mg/mL, which was stirred for 24 h. After centrifugation at 5000 rpm for 30 min, the bottom solution (about 30% in volume) was kept and diluted to 2 mg/mL for the next cycle of centrifugation. After repeating the above process seven times, the gel at bottom was collected and used for investigations in this study.

2.3. Preparation of Ag NW/RLGO Composite Electrodes

The glass substrates were cleaned ultrasonically with acetone, deionized water and ethanol for 15 min in sequence, and then dried with N_2, followed by UV/ozone treatment for 15 min. The pristine Ag NW dispersion was diluted to 1.3 mg/mL before use. Highly conductive Ag NW transparent electrodes were fabricated by spin-coating the Ag NW dispersion at 2500 rpm for 30 s and annealing at 150 °C for 10 min in air. According to our previous research [40], 32 μmol $NaBH_4$ were added to 2.5 mL LGO aqueous dispersion (0.25 mg/mL). Then, the dispersion was kept for 12 h for the partial reduction of LGO. Fifty microliters of partially-reduced LGO dispersion (RLGO) were dropped onto the Ag NW transparent electrode with area of 2.89 cm^2, allowing the liquid to disperse onto the entire substrate uniformly. Finally, the substrate was dried under a steady hot airflow, which could greatly speed up the evaporation of the solution. The "steady hot air flow" was achieved by a commercial adjustable electric hair dryer (FH6618, FLYCO, Shanghai, China) through carefully controlling wind and power.

2.4. Preparation of Perovskite Solar Cells on Ag NW/RLGO Composite Electrodes

For the fabrication of perovskite solar cells, the transparent electrodes of Ag NW/RLGO and commercial ITO glass were used. The Ag NW/RLGO composite transparent electrodes were prepared as described above. The control ITO substrates were ultrasonically cleaned with detergent, deionized water, acetone, and ethanol for 15 min in sequence. To prepare hole transport layer, the as-received PEDOT:PSS solution was spin-coated on the transparent electrodes at a speed of 2000 rpm for 30 s and annealed on a hotplate at 125 °C for 10 min. The samples were then transferred to a glove-box after cooling to room temperature. The perovskite $MAPbI_3$ films were fabricated on the substrates according to a previously reported method [42]. Generally, PbI_2 and MAI with 1:1 ratio were mixed in the mixing solvent of GBL and DMSO (volume ratio = 7:3) to form the precursor with a concentration of 1.2 M. The mixture solution was stirred at 60 °C overnight before use. The perovskite films were fabricated by a successive two step spin-coating process of 1000 rpm for 10 s and 4000 rpm for 30 s. During the second step of spin-coating, 150 μL of toluene were quickly dropped on the substrate, which was then annealed on a hotplate at 100 °C for 10 min. The PCBM layer was deposited on the as-formed perovskite film by spin-coating its chlorobenzene solution (17 mg/mL) at 2000 rpm for 30 s, followed by annealing at 100 °C for 10 min. Subsequently, a thin BCP film, used as buffer layer, was deposited by spin-coating its saturated solution at 3000 rpm for 60 s. Finally, the device was completed by thermal evaporation of 150 nm thick Ag cathode under 3.0×10^{-4} Pa.

2.5. Characterizations

The surface morphology and cross section of the as-prepared samples were examined by field-emission scanning electron microscope (FE-SEM, Zeiss Ultra Plus, Oberkochen, German). GO sheets were measured by optical microscope (XJP-107JX, Pudan Optical Instrument Co., Ltd., Shanghai, China). The diffused transmission spectra of the transparent electrodes were obtained from a UV-vis spectrophotometer (UV2600, Shimadzu, Tokyo, Japan) with an integrating sphere. The sheet resistances of the electrodes were measured using a four-point probe system with a current source-meter (Keithley 2400, Tektronix, Beaverton, OR, USA). To characterize the resistance variation of composite electrode before and after device fabrication, two electrical contacts were formed on the two sides of the electrode using silver paste and a digital multimeter (DT9206, FLUKE Corporation, Elite, WA USA) was used for resistance measurement. The photocurrent density–voltage (J–V) characteristics of all devices were analyzed using an AM1.5G solar simulator (Oriel Sol3A, Newport Corporation, Irvine, CA, USA) and Keithley 2400 sourcemeter (Beaverton, OR, USA), scanning from −0.1 to 1.2 V at a scan rate of 0.1 V s^{-1}. The IPCE (Newport Corporation, Irvine, CA, USA) system was employed to study the quantum efficiency of the solar cells.

3. Results and Discussion

In our previous research [40], we demonstrated that Ag NWs underneath perovskite layer was severely corroded after the formation of perovskite film, which is the most critical issue that hinders their application in PVSCs as bottom electrode. To solve the chemical instability and improve the optical transparency, we used a fully-covered film composed of LGO sheets on silver nanowire electrode as an anti-corrosive barrier. As depicted in Figure 1, three different GO sheets with average sizes of 60 μm, 30 μm and 3 μm were successfully separated by a centrifugal classification method. The obtained GO sheets were diluted to an appropriate concentration, dropped on clean silicon wafer with a 300 nm SiO_2 layer, and then characterized by optical microscope. In Figure 1, we can clearly observe that most GO sheets were thin and well classified, which was beneficial to the transmittance of Ag NW/GO composite electrodes and our later research. Figure 1b,d,f shows the size distributions of the three different GO sheets.

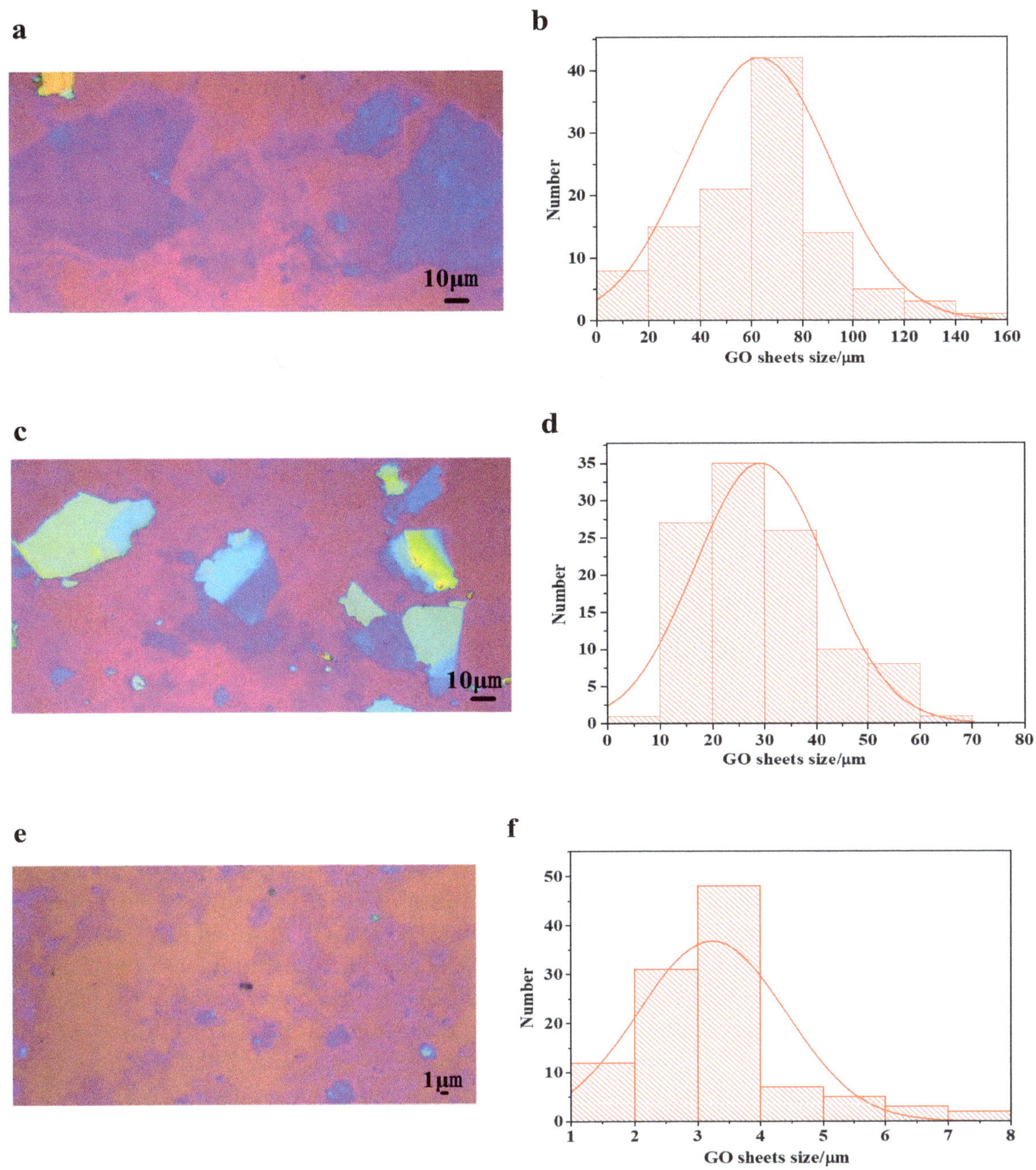

Figure 1. (**a**,**c**,**e**) The optical microscope images of 60 μm, 30 μm and 3 μm GO sheets, respectively;and (**b**,**d**,**f**) the size distributions of 60 μm, 30 μm and 3 μm GO sheets, respectively.

It is also worth mentioning our strategy of forming fully-covered GO film on Ag NW substrate. Our previous method of forming the GO film was obtained through self-assembly and the solvent was dried naturally under ambient condition, which was very time consuming and not suitable for large area film fabrication. Here, a rapid GO film formation approach was tested by using the steady hot airflow for the drying of the solvent. We believe this strategy is compatible with roll-to-roll technology. The preparation procedure of the composite electrode is shown in Figure 2a. To illuminate the difference from two approaches for the formation of composite electrode, a relatively high concentration of GO solution (0.5 mg/mL) was used. Figure 2b shows the photo of Ag NW/LGO composite electrodes that were prepared by the two methods with the large area of 6.25 cm^2. Sample 1 was prepared by a conventional method, in which water was evaporated naturally. Sample 2 was prepared with the assistance of a steady hot airflow, as indicated in Figure 2a. The preparation of Sample 1 was time consuming and resulted in non-uniform film formation, as indicated in Figure 2b. However, with the assistance of steady hot airflow, a flat and uniform GO film could be rapidly fabricated, and the preparation process of the Ag NW/LGO composite was efficient and suitable for large-scale fabrication.

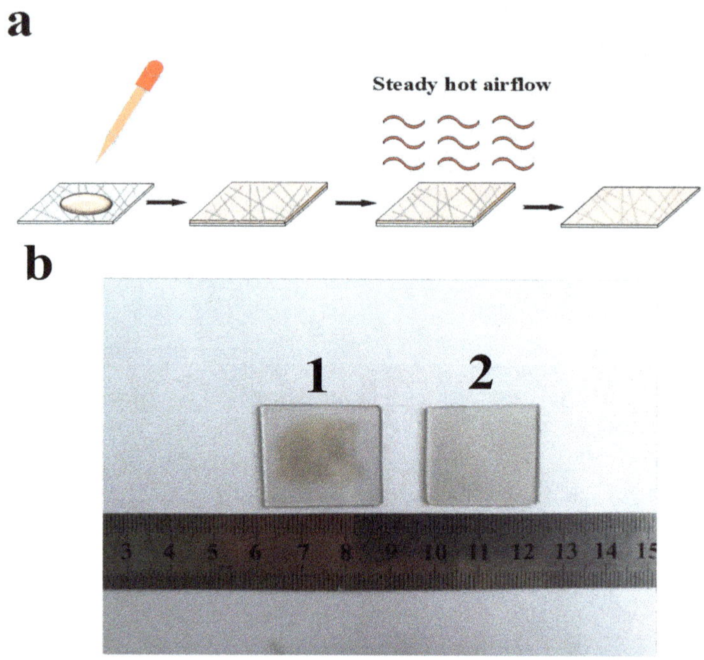

Figure 2. (**a**) Schematic representation of the Ag NW/LGO composite electrode fabrication process; and (**b**) photos of Ag NW/LGO composite electrodes fabricated from relatively high concentration of LGO solution (0.5 mg/mL) on the area of 6.25 cm^2 using two different methods.

To evaluate the anticorrosive effect of the film made of different size of GO sheets, the same volumes of GO aqueous solution with a constant concentration of 0.25 mg/mL were dropped on three identical Ag NW electrodes and allowed to dry under a steady hot airflow, as discussed above. Two electrical contacts using silver paste were formed on the two sides of the electrode to measure its resistance variation before and after device fabrication. PEDOT:PSS layer, perovskite ($MAPbI_3$) layer and PCBM layer were fabricated on the composite electrodes in sequence. The control sample of Ag NW electrode without GO was also prepared following the same procedure. The resistance variations of the Ag NW/GO composite electrodes were measured in a glove-box filled with N_2 (O_2 <0.1 ppm, H_2O <0.1 ppm) after device fabrication. As illustrated in Figure 3a, the resistances of all composite electrodes remained stable for the first 10 h, including the control sample. This was because of the glove-box, in which water and oxygen content were low. The process of decomposition of perovskite layer was relatively slow. Meanwhile, the corrosion of halide species against silver nanowires was also a gradual process in such condition. Over time, the resistance of the control electrode changed dramatically, by exhibiting more than 100 times the original resistance during 8 h. This result directly reflected the severe corrosion of the silver nano-network underneath the perovskite film. In contrast, the resistances of the electrodes protected by GO sheets increased slowly, and the resistance increasing rates decreased as the sizes of GO sheets increased. The inset in Figure 3a clearly indicates that the Ag NW electrode protected by 60 μm GO sheets showed an excellent chemical stability, which can keep stable with negligible variation for 24 h. In addition, we also measured the resistance of the Ag NW electrode protected by 60 μm GO sheets under continuous 0.8 V bias and light irradiation (0.5 sun), as shown in Figure 3b. The result clearly demonstrates that, even under continuous electricity and irradiation, the composite electrode was almost unaffected and maintained a stable resistance, proving the outstanding protective effect of large size GO sheets.

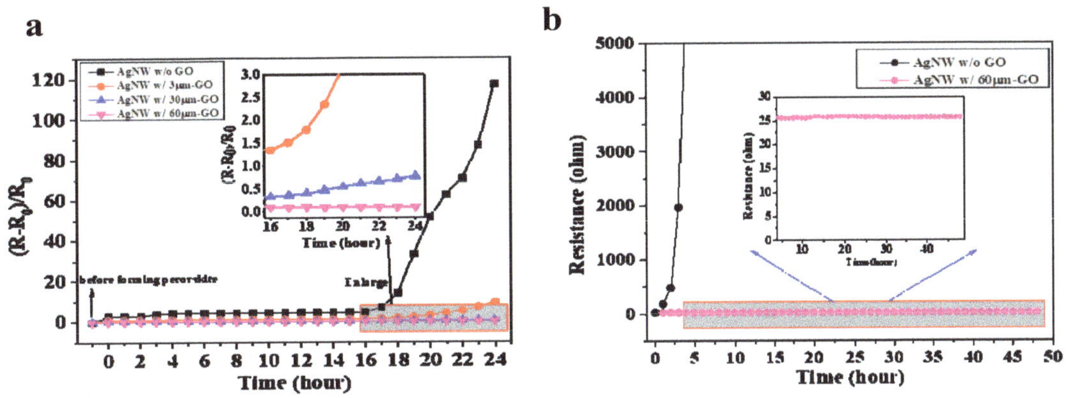

Figure 3. (**a**) The resistance variation of the Ag NW electrodes protected by different sizes of GO sheets (40 μL, 0.25 mg/mL) before and after the fabrication of perovskite devices: and (**b**) the resistance of the Ag NW electrode protected by 60 μm GO sheets under continuous 0.8 V bias and light irradiation (0.5 sun).

To illuminate the enhanced stability of the composite electrode of Ag NW/LGO against perovskite layer, the electrode was characterized by SEM. As evidenced by the SEM images from the silver nano-network coated with 60 μm GO sheets (Figure 4), the silver nanowires were widely covered by single graphene oxide sheet, which acted as anti-corrosive barrier against halide ions from perovskite. Since the average length of the silver nanowires was about 30 μm, it is easy to understand that a larger area of GO sheet can protect more silver nanowires from corrosion. In other words, less GO would be necessary to fully cover the same area of Ag NW electrode than one composed of small-sized GO sheets, as schematically depicted in Figure 5. Due to the physical stacking of the GO sheets, it is also reasonable to believe that the penetration of halide species through the cracks indicated by the dotted lines formed between GO sheets could still happen and caused the deterioration of the electrode. Therefore, the gradual increase of the resistance of electrode composited by 30 μm and 3 μm GO sheets after the fabrication of perovskite layer could be attributed to the formation of more cracks or defects in the anticorrosive film. Stacking more GO sheets on the electrode to block the cracks with additional GO layers might be an efficient method to prevent such penetration; however, this would be harmful to the optical transparency and electrical conductivity of transparent electrode. Solar cells fabricated on the composite electrode based on that strategy would be difficult to optimize to achieve good performance. It should also be noted that, even though the combination of metal nanowire electrode and single- or double-layer graphene in CVD method being able to obtain the best optical, electrical and stability properties, the hydrophobic feature of the film would add difficulty to the fabrication of the charge transporting layer and perovskite film. Therefore, relatively large GO sheets with average size of 60 μm were used for demonstration.

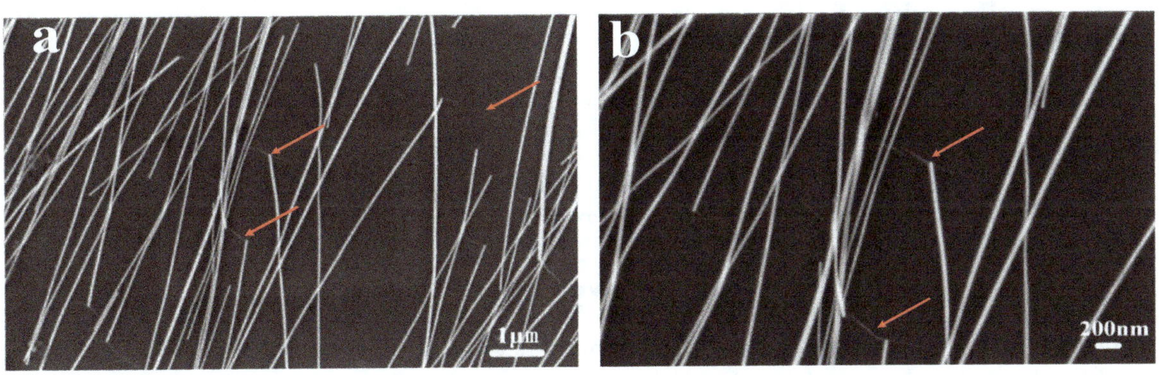

Figure 4. (**a**) Large-scale; and (**b**) enlarged SEM images of silver nano-network under the protection of 60 μm GO sheets. The red arrows inside indicate the crinkles of GO.

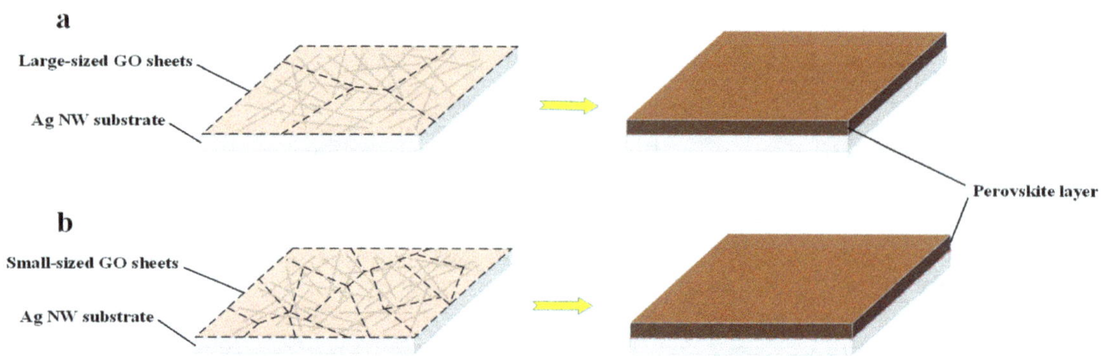

Figure 5. Schematic of Ag NW electrodes protected by: (**a**) large GO sheets; and (**b**) small GO sheets.

Even though the reduced stacking layer from large size GO sheets helped decrease the vertical resistance between bottom Ag NW electrode and charge transporting layer fabricated above, the lateral resistance of GO was poor due to the high degree of oxidation. Partially-reduced LGO was still essential to improve the lateral conductivity, which could help the efficient collection of photo-induced carriers generated at the position away from the metal nanowires. Hence, an appropriate amount of $NaBH_4$ was added into the pristine LGO solution as a reducing agent to enhance the electrical conductivity of LGO film. More details are shown in Experimental Section 2.3. To characterize the optical property of the composite transparent electrodes, diffused transmission spectra of the silver nano-network electrodes coated with 40 μL of LGO (average size of 60 μm) solution with and without partial reduction by optimal amounts of $NaBH_4$ are shown in Figure 6. The spectra indicated the composite electrodes of Ag NW/LGO and Ag NW/RLGO including the glass substrate exhibited transmittances of 83.8% and 81.8%, respectively, at the wavelength of 550 nm. Contributed by the size effect of GO sheet, as discussed above, the quantity of GO sheets forming the anticorrosive film on the electrode surface could be reduced. The optical loss of bare silver nanowire electrode brought by the addition of 60 μm LGO and 60 μm RLGO were minimized to 1.6% and 4.6% of average transmittance (AVT), respectively, at the spectral range of 400–800 nm.

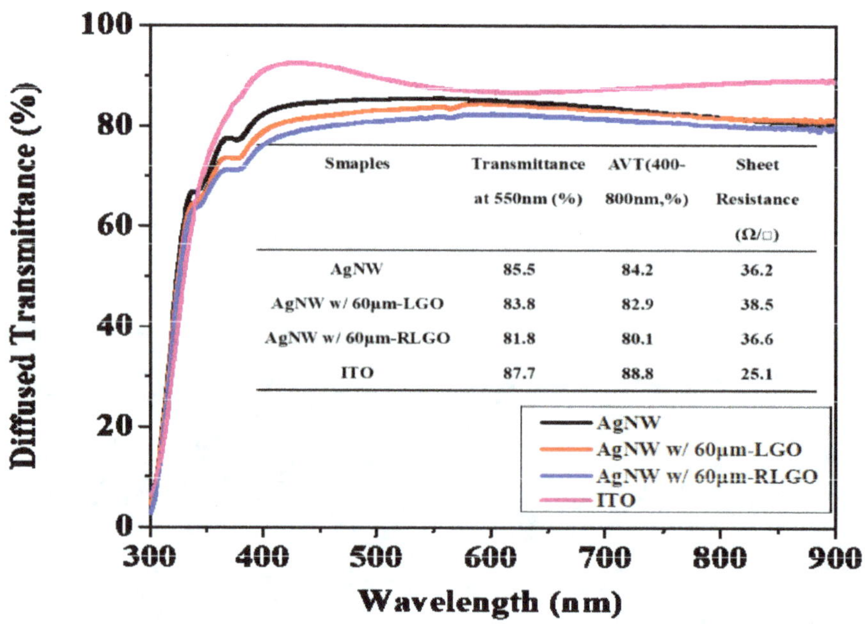

Figure 6. Diffused transmission spectra of silver nano-network electrodes and the network electrode coated by the same amount of 60 μm LGO sheets with and without chemical reduction, and the commercial ITO substrate. The inset is the performance parameters of these samples.

Based on the excellent chemical stability and transmittance of the Ag NW/LGO composite electrodes, we further evaluated the transparent electrodes for perovskite solar cell applications. To fabricate PVSCs based on the Ag NW nano-network transparent electrodes protected by anti-corrosive LGO film, PEDOT:PSS, $MAPBI_3$, PCBM and BCP were deposited as hole transport layer (HTL), absorber layer, electron transport layer (ETL) and buffer layer, respectively. The devices were finally finished after thermal evaporation of the silver counter electrode. The schematic diagram and cross section SEM image are shown in Figure 7a,b. The control devices of perovskite solar cells on commercial ITO electrode were also fabricated for reference. More details on device fabrication can be found in Experiment Section 2.4.

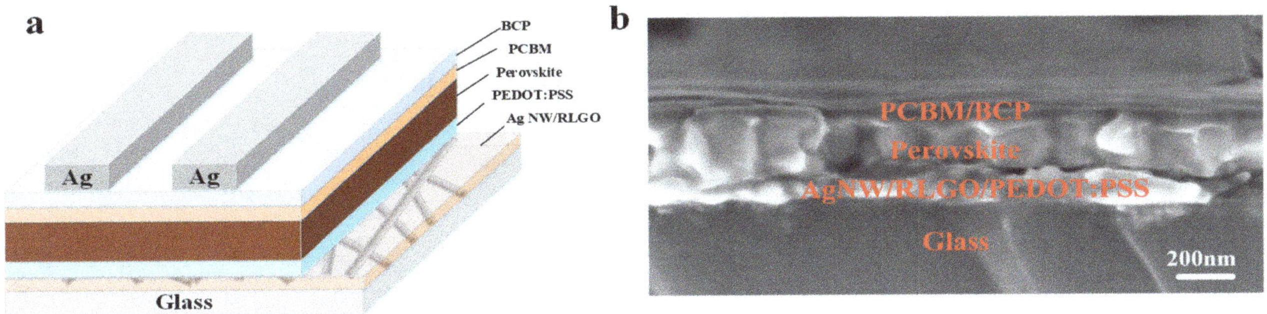

Figure 7. (**a**) Schematic illustration of the perovskite solar cell structure on an Ag NW/LGO composite electrode; and (**b**) cross-sectional SEM image of the perovskite solar cell fabricated on Ag NW/LGO composite electrode.

The current density–voltage (J–V) curves of PVSCs fabricated on ITO glass, the pristine Ag NW electrode without GO and the composite electrode of Ag NW nano-network with 60 μm LGO sheets are shown in Figure 8a, and the performance parameters are summarized in Table 1. The performance of PVSCs was obtained under 1 sun illumination (AM 1.5G, 100 mW/cm^2) irradiated from the bottom electrode side covered with a black shadow mask. The perovskite solar cell using the Ag NW/RLGO electrode showed an open circuit voltage (V_{OC}) of 0.87 V, short-circuit current (J_{SC}) of 15.43 mA/cm^2, fill factor (FF) of 70.9%, and PCE of 9.62%. The incident photon-to-electron conversion efficiency (IPCE) of the device based on the composite electrode is shown in Figure 8b. The calculated J_{SC} from the IPCE spectrum is 15.23 mA/cm^2, which is consistent with the value of 15.43 mA/cm^2 from the J–V curve. Figure 8c exhibits the J–V hysteresis characteristics of the cells based on the optimized Ag NW/RLGO composite electrode using a dwell time of 10 ms for both forward (J_{SC} to V_{OC}) and backward (V_{OC} to J_{SC}) scan directions. The detailed hysteresis performance parameters are summarized in Table 2. As shown in Table 2, a PCE of 9.62% was achieved in the backward scan, whereas the PCE value was 9.08% in the forward scan. The little hysteresis phenomenon might be due to the coating of the PCBM layer [43,44]. The stabilized photocurrent and PCE at the maximum power output point (MPP) under AM1.5 sun illumination are shown in Figure 8d, which exhibited stabilized photocurrent of 12.72 mA/cm^2 and PCE of 9.35% at 0.76 V under 400 s continuing light irradiation, agreeing well with the value measured from J–V curves (Figure 8c). As illustrated in Table 1, the J_{SC} value from Ag NW/RLGO composite electrode accounted for 95% of the value of reference device, which was consistent with the optical transparency ratio of the composite electrode and ITO glass, indicating their qualification of optical transparency and electrical conductivity for transparent bottom electrode. The slight decrease of the FF value from the Ag NW/RLGO composite electrode-based device could be attributed to relative larger surface roughness of the composite electrode. Even though the coating of GO and PEDOT:PSS layers could improve the smoothness of bare silver nanowire electrode, the surface of composite electrode in large scale was still worse than the ITO film possessing nanometer surface roughness. The additional GO layer and thick PEDOT:PSS layer slightly affected the charge transferring between the composite electrode and perovskite layer, as evidenced by the slight increase of the serial resistance in Table 1. Nevertheless, a PCE of 9.62% was achieved, which corresponded to nearly 85% of that of the reference device with an ITO electrode, indicating that it is suitable to use our Ag NW/RLGO based electrode for PVSCs. More importantly, our Ag NW composite electrode protected by large size GO sheets exhibited superior chemical stability over those that are pristine or protected by small size GO sheets, and maintained 95.1% average transmittance of pristine Ag NW electrode. Furthermore, the fast GO film formation method, which was assisted by steady hot airflow, was beneficial to large scale fabrication technology. Future optimization on the surface of composite electrode and perovskite film quality on the composite electrode would undoubtedly boost the efficiency of the devices.

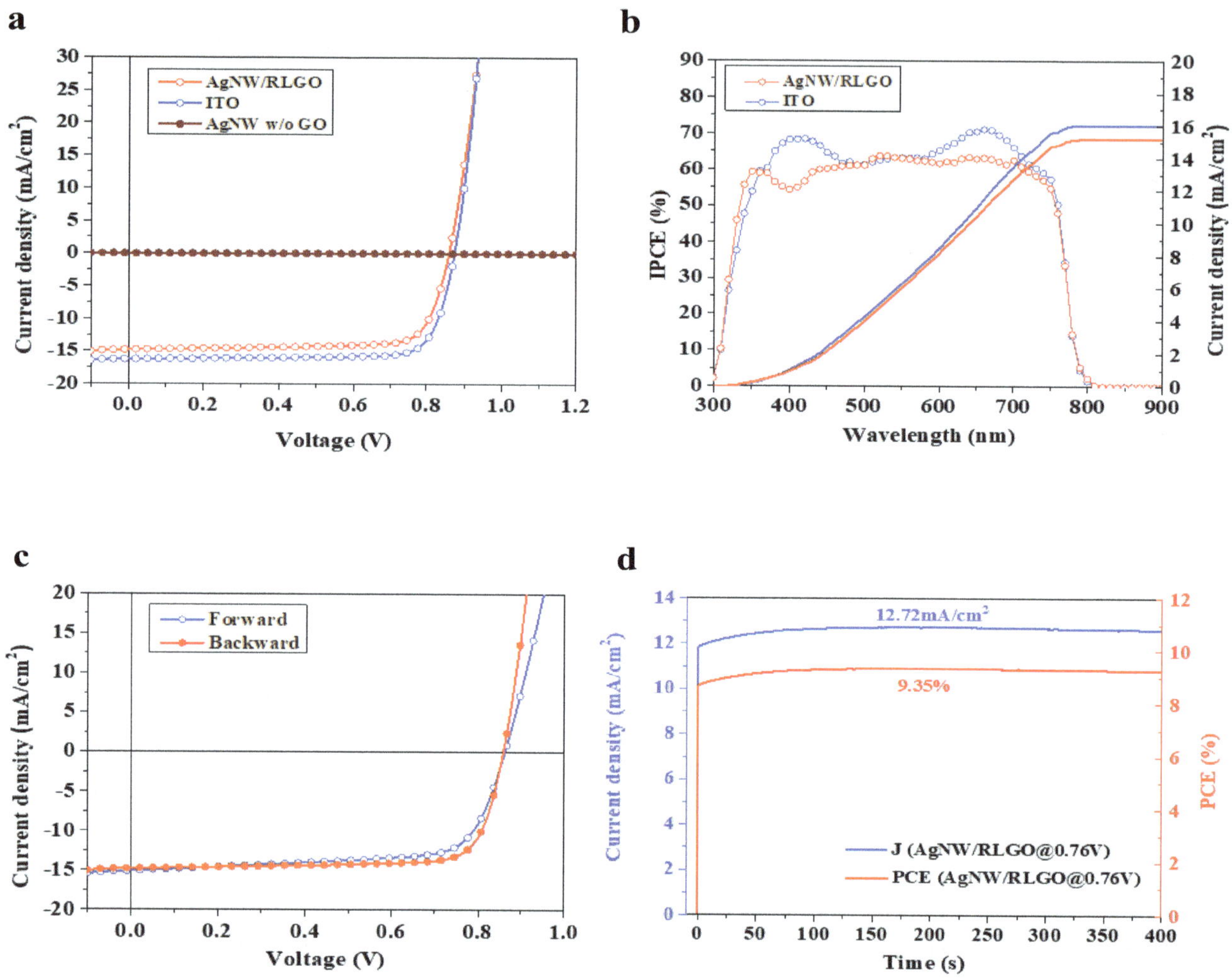

Figure 8. (**a**) J–V curves of the PVSC fabricated on the Ag NW nano-network electrode protected by 60 μm LGO sheets. Data from commercial ITO glass and Ag NW electrode without GO are also presented. (**b**) IPCE performances of the devices fabricated on the nano-composite electrode and ITO substrate. The blue and red solid lines indicate the integrated current of the ITO-based and composite electrode based devices, respectively. (**c**) J–V curves of the PVSC fabricated on the Ag NW nano-network electrode protected by 60 μm LGO sheets measured under forward and backward bias scanning. (**d**) Steady photocurrent (blue) and PCE (red) under 1 sun illumination of the Ag NW/RLGO based PVSC.

Table 1. Summarized photovoltaic parameters of PVSC devices based on Ag NW/RLGO composite electrodes, control ITO substrates and pristine Ag NW electrodes without GO.

Samples	V_{oc} (V)	J_{sc} (mA/cm^2)	FF (%)	PCE (%)	Rs (Ω cm^2)
Ag NW/RLGO	0.87	15.43	70.9	9.62	98.85
Average	0.86 ± 0.01	14.82 ± 0.59	69.7 ± 4.2	8.89 ± 0.67	96.79 ± 24.61
ITO	0.87	16.23	80	11.34	54.49
Average	0.87 ± 0.02	15.98 ± 0.62	78.1 ± 1.5	10.87 ± 0.35	61.54 ± 11.42
Ag NW without GO	0.04185	0.000026	30.66	5.12×10^{-9}	1.38×10^{8}
Average	0.037 ± 0.009	$(21.2 \pm 5.4) \times 10^{-6}$	27.41 ± 2.1	$(3.48 \pm 1.27) \times 10^{-9}$	$(1.58 \pm 0.19) \times 10^{8}$

Table 2. Photovoltaic parameters of PVSC devices based on Ag NW/RLGO composite electrodes scanned forward and scanned backward.

Samples	V_{oc} (V)	J_{sc} (mA/cm^2)	FF (%)	PCE (%)
Forward	0.86	15.08	69.8	9.08
Backward	0.87	15.43	70.9	9.62

4. Conclusions

In summary, different sizes of GO sheets from a centrifugal classification method were used and compared for the formation of fully-covered barrier on silver nanowire electrode with the assistance of steady hot airflow. Experimental results show that the size of GO sheet played an important role in preventing the corrosion of halogen species. By employing large size GO sheets, the anti-corrosive ability of Ag NW/LGO composite electrodes were significantly improved compared to those with small size GO sheets, as proven by monitoring the resistance variation under external bias and light illumination. Moreover, the Ag NW/RLGO composite electrode exhibited excellent transmittance of 81.8% at 550 nm and maintained 95.1% average transmittance (AVT) of the pristine Ag NW electrode. Consequently, the PVSCs fabricated on the Ag NW/RLGO based composite electrodes achieved a power conversion efficiency (PCE) of 9.62%, i.e., nearly 85% of the PCE from the solar cell on commercial ITO. These results clearly demonstrate the potential of Ag NW/RLGO composite as window electrode for PVSCs, which would be beneficial to large-scale low-temperature fabrication and future commercialization of PVSCs.

Author Contributions: H.C. and H.L. conceived the idea. H.C. and H.L. designed the experiment and guided the experiment. M.L., X.W. and W.C.H.C. helped the data analysis. Y.Y. and D.H. helped the device fabrication. H.C. conducted data collection and wrote the manuscript. H.C. and H.L. revised the manuscript.

Acknowledgments: This work was supported by the National Natural Science Foundation of China (Grant No. 11704293), Hubei Provincial Natural Science Foundation of China (Grant No. 2017CFB286) and the Fundamental Research Funds for the Central Universities (WUT: 2018IB008, 2018IVB016).

References

1. Kojima, A.; Teshima, K.; Shirai, Y.; Miyasaka, T. Organometal Halide Perovskites as Visible-Light Sensitizers for Photovoltaic Cells. *J. Am. Chem. Soc.* **2009**, *131*, 6050–6051. [CrossRef] [PubMed]
2. Im, J.H.; Lee, C.R.; Lee, J.W.; Park, S.W.; Park, N.G. 6.5% efficient perovskite quantum-dot-sensitized solar cell. *Nanoscale* **2011**, *3*, 4088. [CrossRef] [PubMed]
3. Lee, M.M.; Teuscher, J.; Miyasaka, T.; Murakami, T.N.; Snaith, H.J. Efficient Hybrid Solar Cells Based on Meso-Superstructured Organometal Halide Perovskites. *Science* **2012**, *338*, 643–647. [CrossRef] [PubMed]
4. Liu, M.; Johnston, M.B.; Snaith, H.J. Efficient planar heterojunction perovskite solar cells by vapour deposition. *Nature* **2013**, *501*, 395–398. [CrossRef]
5. Jeon, N.J.; Lee, J.; Noh, J.H.; Nazeeruddin, M.K.; Grätzel, M.; Seok, S.I. Efficient Inorganic–Organic Hybrid Perovskite Solar Cells Based on Pyrene Arylamine Derivatives as Hole-Transporting Materials. *J. Am. Chem. Soc.* **2013**, *135*, 19087–19090. [CrossRef]

6. Jeon, N.J.; Noh, J.H.; Yang, W.S.; Kim, Y.C.; Ryu, S.; Seo, J.; Seok, S.I. Compositional engineering of perovskite materials for high-performance solar cells. *Nature* **2015**, *517*, 476–480. [CrossRef]
7. Yang, W.S.; Park, B.W.; Jung, E.H.; Jeon, N.J.; Kim, Y.C.; Lee, D.U.; Shin, S.S.; Seo, J.; Kim, E.K.; Noh, J.H.; et al. Iodide management in formamidinium-lead-halide-based perovskite layers for efficient solar cells. *Science* **2017**, *356*, 1376–1379. [CrossRef]
8. Chen, J.Y.; Chueh, C.C.; Zhu, Z.; Chen, W.C.; Jen, A.K.Y. Low-temperature electrodeposited crystalline SnO_2 as an efficient electron-transporting layer for conventional perovskite solar cells. *Sol. Energy Mater. Sol. Cells* **2017**, *164*, 47–55. [CrossRef]
9. Ke, W.; Fang, G.; Liu, Q.; Xiong, L.; Qin, P.; Tao, H.; Wang, J.; Lei, H.; Li, B.; Wan, J.; et al. Low-Temperature Solution-Processed Tin Oxide as an Alternative Electron Transporting Layer for Efficient Perovskite Solar Cells. *J. Am. Chem. Soc.* **2015**, *137*, 6730–6733. [CrossRef]
10. Cai, F.; Yang, L.; Yan, Y.; Zhang, J.; Qin, F.; Liu, D.; Cheng, Y.B.; Zhou, Y.; Wang, T. Eliminated hysteresis and stabilized power output over 20% in planar heterojunction perovskite solar cells by compositional and surface modifications to the low-temperature-processed TiO_2 layer. *J. Mater. Chem. A* **2017**, *5*, 9402–9411. [CrossRef]
11. Kim, J.; Kim, G.; Kim, T.K.; Kwon, S.; Back, H.; Lee, J.; Lee, S.H.; Kang, H.; Lee, K. Efficient planar-heterojunction perovskite solar cells achieved via interfacial modification of a sol–gel ZnO electron collection layer. *J. Mater. Chem. A* **2014**, *2*, 17291–17296. [CrossRef]
12. Konrad, W.; Michael, S.; Tomas, L.; Abate, A.; Snaith, H.J. Sub-150 °C processed meso-superstructured perovskite solar cells with enhanced efficiency. *Energy Environ. Sci.* **2014**, *7*, 1142–1147.
13. Batmunkh, M.; Macdonald, T.J.; Peveler, W.J.; Bati, A.S.R.; Carmalt, C.J.; Parkin, I.P.; Shapter, J.G. Plasmonic Gold Nanostars Incorporated into High-Efficiency Perovskite Solar Cells. *ChemSusChem* **2017**, *10*, 3750–3753. [CrossRef] [PubMed]
14. Song, J.; Zheng, E.; Wang, X.F.; Tian, W.; Miyasaka, T. Low-temperature-processed ZnO–SnO_2 nanocomposite for efficient planar perovskite solar cells. *Sol. Energy Mater. Sol.* **2016**, *144*, 623–630. [CrossRef]
15. Arora, N.; Dar, M.I.; Hinderhofer, A.; Pellet, N.; Schreiber, F.; Zakeeruddi, S.M.; Grätzel, M. Perovskite solar cells with CuSCN hole extraction layers yield stabilized efficiencies greater than 20%. *Science* **2017**, *358*, 768–771. [CrossRef] [PubMed]
16. Zhang, H.; Cheng, J.; Lin, F.; He, H.; Mao, J.; Wong, K.S.; Jen, A.K.Y.; Choy, W.C.H. Pinhole-Free and Surface-Nanostructured NiOx Film by Room-Temperature Solution Process for High-Performance Flexible Perovskite Solar Cells with Good Stability and Reproducibility. *ACS Nano* **2016**, *10*, 1503–1511. [CrossRef] [PubMed]
17. Ouyang, D.; Xiao, J.; Ye, F.; Huang, Z.; Zhang, H.; Zhu, L.; Cheng, J. Strategic Synthesis of Ultrasmall NiCo2O4 NPs as Hole Transport Layer for Highly Efficient Perovskite Solar Cells. *Adv. Funct. Mater.* **2018**, 1702722.
18. Zhang, H.; Wang, H.; Zhu, H.; Chueh, C.C.; Chen, W.; Yang, S.; Jen, A.K.Y. Low-Temperature Solution-Processed $CuCrO_2$ Hole-Transporting Layer for Efficient and Photostable Perovskite Solar Cells. *Adv. Energy Mater.* **2018**, *8*, 1702762. [CrossRef]
19. Sun, W.; Ye, S.; Rao, H.; Li, Y.; Liu, Z.; Xiao, L.; Chen, Z.; Bian, Z.; Huang, C. Room-temperature and solution-processed copper iodide as the hole transport layer for inverted planar perovskite solar cells. *Nanoscale* **2016**, *8*, 15954–15960. [CrossRef] [PubMed]
20. Xu, L.; Wan, F.; Rong, Y.; Chen, H.; He, S.; Xu, X.; Liu, G.; Han, H.; Yuan, Y.; Yang, J.; et al. Stable monolithic hole-conductor-free perovskite solar cells using TiO_2 nanoparticle binding carbon films. *Org. Electron.* **2017**, *45*, 131–138. [CrossRef]
21. Xiong, Y.; Zhu, X.; Mei, A.; Fei, Q.; Liu, S.; Zhang, S.; Jiang, Y.; Zhou, Y.; Han, H. Bifunctional Al_2O_3 Interlayer Leads to Enhanced Open-Circuit Voltage for Hole-Conductor-Free Carbon-Based Perovskite Solar Cells. *Sol. RRL* **2018**, *2*, 1800002. [CrossRef]
22. Tran, V.; Pammi, S.V.N.; Dao, V.D.; Choi, H.S.; Yoon, S.G. Chemical vapor deposition in fabrication of robust and highly efficient perovskite solar cells based on single-walled carbon nanotubes counter electrodes. *J. Alloys Compd.* **2018**, *747*, 703–711. [CrossRef]
23. Zhen, L.; Kulkarni, S.A.; Boix, P.P.; Shi, E.; Cao, A.; Fu, K.; Batabyal, S.K.; Zhang, J.; Xiong, Q.; Wong, L.H.; et al. Laminated Carbon Nanotube Networks for Metal Electrode-Free Efficient Perovskite Solar Cells. *ACS Nano* **2014**, *8*, 6797–6804.

24. Habisreutinger, S.N.; Wenger, B.; Snaith, H.J.; Nicholas, R.J. Dopant-Free Planar n-i-p Perovskite Solar Cells with Steady-State Efficiencies Exceeding 18%. *ACS Energy Lett.* **2017**, *2*, 622–628. [CrossRef]
25. Batmunkh, M.; Macdonald, T.J.; Shearer, C.J.; Bat-Erdene, M.; Wang, Y.; Biggs, M.J.; Parkin, I.P.; Nann, T.; Shapter, J.G. Carbon Nanotubes in TiO_2 Nanofber Photoelectrodes for High-Performance Perovskite Solar Cells. *Adv. Sci.* **2017**, *2017*, 1600504. [CrossRef] [PubMed]
26. Zhou, J.; Ren, Z.; Li, S.; Liang, Z.; Surya, C.; Shen, H. Semi-transparent Cl-doped perovskite solar cells with graphene electrodes for tandem application. *Mater. Lett.* **2018**, *220*, 82–85. [CrossRef]
27. Liu, T.; Liu, W.; Zhu, Y.; Wang, S.; Wu, G.; Chen, H. All solution processed perovskite solar cells with Ag@Au nanowires as top electrode. *Sol. Energy Mater. Sol. Cells* **2017**, *171*, 43–49. [CrossRef]
28. Guo, F.; Azimi, H.; Hou, Y.; Przybilla, T.; Hu, M.; Bronnbauer, C.; Langner, S.; Spiecker, E.; Forberich, K.; Brabec, C.J. High-performance semitransparent perovskite solar cells with solution-processed silver nanowires as top electrodes. *Nanoscale* **2015**, *7*, 1642–1649. [CrossRef] [PubMed]
29. Jeon, I.; Yoon, J.; Ahn, N.; Atwa, M.; Delacou, C.; Anisimov, A.; Kauppinen, E.I.; Choi, M.; Maruyama, S.; Matsuo, Y. Carbon Nanotubes versus Graphene as Flexible Transparent Electrodes in Inverted Perovskite Solar Cells. *J. Phys. Chem. Lett.* **2017**, *8*, 5395–5401. [CrossRef] [PubMed]
30. Heo, J.H.; Shin, D.H.; Song, D.H.; Kim, D.H.; Lee, S.J.; Im, S.H. Super-flexible bis(trifluoromethanesulfonyl)-amide doped graphene transparent conductive electrodes for photo-stable perovskite solar cells. *J. Mater. Chem. A* **2018**, *6*, 8251–8258. [CrossRef]
31. Sung, H.; Ahn, N.; Jang, M.S.; Lee, J.K.; Yoon, H.; Park, N.G.; Choi, M. Transparent Conductive Oxide-Free Graphene-Based Perovskite Solar Cells with over 17% Efficiency. *Adv. Energy Mater.* **2016**, *6*, 1501873. [CrossRef]
32. Sun, K.; Li, P.; Xia, Y.; Chang, J.; Ouyang, J. Transparent Conductive Oxide-Free Perovskite Solar Cells with PEDOT:PSS as Transparent Electrode. *ACS Appl. Mater. Interface* **2015**, *7*, 15314–15320. [CrossRef] [PubMed]
33. Kim, K.; Kwon, H.C.; Ma, S.; Lee, E.; Yun, S.C.; Jang, G.; Yang, H.; Moon, J. All-Solution-Processed Thermally and Chemically Stable Copper–Nickel Core–Shell Nanowire-Based Composite Window Electrodes for Perovskite Solar Cells. *ACS Appl. Mater. Interface* **2018**, *10*, 30337–30347. [CrossRef] [PubMed]
34. Hwang, H.; Ahn, J.; Lee, E.; Kim, K.; Kwon, H.C.; Moon, J. Enhanced compatibility between a copper nanowire-based transparent electrode and a hybrid perovskite absorber by poly(ethylenimine). *Nanoscale* **2017**, *9*, 17207–17211. [CrossRef] [PubMed]
35. Lu, H.; Zhang, D.; Cheng, J.; Liu, J.; Mao, J.; Choy, W.C.H. Locally Welded Silver Nano-Network Transparent Electrodes with High Operational Stability by a Simple Alcohol-Based Chemical Approach. *Adv. Funct. Mater.* **2015**, *25*, 4211–4218. [CrossRef]
36. Lu, H.; Zhang, D.; Ren, X.; Liu, J.; Choy, W.C.H. Selective Growth and Integration of Silver Nanoparticles on Silver Nanowires at Room Conditions for Transparent Nano-Network Electrode. *ACS Nano* **2014**, *8*, 10980–10987. [CrossRef] [PubMed]
37. Lee, E.; Ahn, J.; Kwon, H.C.; Ma, S.; Kim, K.; Yun, S.; Moon, J. All-Solution-Processed Silver Nanowire Window Electrode-Based Flexible Perovskite Solar Cells Enabled with Amorphous Metal Oxide Protection. *Adv. Energy Mater.* **2018**, *8*, 1702182. [CrossRef]
38. Han, J.; Yuan, S.; Liu, L.; Qiu, X.; Gong, H.; Yang, X.; Li, C.; Hao, Y.; Cao, B. Fully indium-free flexible Ag nanowires/ZnO:F composite transparent conductive electrodes with high haze. *J. Mater. Chem. A* **2015**, *3*, 5375–5384. [CrossRef]
39. Kim, A.; Lee, H.; Kwon, H.C.; Jung, H.S.; Park, N.G.; Jeong, S.; Moon, J. Fully solution-processed transparent electrodes based on silver nanowire composites for perovskite solar cells. *Nanoscale* **2016**, *8*, 6308–6316. [CrossRef]
40. Lu, H.; Sun, J.; Zhang, H.; Lu, S.; Choy, W.C.H. Room-temperature solution-processed and metal oxide-free nano-composite for the flexible transparent bottom electrode of perovskite solar cells. *Nanoscale* **2016**, *8*, 5946–5953. [CrossRef]
41. Peng, L.; Xu, Z.; Liu, Z.; Guo, Y.; Li, P.; Gao, C. Ultrahigh Thermal Conductive yet Superflexible Graphene Films. *Adv. Mater.* **2017**, *29*, 1700589. [CrossRef] [PubMed]
42. Seo, J.W.; Park, S.; Kim, Y.C.; Jeon, N.J.; Noh, J.H.; Yoon, S.C.; Seok, S.I. Benefits of very thin PCBM and LiF layers for solution-processed p–i–n perovskite solar cells. *Energy Environ. Sci.* **2014**, *7*, 2642–2646. [CrossRef]

43. Namkoong, G.; Mamun, A.A.; Ava, T.T. Impact of PCBM/C60 electron transfer layer on charge transports on ordered and disordered perovskite phases and hysteresis-free perovskite solar cells. *Org. Electron.* **2018**, *56*, 163–169. [CrossRef]

44. Wei, J.; Li, H.; Zhao, Y.; Zhou, W.; Fu, R.; Leprince-Wang, Y.; Yu, D.; Zhao, Q. Suppressed hysteresis and improved stability in perovskite solar cells with conductive organic network. *Nano Energy* **2016**, *26*, 139–147. [CrossRef]

14

Solution-Processed PEDOT:PSS/MoS_2 Nanocomposites as Efficient Hole-Transporting Layers for Organic Solar Cells

Madeshwaran Sekkarapatti Ramasamy, Ka Yeon Ryu, Ju Won Lim, Asia Bibi, Hannah Kwon, Ji-Eun Lee, Dong Ha Kim * and Kyungkon Kim *

Deprtment of Chemistry and Nanoscience, Ewha Womans University, 52 Ewhayeodae-gil, Seodaemun-gu, Seoul 03760, Korea

* Correspondence: dhkim@ewha.ac.kr (D.H.K.); kimkk@ewha.ac.kr (K.K.)

Abstract: An efficient hole-transporting layer (HTL) based on functionalized two-dimensional (2D) MoS_2-poly(3,4-ethylenedioxythiophene):poly(styrenesulfonate) (PEDOT:PSS) composites has been developed for use in organic solar cells (OSCs). Few-layer, oleylamine-functionalized MoS_2 ($FMoS_2$) nanosheets were prepared via a simple and cost-effective solution-phase exfoliation method; then, they were blended into PEDOT:PSS, a conducting conjugated polymer, and the resulting hybrid film (PEDOT:PSS/$FMoS_2$) was tested as an HTL for poly(3-hexylthiophene):[6,6]-phenyl-C_{61}-butyric acid methyl ester (P3HT:PCBM) OSCs. The devices using this hybrid film HTL showed power conversion efficiencies up to 3.74%, which is 15.08% higher than that of the reference ones having PEDOT:PSS as HTL. Atomic force microscopy and contact angle measurements confirmed the compatibility of the PEDOT:PSS/$FMoS_2$ surface for active layer deposition on it. The electrical impedance spectroscopy analysis revealed that their use minimized the charge-transfer resistance of the OSCs, consequently improving their performance compared with the reference cells. Thus, the proposed fabrication of such HTLs incorporating 2D nanomaterials could be further expanded as a universal protocol for various high-performance optoelectronic devices.

Keywords: organic solar cells; MoS_2; hole-transporting layer; oleylamine

---

1. Introduction

Organic solar cells (OSCs) have many striking properties such as flexibility, solution processability, light weight, and simple manufacturing, especially if compared with their inorganic counterparts. To enhance their performance, numerous strategies have been proposed, including novel photoactive materials, morphology control, interfacial engineering, plasmonic nanoparticles incorporation, and alternative buffer layers and electrodes [1–6]. Their power conversion efficiency (PCE) has been recently improved up to >13% with rapid advances in new photovoltaic materials [7]. In the typical bulk heterojunction (BHJ) OSCs configuration, a photoactive blend layer consisting of acceptor/donor pairs is sandwiched between a bottom transparent anode and a top low-work-function cathode, combined with the corresponding interlayers. Such interlayers are crucial for determining the overall PCE and stability of OSCs because they reduce the potential energy barrier between photoactive layer and electrodes, enhancing the extraction of holes and electrons at the anode and cathode, respectively.

Until now, many hole-transporting layer (HTL) materials, such as conducting conjugated polymers [8–10], conjugated polyelectrolytes [11,12] metal oxides/sulfides [13–17], and graphene oxide and its hybrid films [18–21], have been explored for use in OSCs. Among them, the conjugated polymer poly(3,4 ethylenedioxythiophene):poly(styrenesulfonate) (PEDOT:PSS) has been the most widely used due to its adequate work function for creating a good ohmic contact between active layer and anode, solution processability, and high conductivity. However, its hygroscopic and acidic nature often induces chemical instability between active layers and indium tin oxide (ITO) anodes, affecting the device stability and efficiency [22,23]. Moreover, there is a clear surface energy mismatch between PEDOT:PSS (hydrophilic nature) and the active layer (hydrophobic and made of, e.g., poly(3-hexylthiophene) (P3HT)) [24,25]. To overcome such drawbacks, various PEDOT:PSS modification strategies, such as incorporating metal nanoparticles [26–28], modification by metal salts [29,30], polymer doping [31,32], and hybridization with graphene [33,34], have been developed. Interfacial engineering with long alkyl chains is an alternative but attractive method to reduce the surface energy mismatch between HTL and active layer and also to accomplish desirable molecular orientation in the active layer for enhancing the charge transport in OSCs [35].

Single and few-layer molybdenum disulfide, a two-dimensional (2D) transition metal dichalcogenide (TMDC), has recently received much interest in electronics and optoelectronics research due to its excellent optical (bandgap: 1.8 eV), electrical (device mobility: 10–130 $cm^2\ V^{-1}\ S^{-1}$), and mechanical (Young modulus: 270 GPa) properties [36,37]. Among the key preparation/exfoliation methods for TMDCs, namely, micromechanical cleavage [38], chemical vapor deposition [39], and liquid-phase exfoliation (LPE) [40], the latter is more attractive because it is scalable and cost-effective. MoS_2 has been tested as HTL for OSCs [41–43] to exploit its extraordinary optical and electrical properties in photovoltaics; nevertheless, the results have revealed that neat MoS_2 is not sufficient to replace PEDOT:PSS as OSC HTL, possibly because of its work function mismatch and unexpected phase transition. Hence, Xing et al. fabricated PEDOT:PSS/WS_2 hybrid films and demonstrated their applicability as effective OSC HTLs [44]. However, the long-time (48 h) sonication they adopted for TMDC exfoliation in the PEDOT:PSS aqueous dispersion may affect the structure of both PEDOT:PSS and MoS_2 in the final product; therefore, innovative strategies for effectively integrating these materials in OSCs are still highly demanded.

Here, we report the fabrication of oleylamine-functionalized MoS_2 ($FMoS_2$) combined with PEDOT:PSS as an effective hybrid HTL (PEDOT:PSS/$FMoS_2$) for use in conventional P3HT:[6,6]-phenyl-C_{61}-butyric acid methyl ester (PCBM)-based OSCs. The so-obtained OSCs exhibited better PCE and short-circuit current density (J_{sc}) values compared with the reference cell having simple PEDOT:PSS as HTL. $FMoS_2$ was characterized by various spectroscopic techniques including Raman spectroscopy, ultraviolet–visible (UV-Vis) absorption and transmittance, photoluminescence (PL), and transmission electron microscopy (TEM); the active layer microstructure and the surface properties of the hybrid HTL were analyzed by grazing-incidence wide-angle X-ray scattering (GIWAXS), atomic force microscopy (AFM), and contact angle measurements. Electrochemical impedance spectroscopy (EIS) measurements were carried out using an electrochemical analyzer (IVIUMSTAT.XR, IVIUM Technologies) under illumination at 0.1 V.

2. Experimental

2.1. Materials and Methods

The following chemicals were used in our experiment: molybdenum (IV) sulfide (<2 μm, 99%) and oleylamine (Sigma-Aldrich, Gyeonggi-do, Korea), P3HT (1-Material, Gyeonggi-do, Korea), PEDOT:PSS (Heraeus Deutschland GmbH & Co., Leverkusen, Germany), isopropyl alcohol (IPA) (Dae-Jung Chemicals & Metals Co., Ltd., Gyeonggi-do, Korea), and methanol (Samchun Chemicals, Seoul, Korea).

2.2. *Synthesis of $FMoS_2$ Nanosheets and PEDOT:PSS/$FMoS_2$ Hybrids*

$FMoS_2$ nanosheets were synthesized according to the liquid-phase exfoliation method reported in literature [45], with small modifications. Briefly, bulk MoS_2 powder (200 mg) was bath-sonicated in oleylamine (2 mL) by using a Branson ultrasonic bath for 20 min and successively stirred at 60 °C for 12 h in an N_2-filled glove box. Then, 1,2-dichlorobenzene (DCB) (18 mL) was added, and the dispersion was further bath-sonicated for 5 h. The resulting suspension was centrifuged at 4000 rpm, and the top 80% dark-green color supernatant, which contains excess oleylamine, DCB, and $FMOS_2$ was collected. Then, the $FMoS_2$ nanosheets were separated by adding excess acetone, followed by sonication for 2 min and high-speed centrifugation (10000 rpm). The separated $FMoS_2$ nanosheets were settled at the bottom of the centrifuge tube, which was re-dispersed in a small amount of IPA by mild sonication, and different concentrations (5, 20, and 50 μL) of the resulting dispersion were added into PEDOT:PSS:methanol (1:1 V%) aqueous solutions, which were successively ultrasonicated for 30 min to obtain PEDOT:PSS/$FMoS_2$ hybrid solutions.

2.3. Fabrication of OSCs

The OSCs having device architectures of ITO/PEDOT:PSS/P3HT:PCBM/LiF/Al and ITO/(PEDOT:PSS/$FMoS_2$)/P3HT:PCBM/LiF/Al were fabricated as follows. ITO-coated glass substrates were cleaned via sequential ultrasonication in acetone, IPA, and distilled water, followed by oxygen plasma treatment for 10 min; then, they were spin-coated with a PEDOT:PSS (Clevios P VP Al 4083) or PEDOT:PSS/$FMoS_2$ solution at 4000 rpm for 40 s and dried at 130 °C for 30 min to complete the HTL deposition. Next, an active layer consisting of a P3HT:PCBM (1:0.6 wt%) binary blend solution was spin-coated on the resulting HTL layer at 2500 rpm for 40 s inside an N_2-filled glove box and annealed at 150 °C. Finally, LiF and Al layers were deposited by thermal evaporation. The active area of the fabricated OSCs was 0.06 cm^2.

2.4. Characterization

The absorption properties of the samples were analyzed using a UV-Vis absorption spectrometer (Cary 5000, Varian, Inc.). Raman spectra were recorded on a Horiba Jobin-Yvon spectrometer. The emission properties were investigated with a luminescence spectrometer (LS55 Perkin Elmer). The TEM measurements were carried out on a JEOL JSM-2100-F system. The surface morphologies were investigated using a tapping-mode atomic force microscope (Veeco D3100). The water contact angles of the samples were measured with a KSV CAM 101 instrument. The GIWAXS analysis was conducted at the PLS-II 9A U-SAXs beamline of the Pohang Accelerator Laboratory (Korea) at the following operating conditions: incidence angle of ~0.12°, wavelength of 1.12 Å, and sample-to-detector distance of 224 nm. The GIWAXS patterns were recorded using a 2D charge-coupled device camera (Rayonix, SX-165, USA) with an exposure time of 10–30 s. The JV properties of the solar cells were measured with a Keithley 2400 solar cell IV measurement system under AM 1.5 G illumination at 100 mW cm^{-2}.

3. Results and Discussion

OSCs having two different device architectures, ITO/(PEDOT:PSS/$FMoS_2$)/P3HT:PCBM/LiF/Al and ITO/PEDOT:PSS/P3HT:PCBM/LiF/Al (for comparison), were fabricated as schematized in Figure 1.

First, we synthesized $FMoS_2$ nanosheets via the solution-phase ultrasonic exfoliation of bulk MoS_2 in the presence of oleylamine and 1,2-dichlorobenzene as a solvent; then, they were incorporated in different concentrations (5, 20, and 50 μL) into PEDOT:PSS, and the resulting PEDOT:PSS/$FMoS_2$ (denoted as PEDOT:PSS/$FMoS_2$(5), PEDOT:PSS/$FMoS_2$(20), and PEDOT:PSS/$FMoS_2$(50) according to the $FMoS_2$ loading) was used as HTL for conventional OSCs.

Raman spectroscopy is a powerful nondestructive technique for monitoring structural changes in 2D materials [46]. The Raman spectrum of bulk MoS_2 showed two characteristic peaks at 374.83 and 402.05 cm^{-1} corresponding, respectively, to the E^1_{2g} and A_{1g} vibrational modes (Figure 2a); the first arose from the in-plane vibration of Mo and S atoms, while the second resulted from the out-of-plane vibrations of sulfur [47,48]. As regards $FMoS_2$, the peaks for both the E^1_{2g} and A_{1g} vibrational modes were blue-shifted toward higher wavenumbers (respectively, 382.66 and 405.66 cm^{-1}), suggesting interactions between oleylamine and MoS_2. Moreover, the wavenumber difference between these two vibrational modes is closely related to the layer number present in the MoS_2 nanosheets [49], and in our case, this difference decreased from 27.2 cm^{-1} for bulk MoS_2 to 23 cm^{-1} for $FMoS_2$ nanosheets, demonstrating the successful exfoliation of MoS_2 nanosheets during the oleylamine treatment.

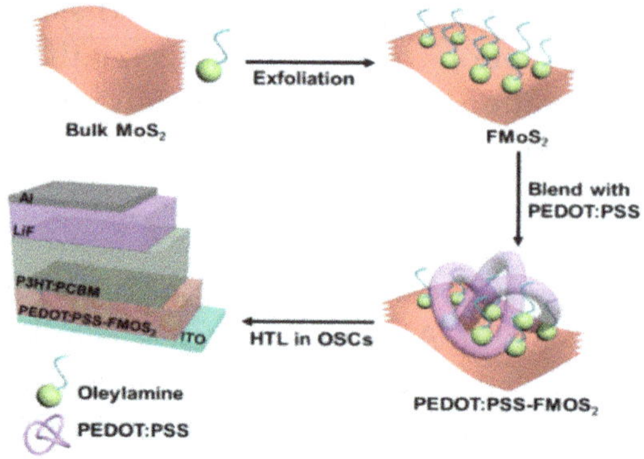

Figure 1. Fabrication process for poly(3,4-ethylenedioxythiophene):poly(styrenesulfonate) (PEDOT:PSS)/oleylamine-functionalized MoS_2 ($FMoS_2$) hybrid hole-transporting layer (HTL) for organic solar cells.

The absorption properties of the $FMoS_2$ nanosheets were further investigated via UV-Vis absorption spectroscopy; their spectrum (Figure 2b) clearly showed two characteristic absorption peaks of MoS_2 at 618 and 677 cm^{-1} corresponding, respectively, to the A1 and B1 direct excitonic transitions with the energy split from valence band spin–orbital coupling [50]. Furthermore, unlike bulk MoS_2, $FMoS_2$ yielded dark-greenish dispersion in 1,2-dichlorobenzene. These results clearly indicate some alteration in the surface properties of MoS_2 due to the oleylamine treatment [51].

Bulk MoS_2 is an indirect bandgap semiconductor that does not exhibit any photoluminescence; however, upon exfoliation, its luminescence increases with decreasing its layer thickness, so that single-layer MoS_2 shows the highest photoluminescence due to its transition into a direct bandgap semiconductor [52,53]. As expected, $FMoS_2$ exhibited significant photoluminescence (see the PL spectra in Figure S1, Electronic Supporting Information (ESI)), which clearly proves the successful layer thinning of MoS_2 during the functionalization process.

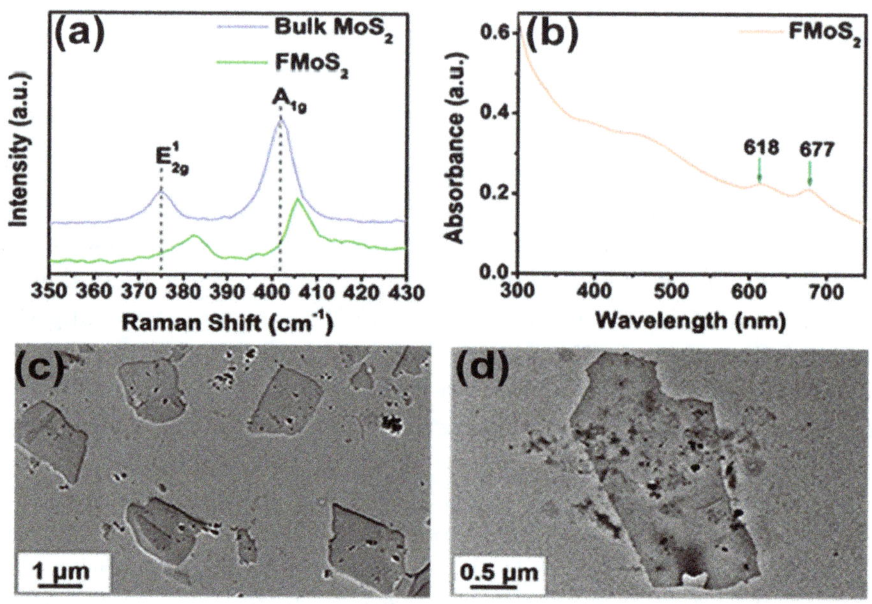

Figure 2. (**a**) Raman spectra of bulk and oleylamine-functionalized MoS_2 ($FMoS_2$). (**b**) Ultraviolet–visible light absorption spectrum of $FMoS_2$. (**c**,**d**) Transmission electron microscopy images of $FMoS_2$ nanosheets.

The TEM images of the $FMoS_2$ nanosheets are displayed in Figure 2c,d, showing a thin nanosheet morphology with sizes of several hundred nanometers. A careful observation of the nanosheet edges reveals the presence of few-layer nanosheets, confirming the effectiveness of the liquid-based exfoliation with oleylamine. AFM measurements were carried out (Figure S2, ESI) to further evaluate the layer thickness; that of $FMoS_2$ was ~6.7 nm, suggesting the existence of few-layer nanosheets, while the reported thickness of monolayer MoS_2 ranges between 0.9 and 1.2 nm [54].

To improve the performance of conventional PEDOT:PSS-based HTL for OSCs, we incorporated it with $FMoS_2$ via a simple solution-blending method because we believed that the introduction of 2D sheet-like MoS_2 functionalized with a long-chain primary alkyl amine (oleylamine) would have made the PEDOT:PSS surface more hydrophobic, facilitating the following deposition of the hydrophobic active layer. In addition, the amine group of oleylamine tends to be located near Mo atoms in MoS_2 due to metal–amine interactions, while its long alkyl chain with $-CH_3$ groups is oriented toward the active layer, and this kind of configuration should enforce the active layer with a desirable molecular orientation for efficient charge transport in OSCs; P3HT thin films deposited on insulator substrates modified with $-CH_3$ groups formed face-on orientation because of π–H interactions [55,56].

The contact angles of ITO with PEDOT:PSS and PEDOT:PSS/$FMoS_2$ containing 5, 20, and 50 μL of $FMoS_2$ were 30°, 47°, 54°, and 56°, respectively (Figure S3, ESI), which indicates that the hydrophobicity of PEDOT:PSS was slightly increased by the $FMoS_2$ addition and, hence, the hydrophobic active layer solution was more compatible on hybrid HTL than that of the hydrophilic PEDOT:PSS one.

The surface morphology of the various samples was compared via tapping-mode AFM analysis (Figure S4, ESI); the root-mean-square (rms) roughness value of PEDOT:PSS was 1 nm and decreased down to 0.69 nm for PEDOT:PSS/$FMoS_2$(5), suggesting a smooth surface morphology in the hybrid HTL. However, PEDOT:PSS/$FMoS_2$(50) exhibited an rms roughness value of 0.97 nm, indicating that the addition of higher $FMoS_2$ concentrations would decrease the film smoothness.

All the synthesized PEDOT:PSS and PEDOT:PSS/$FMoS_2$ hybrid films exhibited similar UV-Vis transmittance values (Figure S5a, ESI), showing that the $FMoS_2$ addition did not affect any absorption property of the PEDOT:PSS matrix. As regards the P3HT:PCBM (active layer) films spin-coated on glass substrates predeposited with PEDOT:PSS or PEDOT:PSS/$FMoS_2$ HTLs (Figure S5b), for all the samples, their absorbance ranged from 400 to 650 nm, with a maximum at 512 nm, and two shoulders around 550 and 600 nm. The existence of vibronic feature at 600 nm suggests that the P3HT film existed in a high degree of ordered crystalline lamella due to strong interchain interactions [57].

The current–voltage (JV) characteristics of the fabricated P3HT:PCBM OSCs having PEDOT:PSS/$FMoS_2$ as HTL are shown in Figure 3a. Their performance is compared with that of reference devices having PEDOT:PSS as HTL in Table 1. The reference cells showed PCE = 3.25%, J_{sc} = 7.92 mA cm^{-2}, V_{oc} = 0.671 V, and FF = 0.61. The $FMoS_2$ incorporation led to significant PCE and J_{sc} improvements; in particular, the device based on PEDOT:PSS/$FMoS_2$(5) exhibited the highest PCE, J_{sc}, and FF.

The external quantum efficiency (EQE) measurements (Figure 3b) showed improved EQE for the hybrid HTL-based OSCs compared with the reference cells and confirmed also their increased J_{sc}, demonstrating the enhanced charge extraction at the HTL/active layer interface and the charge collection at the electrodes [58,59]. The photovoltaic parameters such as PCE, J_{sc}, FF and V_{oc} as a function of $FMoS_2$ in PEDOT:PSS HTLs are plotted in Figure 3c, d, e and f respectively.

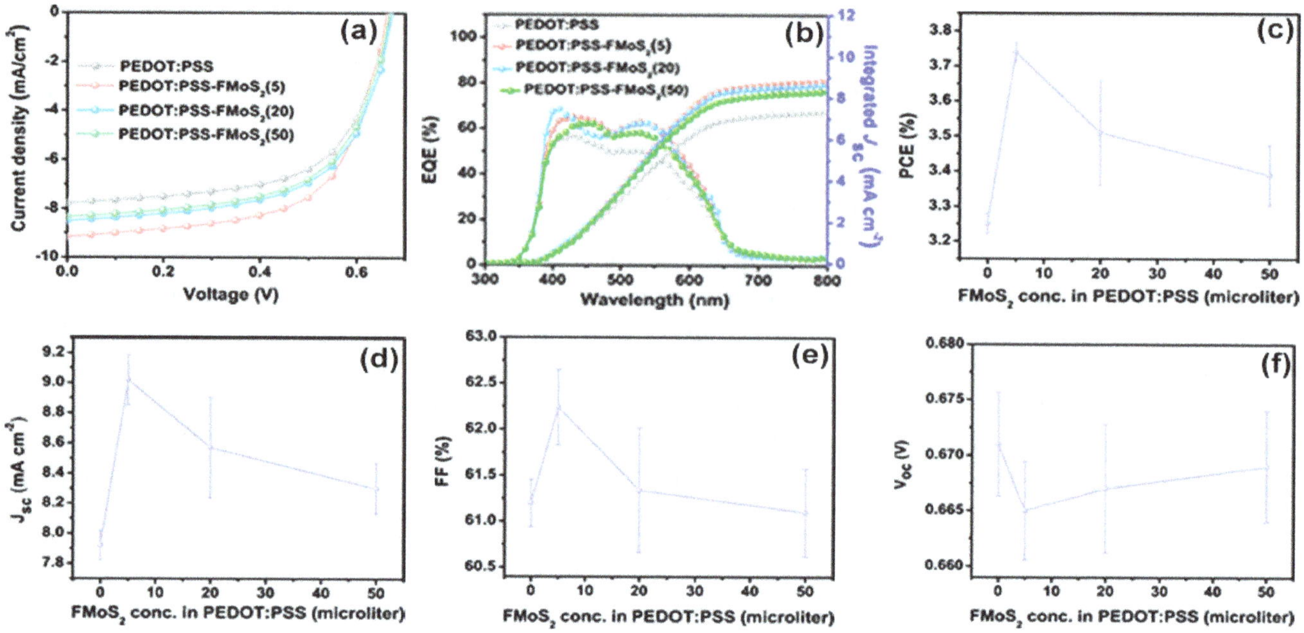

Figure 3. (**a**) Current density–voltage curves, (**b**) external quantum efficiency (EQE) profiles, (**c**) power conversion efficiencies (PCE), (**d**) short-circuit current density (J_{SC}), (**e**) fill factor, and (**f**) open-circuit voltage (V_{OC}) values of organic solar cells based on poly(3,4-ethylenedioxythiophene):poly(styrenesulfonate) (PEDOT:PSS) and PEDOT:PSS/oleylamine-functionalized MoS_2 ($FMoS_2$) as hole-transporting layers. The reported average PCE values are extracted from nine identical cells for each sample.

Table 1. Photovoltaic performance of poly(3-hexylthiophene):[6,6]-phenyl-C61-butyric acid methyl ester-based organic solar cells having poly(3,4-ethylenedioxythiophene):poly(styrenesulfonate) (PEDOT:PSSS) and PEDOT:PSS/oleylamine-functionalized MoS_2 ($FMoS_2$) as hole-transportation layers.

$FMoS_2$ Concentration (μL) in PEDOT:PSS	**PCE (%)**	**V_{oc} (V)**	**J_{sc} (mA cm^{-2})**	**FF (%)**
0 (Reference)	3.25 ± 0.03	0.671 ± 0.004	7.92 ± 0.09	61.2 ± 0.26
5	3.74 ± 0.02	0.665 ± 0.004	9.02 ± 0.17	62.24 ± 0.41
20	3.51 ± 0.15	0.667 ± 0.006	8.57 ± 0.33	61.34 ± 0.68
50	3.39 ± 0.08	0.669 ± 0.005	8.30 ± 0.17	61.10 ± 0.48

To understand the charge transport, we analyzed the microstructure (chain-orientation and crystallinity) of the active layer (P3HT:PCBM) on both the PEDOT:PSS and PEDOT:PSS/$FMoS_2$ samples by GIWAXS (Figures 4 and 5). Charge transport in conjugated polymers occurs either in the π–π staking direction or the chain backbone one, which is the fastest but its vertical alignment of chains backbones along the z direction is rarely observed [60,61]. In general, P3HT crystallizes into two main configurations, namely, edge-on and face-on orientations; in the former, both chain backbone and π–π staking directions lie parallel to the substrate; in the latter, π–π staking occurs perpendicular to the substrate, which is a desirable orientation in OSCs for vertical charge transport [62]. Figure 4 shows the GIWAXS diffraction patterns of P3HT:PCBM thin films deposited on PEDOT:PSS and PEDOT:PSS/$FMoS_2$ HTLs. In both cases, the thin films exhibited strong (100), (200), and (300) diffractions along the z axis, confirming the existence of the strong edge-on lamellae configuration of P3HT [63]. In addition, the absence of π–π staking peak (010), corresponding to the face-on orientation near the z axis, indicates that P3HT preferentially adopted the edge-on configuration in both PEDOT:PSS and PEDOT:PSS/$FMoS_2$ HTLs. Since the use of $-CH_3$ group-functionalized substrates tends to promote the face-on orientation of P3HT [55,56], we aimed to improve such configuration of the active layer by incorporating the described oleylamine (having $-CH_3$ groups)-functionalized

MoS_2 into PEDOT:PSS, but we did not observe any significant difference in its molecular orientation, maybe because the low $FMoS_2$ concentrations used were not sufficient for such change. Thus, we can conclude that the PCE and J_{sc} enhancement in the OCSs having PEDOT:PSS/$FMoS_2$ as HTL may be due to its surface compatibility for the active layer deposition, as observed in the AFM and contact angle measurements.

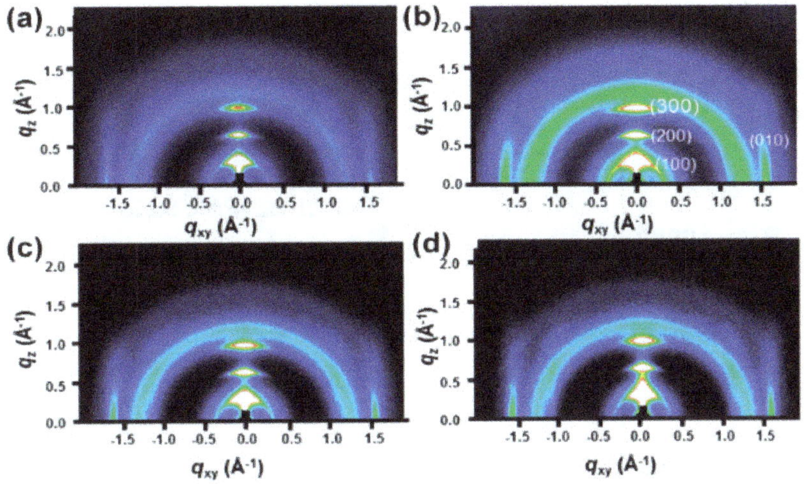

Figure 4. Grazing-incidence wide-angle X-ray scattering diffraction patterns of poly(3-hexylthiophene):[6,6]-phenyl-C_{61}-butyric acid methyl ester thin films deposited on (**a**) poly(3,4-ethylenedioxythiophene):poly(styrenesulfonate) (PEDOT:PSS) and PEDOT:PSS combined with (**b**) 5, (**c**) 20, and (**d**) 50 μL of oleylamine-functionalized MoS_2.

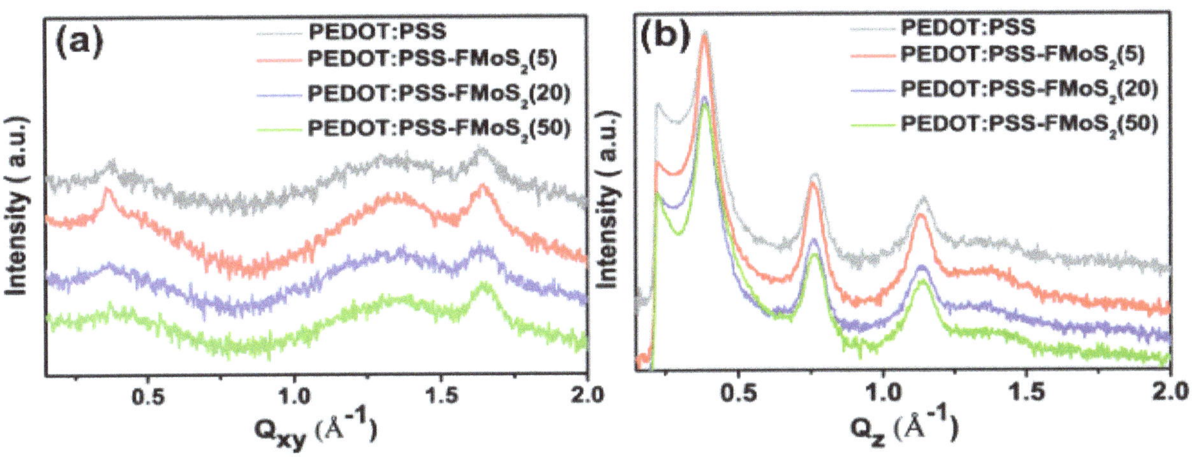

Figure 5. (**a**) In-plane and (**b**) out-of-plane spectra of poly(3-hexylthiophene):[6,6]-phenyl-C61-butyric acid methyl ester thin films deposited on poly (3,4-ethylendioxythiophene): poly(styrenesulfonate) (PEDOT:PSS) and PEDOT:PSS combined with 5, 20, and 50 μL of oleylamine-functionalized MoS_2 samples obtained from grazing-incidence wide-angle X-ray scattering.

Electrical impedance spectroscopy (EIS) was performed to investigate the charge transport dynamics of the OSCs fabricated with PEDOT:PSS and PEDOT:PSS/$FMoS_2$(5) as HTL (Figure 6). This analysis allowed us to observe the current response by applying alternating current voltage as a function of frequency; the OSCs with PEDOT:PSS/$FMoS_2$(5) demonstrated slightly lower charge transfer resistance, revealing that the holes were effectively transported from the active layer to the anode (ITO). In order to elucidate the origin of the improvement in the photovoltaic performance, especially both FF and J_{sc} for PEDOT:PSS/$FMoS_2$(5), we further calculated the resistance of the devices. In general, it is well known that lower series resistance (R_S) and higher shunt resistance (R_{SH}) are required to achieve higher FF in the solar cell device [64]. Based on the J–V curves obtained from the devices, it is clearly revealed that the device with PEDOT:PSS/FMoS$_2$(5) as HTL showed the lowest R_S while maintaining higher R_{SH}, leading to enhancement in charge extraction. The corresponding R_S value of cells employing PEDOT:PSS/FMoS$_2$(5) as HTL was 134.8 $\Omega \cdot cm^2$, while the reference showed 180.0 $\Omega \cdot cm^2$. Lower R_S indicates that better interfacial contact and charge collection efficiency were obtained due to the addition of the conducting FMoS$_2$ layer. In the case of the R_{SH}, no significant changes in the shunt resistance were observed for the devices. In the point of view of the identical R_{SH}, barrier resistance at the interface and the leakage current level flowing across the photoactive layer is similar. Therefore, the addition of FMoS$_2$ might contribute to extract photoexcited charges efficiently by lowering the R_S.

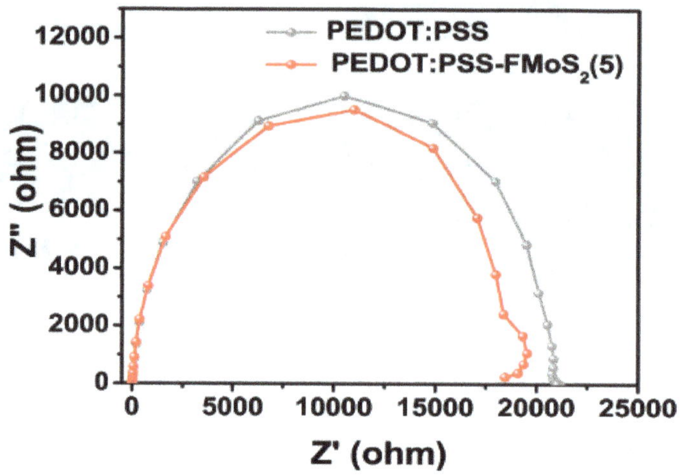

Figure 6. Electrical impedance spectra of organic solar cells based on poly(3,4-ethylenedioxythiophene):poly(styrenesulfonate) (PEDOT:PSS) and PEDOT:PSS/oleylamine-functionalized MoS$_2$ (5 μL) (PEDOT:PSS/FMoS$_2$(5)) as hole-transportation layer.

4. Conclusions

The application of solution-processed PEDOT:PSS/FMoS$_2$ hybrids as effective HTLs for OSCs has been successfully demonstrated. Raman, UV-Vis, PL, TEM, and AFM analyses confirmed the successful exfoliation of bulk MoS$_2$ into few-layer nanosheets in the presence of oleylamine via a simple and cost-effective solution-based method. The OSCs fabricated with the synthesized PEDOT:PSS/FMoS$_2$ hybrids as HTL exhibited PCE values up to 3.74%, which is 15.08% higher than that of the reference cells having simple PEDOT:PSS as HTL. The hybrid HTL films showed better surface properties for the deposition of the hydrophobic active layer, consequently, the charge-transfer resistance was minimized for OSCs fabricated with hybrid HTL compared with reference cells, improving the OSC performance. Due to their simple preparation method, 2D FMoS$_2$-incorporated PEDOT:PSS-based HTL provides valuable alternative HTL for OSCs.

Author Contributions: M.S.R., prepared the content of this research, carried out device fabrication, analysis of results and wrote the manuscript. K.Y.R. performed GIWAXS measurements and interpreted the data. J.W.L. performed IPCE measurements and guided for manuscript revisions. A.B., H.K., J.-E.L. performed the design of device configuration, performed AFM and PL measurements. D.H.K. and K.K. guided the overall scope of this research and wrote the manuscript.

References

1. Jang, Y.H.; Jang, Y.J.; Kim, S.; Quan, L.N.; Chung, K.; Kim, D.H. Plasmonic solar cells: From rational design to mechanism overview. *Chem. Rev.* **2016**, *116*, 14982–15034. [CrossRef] [PubMed]
2. Oh, Y.; Lim, J.W.; Kim, J.G.; Wang, H.; Kang, B.H.; Park, Y.W.; Kim, H.; Jang, Y.J.; Kim, J.; Kim, D.H.; et al.

Plasmonic periodic nanodot arrays via laser interference lithography for organic photovoltaic cells with >10% efficiency. *ACS Nano* **2016**, *10*, 10143–10151. [CrossRef] [PubMed]

3. Xie, L.; Lee, J.S.; Jang, Y.; Ahn, H.; Kim, Y.H.; Kim, K. Organic photovoltaics utilizing a polymer nanofiber/fullerene interdigitated bilayer prepared by sequential solution deposition. *J. Phys. Chem. C* **2016**, *120*, 12933–12940. [CrossRef]
4. Zhu, M.; Kim, H.; Jang, Y.J.; Park, S.; Ryu, D.Y.; Kim, K.; Tang, P.; Qiu, F.; Kim, D.H.; Peng, J. Toward high efficiency organic photovoltaic devices with enhanced thermal stability utilizing P3HT-b-P3PHT block copolymer additives. *J. Mater. Chem. A* **2016**, *4*, 18432–18443. [CrossRef]
5. Kim, Y.; Son, J.; Shafian, S.; Kim, K.; Hyun, J.K. Semitransparent blue, green, and red organic solar cells Using Color Filtering Electrodes. *Adv. Opt. Mater.* **2018**, *6*, 1800051. [CrossRef]
6. Xie, L.; Yoon, S.; Cho, Y.J.; Kim, K. Effective protection of sequential solution-processed polymer/fullerene bilayer solar cell against charge recombination and degradation. *Org. Electron.* **2015**, *25*, 212–218. [CrossRef]
7. Zhao, W.; Li, S.; Yao, H.; Zhang, S.; Zhang, Y.; Yang, B.; Hou, J. Molecular optimization enables over 13% efficiency in organic solar cells. *J. Am. Chem. Soc.* **2017**, *139*, 7148–7151. [CrossRef] [PubMed]
8. Ahn, S.; Jeong, S.H.; Han, T.H.; Lee, T.W. Conducting polymers as anode buffer materials in organic and perovskite optoelectronics. *Adv. Opt. Mater.* **2017**, *5*, 1600512. [CrossRef]
9. Zhang, S.; Yu, Z.; Li, P.; Li, B.; Isikgor, F.H.; Du, D.; Sun, K.; Xia, Y.; Ouyang, J. Poly(3,4-ethylenedioxythiophene): Polystyrene sulfonate films with low conductivity and low Acidity through a treatment of their solutions with probe ultrasonication and their application as hole transport layer in polymer solar cells and perovskite solar cells. *Org. Electron.* **2016**, *32*, 149–156.
10. Lee, J.J.; Lee, S.H.; Kim, F.S.; Choi, H.H.; Kim, J.H. Simultaneous enhancement of the efficiency and stability of organic solar cells using PEDOT: PSS grafted with a PEGME buffer layer. *Org. Electron.* **2015**, *26*, 191–199. [CrossRef]
11. Zhou, H.; Zhang, Y.; Mai, C.K.; Seifter, J.; Nguyen, T.Q.; Bazan, G.C.; Heeger, A.J. Solution-processed pH-neutral conjugated polyelectrolyte improves interfacial contact in organic solar cells. *ACS Nano* **2015**, *9*, 371–377. [CrossRef] [PubMed]
12. Zhou, H.; Zhang, Y.; Mai, C.K.; Collins, S.D.; Nguyen, T.Q.; Bazan, G.C.; Heeger, A.J. Conductive conjugated polyelectrolyte as hole-transporting layer for organic bulk heterojunction solar cells. *Adv. Mater.* **2014**, *26*, 780–785. [CrossRef] [PubMed]
13. Jasieniak, J.J.; Seifter, J.; Jo, J.; Mates, T.; Heeger, A.J. A solution-processed MoOx anode interlayer for use within organic photovoltaic devices. *Adv. Funct. Mater.* **2012**, *22*, 2594–2605. [CrossRef]
14. Xie, F.; Choy, W.C.H.; Wang, C.; Li, X.; Zhang, S.; Hou, J. Low-temperature solution-processed hydrogen molybdenum and vanadium bronzes for an efficient hole-transport layer in organic electronics. *Adv. Mater.* **2013**, *25*, 2051–2055. [CrossRef] [PubMed]
15. Bao, X.; Zhu, Q.; Wang, T.; Guo, J.; Yang, C.; Yu, D.; Wang, N.; Chen, W.; Yang, R. Simple O_2 plasma-processed V_2O_5 as an anode buffer layer for high performance polymer solar cells. *ACS Appl. Mater. Interfaces* **2015**, *7*, 7613–7618. [CrossRef] [PubMed]
16. Bai, S.; Jin, Y.; Dai, X.; Liang, X.; Ye, Z.; Li, M.; Cheng, J.; Xiao, X.; Wu, Z.; Xia, Z.; et al. Low-temperature combustion-synthesized nickel oxide thin films as hole-transport interlayers for solution processed optoelectronic devices. *Adv. Energy Mater.* **2014**, *4*, 1301460. [CrossRef]
17. Yin, Z.; Wei, J.; Zheng, Q. Interfacial materials for organic solar cells: Recent advances and perspectives. *Adv. Sci.* **2016**, *3*, 1500362. [CrossRef] [PubMed]
18. Liu, J.; Xue, Y.; Dai, L. Sulfated graphene oxide as a hole-extraction layer in high-performance polymer solar cells. *J. Phys. Chem. Lett.* **2012**, *3*, 1928–1933. [CrossRef]
19. Li, S.S.; Tu, K.H.; Lin, C.C.; Chen, C.W.; Chhowalla, M. Solution-processable graphene oxide as an efficient hole transport layer in polymer solar cells. *ACS Nano* **2010**, *4*, 3169–3174. [CrossRef]
20. Yun, J.M.; Yeo, J.S.; Kim, J.; Jeong, H.G.; Kim, D.Y.; Noh, Y.J.; Kim, S.S.; Ku, B.C.; Na, S.I. Solution-processable reduced graphene oxide as a novel alternative to PEDOT:PSS hole transport layers for highly efficient and stable polymer solar cells. *Adv. Mater.* **2011**, *23*, 4923–4928. [CrossRef]
21. Niu, J.; Yang, D.; Ren, X.; Yang, Z.; Liu, Y.; Zhu, X.; Zhao, W.; Liu, S. Graphene-oxide doped PEDOT: PSS as a superior hole transport material for high-efficiency perovskite solar cell. *Org. Electron.* **2017**, *48*, 165–171. [CrossRef]

22. Drakonakis, V.M.; Savva, A.; Kokonou, M.; Choulis, S.A. Investigating electrodes degradation in organic photovoltaics through reverse engineering under accelerated humidity lifetime conditions. *Sol. Energy Mater. Sol. Cells* **2014**, *130*, 544–550. [CrossRef]
23. Kawano, K.; Pacios, R.; Poplavskyy, D.; Nelson, J.; Bradley, D.D.C.; Durrant, J.R. Degradation of organic solar cells due to air exposure. *Sol. Energy Mater. Sol. Cells* **2006**, *90*, 3520–3530. [CrossRef]
24. Lim, F.J.; Ananthanarayanan, K.; Luther, J.; Ho, G.W. Influence of a novel fluorosurfactant modified PEDOT: PSS hole transport layer on the performance of inverted organic solar cells. *J. Mater. Chem.* **2012**, *22*, 25057–25064. [CrossRef]
25. Arulkashmir, A.; Krishnamoorthy, K. Disassembly of micelles to impart donor and acceptor gradation to enhance organic solar cell efficiency. *Chem. Commun.* **2016**, *52*, 3486–3489. [CrossRef]
26. Lim, D.C.; Kim, K.D.; Park, S.Y.; Hong, E.M.; Seo, H.O.; Lim, J.H.; Lee, K.W.; Jeong, Y.J.; Song, C.; Lee, E.; et al. Towards fabrication of high-performing organic photovoltaics: New donor polymer, atomic layer deposited thin buffer layer and plasmonic effects. *Energy Environ. Sci.* **2012**, *5*, 9803–9807. [CrossRef]
27. Wang, K.; Yi, C.; Hu, X.; Liu, C.; Sun, Y.; Hou, J.; Li, Y.; Zheng, J.; Chuang, S.; Karim, A.; et al. Enhanced performance of polymer solar cells using PEDOT:PSS doped with Fe_3O_4 magnetic nanoparticles aligned by an external magnetostatic field as an anode buffer layer. *ACS Appl. Mater. Interfaces* **2014**, *6*, 13201–13208. [CrossRef]
28. Yao, K.; Salvador, M.; Chueh, C.C.; Xin, X.K.; Xu, Y.X.; de Quilettes, D.W.; Hu, T.; Chen, Y.; Ginger, D.S.; Jen, A.K.Y. A general route to enhance polymer solar cell performance using plasmonic nanoprisms. *Adv. Energy Mater.* **2014**, *4*, 1400206. [CrossRef]
29. Zhao, Z.; Wu, Q.; Xia, F.; Chen, X.; Liu, Y.; Zhang, W.; Zhu, J.; Dai, S.; Yang, S. Improving the conductivity of PEDOT: PSS hole transport layer in polymer solar cells via copper(II) bromide salt doping. *ACS Appl. Mater. Interfaces* **2015**, *7*, 1439–1448. [CrossRef]
30. Kadem, B.; Cranton, W.; Hassan, A. Metal salt modified PEDOT: PSS as anode buffer layer and its effect on power conversion efficiency of organic solar cells. *Org. Electron.* **2015**, *24*, 73–79. [CrossRef]
31. Kang, D.J.; Cho, H.H.; Lee, I.; Kim, K.H.; Kim, H.J.; Liao, K.; Kim, T.S.; Kim, B.J. Enhancing mechanical properties of highly efficient polymer solar cells using size-tuned polymer nanoparticles. *ACS Appl. Mater. Interfaces* **2015**, *7*, 2668–2676. [CrossRef]
32. Hossain, J.; Liu, Q.; Miura, T.; Kasahara, K.; Harada, D.; Ishikawa, R.; Ueno, K.; Shirai, H. Nafion-modified PEDOT: PSS as a transparent hole-transporting layer for high-performance crystalline-Si/organic heterojunction solar cells with improved light soaking stability. *ACS Appl. Mater. Interfaces* **2016**, *8*, 31926–31934. [CrossRef]
33. Hilal, M.; Han, J.I. Significant improvement in the photovoltaic stability of bulk heterojunction organic solar cells by the molecular level interaction of graphene oxide with a PEDOT: PSS composite hole transport layer. *Sol. Energy* **2018**, *167*, 24–34. [CrossRef]
34. HDehsari, S.; Shalamzari, E.K.; Gavgani, J.N.; Taeomi, F.A.; Ghanbary, S. Efficient preparation of ultralarge graphene oxide using a PEDOT: PSS/GO composite layer as hole transport layer in polymer-based optoelectronic devices. *RSC Adv.* **2014**, *4*, 55067–55076. [CrossRef]
35. Walker, B.; Choi, H.; Kim, J.Y. Interfacial engineering for highly efficient organic solar cells. *Curr. Appl. Phys.* **2017**, *17*, 370–391. [CrossRef]
36. Pham, V.P.; Yeom, G.Y. Recent advances in doping of molybdenum disulfide: Industrial applications and future prospects. *Adv. Mater.* **2016**, *28*, 9024–9059. [CrossRef]
37. Chhowalla, M.; Shin, H.S.; Eda, G.; Li, L.J.; Loh, K.P.; Zhang, H. The chemistry of two-dimensional layered transition metal dichalcogenide nanosheets. *Nat. Chem.* **2013**, *5*, 263–275. [CrossRef]
38. Li, H.; Wu, J.; Yin, Z.; Zhang, H. Preparation and Applications of Mechanically Exfoliated Single-Layer and Multilayer MoS_2 and WSe_2 Nanosheets. *Acc. Chem. Res.* **2014**, *47*, 1067–1075. [CrossRef]
39. Lee, Y.H.; Zhang, X.Q.; Zhang, W.; Chang, M.T.; Lin, C.T.; Chang, K.D.; Yu, Y.C.; Wang, J.T.W.; Chang, C.S.; Li, L.J.; et al. Synthesis of large-area MoS_2 atomic layers with chemical vapor deposition. *Adv. Mater.* **2012**, *24*, 2320–2325. [CrossRef]
40. Zhang, X.; Lai, Z.; Tan, C.; Zhang, H. Solution-processed two-Dimensional MoS_2 nanosheets: Preparation, hybridization, and applications. *Angew. Chem. Int. Ed.* **2016**, *55*, 8816–8838. [CrossRef]

41. Le, Q.V.; Nguyen, T.P.; Jang, H.W.; Kim, S.Y. The use of UV/ozone-treated MoS_2 nanosheets for extended air stability in organic photovoltaic cells. *Phys. Chem. Chem. Phys.* **2014**, *16*, 13123–13128. [CrossRef]
42. Yun, J.M.; Noh, Y.J.; Yeo, J.S.; Go, Y.J.; Na, S.I.; Jeong, H.G.; Kim, J.; Lee, S.; Kim, S.S.; Koo, H.Y.; et al. Efficient work-function engineering of solution-processed MoS_2 thin-films for novel hole and electron transport layers leading to high-performance solar cells. *J. Mater. Chem. C* **2013**, *1*, 3777–3783. [CrossRef]
43. Yang, X.; Fu, W.; Liu, W.; Hong, J.; Cai, Y.; Jin, C.; Xu, M.; Wang, H.; Yang, D.; Chen, H. Engineering crystalline structures of two-dimensional MoS_2 sheets for high-performance organic solar cells. *J. Mater. Chem. A* **2014**, *2*, 7727–7733. [CrossRef]
44. Xing, W.; Chen, Y.; Wu, X.; Xu, X.; Ye, P.; Zhu, T.; Guo, Q.; Yang, L.; Li, W.; Huang, H. PEDOT: PSS-assisted exfoliation and functionalization of 2D nanosheets for high-performance organic solar cells. *Adv. Funct. Mater.* **2017**, *27*, 1701622. [CrossRef]
45. Ahmad, R.; Srivastava, R.; Yadav, S.; Chand, S.; Sapra, S. Functionalized 2D-MoS_2-incorporated polymer ternary solar cells: Role of nanosheet-induced long-range ordering of polymer chains on charge transport. *ACS Appl. Mater. Interfaces* **2017**, *9*, 34111–34121. [CrossRef]
46. Wu, J.B.; Lin, M.L.; Cong, X.; Liu, H.N.; Tan, P.H. Raman spectroscopy of graphene-based materials and its applications in related devices. *Chem. Soc. Rev.* **2018**, *47*, 1822–1873. [CrossRef]
47. Cho, K.; Min, M.; Kim, T.Y.; Jeong, H.; Pak, J.; Kim, J.K.; Jang, J.; Yun, S.J.; Lee, Y.H.; Hong, W.K.; et al. Electrical and optical characterization of MoS_2 with Sulfur vacancy passivation by treatment with alkanethiol molecules. *ACS Nano* **2015**, *9*, 8044–8053. [CrossRef]
48. Yao, Y.; Tolentino, L.; Yang, Z.; Song, X.; Zhang, W.; Chen, Y.; Wong, C. High-concentration aqueous dispersions of MoS_2. *Adv. Funct. Mater.* **2013**, *23*, 3577–3583. [CrossRef]
49. Luo, S.; Qi, X.; Ren, L.; Hao, G.; Fan, Y.; Liu, Y.; Han, W.; Zang, C.; Li, J.; Zhong, J. Photoresponse properties of large-area MoS_2 atomic layer synthesized by vapor phase deposition. *J. Appl. Phys.* **2014**, *116*, 164304. [CrossRef]
50. Zhou, K.G.; Mao, N.N.; Wang, H.X.; Peng, Y.; Zhang, H.L. A Mixed-solvent strategy for efficient exfoliation of inorganic graphene analogues. *Angew. Chem. Int. Ed.* **2011**, *50*, 10839–10842. [CrossRef]
51. Chen, X.; Berner, N.C.; Backes, C.; Duesberg, G.S.; McDonald, A.R. Functionalization of two-dimensional MoS_2: On the reaction between MoS_2 and organic thiols. *Angew. Chem. Int. Ed.* **2016**, *55*, 5803–5808. [CrossRef]
52. Splendiani, A.; Sun, L.; Zhang, Y.; Li, T.; Kim, J.; Chim, C.Y.; Galli, G.; Wang, F. Emerging photoluminescence in monolayer MoS_2. *Nano Lett.* **2010**, *10*, 1271–1275. [CrossRef]
53. Eda, G.; Yamaguchi, H.; Voiry, D.; Fujita, T.; Chen, M.; Chhowalla, M. Photoluminescence from chemically exfoliated MoS_2. *Nano. Lett.* **2011**, *11*, 5111–5116. [CrossRef]
54. Qi, Y.; Wang, N.; Xu, Q.; Li, H.; Zhou, P.; Lu, X.; Zhao, G. A green route to fabricate MoS_2 nanosheets in water-ethanol-CO_2. *Chem. Commun.* **2015**, *51*, 6726–6729. [CrossRef]
55. Kim, D.H.; Park, Y.D.; Jang, Y.; Yang, H.; Kim, Y.H.; Han, J.I.; Moon, D.G.; Park, S.; Chang, T.; Chang, C.; et al. Enhancement of field-effect mobility due to surface-mediated molecular ordering in regioregular polythiophene thin film transistors. *Adv. Funct. Mater.* **2005**, *15*, 77–82. [CrossRef]
56. Kim, D.H.; Jang, Y.; Park, Y.D.; Cho, K. Surface-induced conformational changes in poly(3-hexylthiophene) monolayer films. *Langmuir* **2005**, *21*, 3203–3206. [CrossRef]
57. Huang, J.S.; Goh, T.; Li, X.; Sfeir, M.Y.; Bielinski, E.A.; Tomasulo, S.; Lee, M.L.; Hazari, N.; Taylor, A.D. Polymer bulk heterojunction solar cells employing Forster resonance energy transfer. *Nat. Photonics* **2013**, *7*, 479–485. [CrossRef]
58. Liu, Y.; Renna, L.A.; Page, Z.A.; Thompson, H.B.; Kim, P.Y.; Barnes, M.D.; Emrick, T.; Venkataramanan, D.; Russell, T.P. A polymer hole extraction layer for inverted perovskite solar cells from aqueous solutions. *Adv. Energy Mater.* **2016**, *6*, 1600664. [CrossRef]
59. Choi, H.; Kim, H.B.; Ko, S.J.; Kim, J.Y.; Heeger, A.J. An organic surface modifier to produce a high work function transparent electrode for high performance polymer solar cells. *Adv. Mater.* **2015**, *27*, 892–896. [CrossRef]
60. Noriega, R.; Rivnay, J.; Vandewal, K.; Koch, F.P.V.; Stingelin, N.; Smith, P.; Toney, M.F.; Salleo, A. A general

relationship between disorder, aggregation and charge transport in conjugated polymers. *Nat. Mater.* **2013**, *12*, 1038–1044. [CrossRef]

61. Lan, Y.K.; Huang, C. A theoretical study of the charge transfer behavior of the highly regioregular poly-3-hexylthiophene in the ordered state. *J. Phys. Chem. B* **2008**, *112*, 14857–14862. [CrossRef]
62. Skrypnychuk, V.; Boulanger, N.; Yu, V.; Hilke, M.; Mannsfeld, S.C.B.; Toney, M.F.; Barbero, D.R. Enhanced vertical charge transport in a semiconducting P3HT thin Film on single layer grapheme. *Adv. Funct. Mater.* **2015**, *25*, 664–670. [CrossRef]
63. Boulanger, N.; Yu, V.; Hilke, M.; Toney, M.F.; Barbero, D.R. In situ probing of the crystallization kinetics of rr-P3HT on single layer graphene as a function of temperature. *Phys. Chem. Chem. Phys.* **2017**, *19*, 8496–8503. [CrossRef]
64. Lim, J.W.; Wang, H.; Choi, C.H.; Quan, L.N.; Chung, K.; Park, W.T.; Noh, Y.Y.; Kim, D.H. Polyethylenimine ethoxylated interlayer-mediated ZnO interfacial engineering for high-performance and low-temperature processed flexible perovskite solar cells: A simple and viable route for one-step processed $CH_3NH_3PbI_3$. *J. Power Sources* **2019**, *438*, 226956. [CrossRef]

15

A Novel Method to Achieve Selective Emitter using Surface Morphology for PERC Silicon Solar Cells

Minkyu Ju [1], Jeongeun Park [1], Young Hyun Cho [2], Youngkuk Kim [2], Donggun Lim [1], Eun-Chel Cho [2,*] and Junsin Yi [2,*]

1 School of Electronic, Electrical Engineering, Korea National University of Transportation, Chungju 27469, Korea; mkju@ut.ac.kr (M.J.); ac1331@ut.ac.kr (J.P.); dglim@ut.ac.kr (D.L.)
2 School of Information and Communication Engineering, Sungkyunkwan University, Suwon 16419, Korea; yhcho64@skku.edu (Y.H.C.); bri3tain@skku.edu (Y.K.)
* Correspondence: echo0211@skku.edu (E.-C.C.); junsin@skku.edu (J.Y.)

Abstract: Recently, selective emitter (SE) technology has attracted renewed attention in the Si solar cell industry to achieve an improved conversion efficiency of passivated-emitter rear-contact (PERC) cells. In this study, we presented a novel technique for the SE formation by controlling the surface morphology of Si wafers. SEs were formed simultaneously, that is, in a single step for the doping process on different surface morphologies, nano/micro-surfaces, which were formed during the texturing processes; in the same doping process, the nano- and micro-structured areas showed different sheet resistances. In addition, the difference in sheet resistance between the heavily doped and shallow emitters could be controlled from almost 0 to 60 Ω/sq by changing the doping process conditions, pre-deposition and driving time, and temperature. Regarding cell fabrication, wafers simultaneously doped in the same tube were used. The sheet resistance of the homogeneously doped-on standard micro-pyramid surface was approximately 82 Ω/sq, and those of the selectively formed nano/micro-surfaces doped on were on 62 and 82 Ω/sq, respectively. As a result, regarding doped-on selectively formed nano/micro-surfaces, SE cells showed a J_{SC} increase (0.44 mA/cm^2) and a fill factor (FF) increase (0.6%) with respect to the homogeneously doped cells on the micro-pyramid surface, resulting in about 0.27% enhanced conversion efficiency.

Keywords: selective emitter; surface morphology; doping process; PERC; solar cell

---

1. Introduction

Presently, crystalline silicon (c-Si) solar cells are the leading product in the solar cell market, owing to their efficiency and low production cost [1,2], and this trend is expected to continue [3]. In terms of cell fabrication technology, passivated-emitter rear-contact (PERC) cells have become the mainstream product in the mass production of c-Si solar cells, owing to their high efficiency and cost-effectiveness [4–6]. According to an ITRPV (International Technology Roadmap for Photovoltaics) report, the market share of PERC technology in 2019 was over 30% and, according to estimates, it will exceed 60% by 2030 [3]. With continuous development over the past few years, PERC cell technology has already achieved a mass-production efficiency exceeding 21% [6]. Therefore, to further improve the efficiency, most solar cell companies are trying to develop new technologies and apply them to their production lines. The selective emitter (SE) technology has recently again attracted renewed attention, owing to its high efficiency [3–10].

SE technology has the following advantages. First, it has low sheet resistance because the high doping concentration under the electrodes lowers the contact resistance, leading to an improvement in the fill factor (FF). Second, better surface passivation on the high-sheet-resistance area enhances the short-circuit current density (J_{SC}) and open-circuit voltage (V_{OC}) [6–10]. There are several methods to create an SE, such as dopant-paste printing [11,12], inkjet printing [13,14], etch-back [15,16], and laser techniques [17–24]. However, most of these techniques are applied immediately before or after the doping process, so there are risks to the solar cell manufacturing process. As shown in Figure 1, SE technologies such as dopant-paste printing and inkjet printing are susceptible to contamination; in other words, there is a risk of contaminants spreading into the silicon during the high-temperature doping process [11–14]. Etch-back and etch-paste techniques result in the loss of reflectivity due to smoothing of the texture surfaces [15–17]. The laser doping method can damage the silicon surface, which is detrimental to the open-circuit voltage (V_{OC}) [18,19]. In addition, most SE technologies are likely to cause misalignment, which results in over-alignment losses in the alignment with the front electrode during the printing process [8–23]. Recently, as a new technology to overcome this, the development of a new SE to mitigate the recombination loss by the heavily doped epitaxial layer selectively grown after the passivation process, has been reported [24].

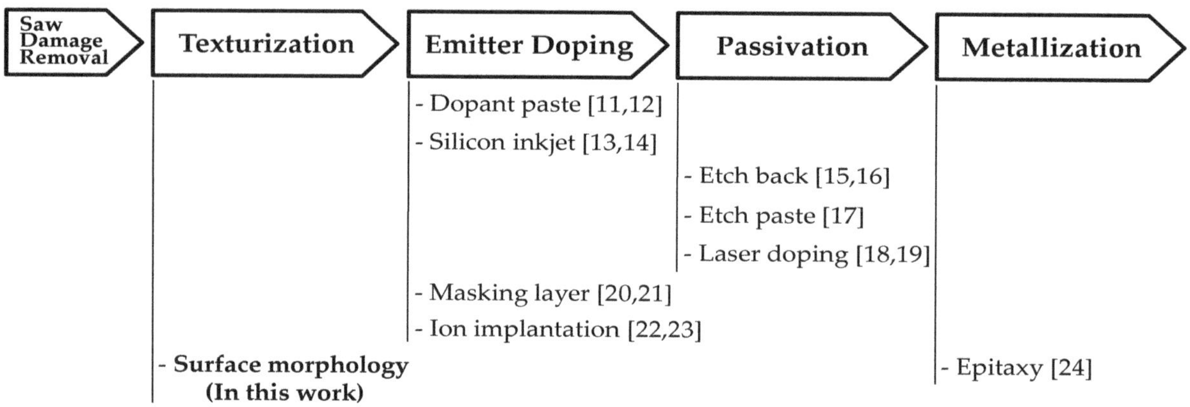

Figure 1. Comparison between the processes for selective emitter (SE) technologies used in our study and in previous studies [11–24].

In this study, a novel SE technique that uses selective nanosurface morphologies for a heavy emitter area was proposed before the pyramid texturing process [3–24]. A new SE technique that uses each different surface morphology was proposed. Even though the nano/micro-structured areas were doped in a single doping process, they exhibited different sheet resistances. In addition, the differences in sheet resistance could be controlled during the doping process by altering the temperature, pre-deposition time, and drive-in time. Importantly, using this SE technique could perform a sufficient cleaning process before the doping process, so there was less contamination risk to the solar cell manufacturing process. Furthermore, the loss from the mismatch between the heavily doped region and the front electrodes by screen-printing could be drastically reduced. Therefore, compared to conventional SE, there was less risk such as surface damage, return loss, and over-alignment loss.

2. Experimental Details

In this experiment, p-type boron-doped 6-inch single c-Si wafers with a resistivity of 1–1.5 ohm-cm and a thickness of 180 ± 20 μm were used. Initially, the c-Si wafers were immersed for 10 min at approximately 75 °C, in a mixed etching solution with a concentration of 5 wt% sodium hydroxide (NaOH) and 0.75 wt% sodium hypochlorite (NaOCl), to ensure saw damage removal (SDR). The wafers were rinsed in flowing DI water, followed by hydrochloric acid (HCl) and hydrofluoric acid (HF), to remove the metal ions and native oxides on the surface. Finally, the wafers were rinsed again in the flowing DI water and then dried.

2.1. Characteristics of Phosphorus Doping According to Nanosurface Morphology

To obtain the nanostructure on one side, two cleaned SDR wafers in contact with each other were placed on the etching carrier. The nanostructure texture was obtained through vapor-texturing using an etching solution of $HF:HNO_3$ = 7:3 volume ratio to form a uniform nanostructure on the surface [25]. The morphologies of the samples were controlled, and the shape and size of the surface nanostructure were modified during etching by varying the post-etching time between 0 and 3 min in an $HF:HNO_3:CH_3COOH$ isotropic etching solution with a volume ratio of (1:100:50), at room temperature. After vapor texturing, the samples were immersed for 10 min at 75 °C in the standard clean 1 (SC1) solution of $NH_4OH:H_2O_2:H_2O$ (1:1:5 volume ratio) to remove the chemicals remaining on the surface. Then, the samples were rinsed in flowing DI water followed by HF, to remove the native oxides on the surface. Finally, the wafers were rinsed again in flowing DI water and were dried. To form the micro-pyramid structure on the other surface of the samples, a 160-nm thickness silicon nitride (SiN_X) texturing barrier was deposited on the nanosurface using plasma-enhanced chemical vapor deposition (PECVD). Then, the remaining silicon particles on the surface were removed from the samples through SC1 cleaning. The samples were immersed for 30 min at 81–83 °C in a mixed solution of 2 wt% sodium hydroxide (NaOH), 6.25 wt% sodium silicate (Na_2SiO_3), and 12.5 vol% isopropyl alcohol (IPA) for micro-pyramid texturization [26]. For removal of the SiN_X texture-barrier film formed on the nanostructure, the samples were immersed for 10 min at room temperature in buffered HF (BHF, NH_4F:HF = 6:1). The samples were rinsed in flowing DI water, followed by HCl and HF, to remove the metal ions and native oxides on the surface. Finally, the wafers were rinsed again in flowing DI water and dried.

Phosphorus doping was performed at a temperature of 825 °C for 10 min in a quartz-tube furnace using $POCl_3$ for the formation of an emitter layer on the surface of the samples, with each sample having a different surface. To remove the phosphorus silicate glass (PSG) layer deposited on the surface, the sample was immersed in an HF solution for 30 s, and then washed with DI water and dried. The sheet resistances of the emitters formed according to the varied surface structure were measured at similar positions, at a uniform distance apart, through a four-point probe. The surface morphology of samples with varied nanostructure was observed using scanning electron microscope (SEM) images.

2.2. *Characteristics of Nanosurface Emitter Layer According to Doping Condition*

To observe the variation in sheet resistance according to the phosphorus doping condition in the variable surface morphology, wafers formed on both sides with a nanostructure and a micro-pyramid structure were used. The prepared samples were cleaned with HCl and HF, rinsed with flowing DI water, and dried. The cleaned samples were processed in a quartz-tube furnace using $POCl_3$ at a temperature range of 805–860 °C for 10 min, to ensure a varied sheet resistance for the shallow emitter on the surface. Subsequently, the samples were processed in a quartz-tube furnace using $POCl_3$ for 10–30 min at 860 °C, to ensure a varied sheet resistance for the deep emitter on the surface. After removing the PSG layer by using HF, the sheet resistances of the emitter formed on the nano- and micro-pyramid structures were measured and analyzed. The analysis of the number of phosphorus atoms in the emitter with varied surface morphology was performed using a CAMECA IMS 7f magnetic-sector secondary ion mass spectroscope (SIMS) (CAMECA in Gennevilliers, France).

2.3. *Fabrication of Novel SE Solar Cells Using Surface Morphology*

To analyze the characteristics of the solar cells according to their surface morphology, they were designed to have a nanostructure (nano), micro-pyramid structure (pyramid), and nano-micro-SE structure (selective). Initially, all the fabricated samples had a uniform nanostructure on the front side from the cleaned SDR by vapor-texturing [25]. For the sample with the nano-micro-SE structure, a SiN_X texturing barrier was deposited on the front side using PECVD. The front texture barrier was patterned using the acid etching resistance through screen-printing using a front-grid mask with a line width of about 100 μm, and spacing of 1.312 mm. Then, the sample was immersed for 10 min at room temperature, and BHF was used to pattern the surface by removing the SiN_X texture barrier. All the samples were immersed for 10 min at 75 °C in an SC1 solution to remove the remaining chemicals on the surface. Then, the acid etching resistance on the surface was removed through SC1 cleaning. The two types of samples were immersed for 30 min at 81–83 °C in a solution of 2 wt% NaOH, 6.25 wt% Na_2SiO_3, and 12.5 vol% IPA for micro-pyramid texturization [26]. Two types of samples were patterned with the nano–micro-SE structure and micro-pyramid structure. To remove the SiN_X texture barrier film of the patterned nano–micro-SE structure of the samples, these were immersed in BHF for 10 min at room temperature. All the samples were rinsed in flowing DI water, followed by HCl and HF, to remove the metal ions and native oxides on the surface. Finally, the wafers were rinsed again in flowing DI water and dried. The surface morphology of the patterned nano- and micro-pyramid structures was observed using SEM. The emitter layer was formed in a quartz tube using $POCl_3$ at 830 °C for 10 min. The PSG layer was removed and etched on the rear side using the InOxide facility of RENA Technologies GmbH.

For the formation of standardized PERC structures, the front side was passivated using PECVD-SiN_X (single layer with refractive index 2.05), and the rear side was passivated with a stacked layer of atomic layer deposition aluminum oxide (ALD-Al_2O_3) and PECVD-SiN_X. Line-shaped laser contact openings (LCO) were formed on the rear side using a picosecond laser. The Ag front grid metal was printed with an aligned nano-micro-SE patterned with Ag paste. The grid pattern design of the front electrode had 40-μm finger width and 1.312-mm spacing. The rear side was screen-printed on the whole area with a commercially available Al paste, and a furnace firing step completed the PERC cell process. To confirm the Ag printing characteristics of the front electrode formed by each surface structure, the shape of the electrode was observed using an optical microscope (OM), and the line width was measured. The current density-voltage (J–V) measurement of the solar cell fabricated by each surface structure was analyzed under the standard condition of AM 1.5 G (100 mW/cm^2) at 25 °C. The internal quantum efficiency was measured in the wavelength range of 300 to 1100 nm using the QEX7 IPCE(Incident-Photon-to-electron Conversion Efficiency) system (PV Measurements, Inc. in Washington, WA, USA).

3. Results

3.1. *Characteristics of Phosphorus Doping According to Nanosurface Morphology*

A nanostructured surface with an approximate size of 50 nm was formed through chemical vapor texturing of the cleaned SDR wafers, as shown in Figure 2a,e. Further post-etching of the nanostructured surface for immersion durations of 1, 2, and 3 min changed the sizes of the nanostructures to approximately 100, 200, and 300 nm, respectively. The SEM images in Figure 2a–d show the changed size of the nanostructures formed after vapor-texturing with post-etching. Figure 2e,f shows the cross-section of nano-vapor texturing and micro-pyramid texturing.

In Figure 3, it can be observed that the average emitter sheet resistance for the micro-pyramid structure surface was 98.5 Ω/sq. At this time, for the 50-nm nanostructure surface (shown in Figure 2a,e), the emitter sheet resistance was 61.1 Ω/sq. By changing the nanostructure size to approximately 100, 200, and 300 nm, the emitter sheet resistance was observed to increase to 82.0, 86.4, and 88.8 Ω/sq. As can be observed in Figure 3, owing to the different factors, the sheet resistance for the micro-pyramid structure was 0.64 (nano-vapor texture) for the 50-nm size, 0.81 (post-etching: 1 min) for the 100-nm size, 0.85 (post-etching: 2 min) for the 200-nm size, and 0.88 (post-etching: 3 min) for the 300-nm size. The difference factor index was gradually increased to 1, which was the reference value for the micro-pyramid structure as the increased size of the nanostructure. As a result, the difference factor index for the novel nano-micro-SE structure, being lower than 1, was better.

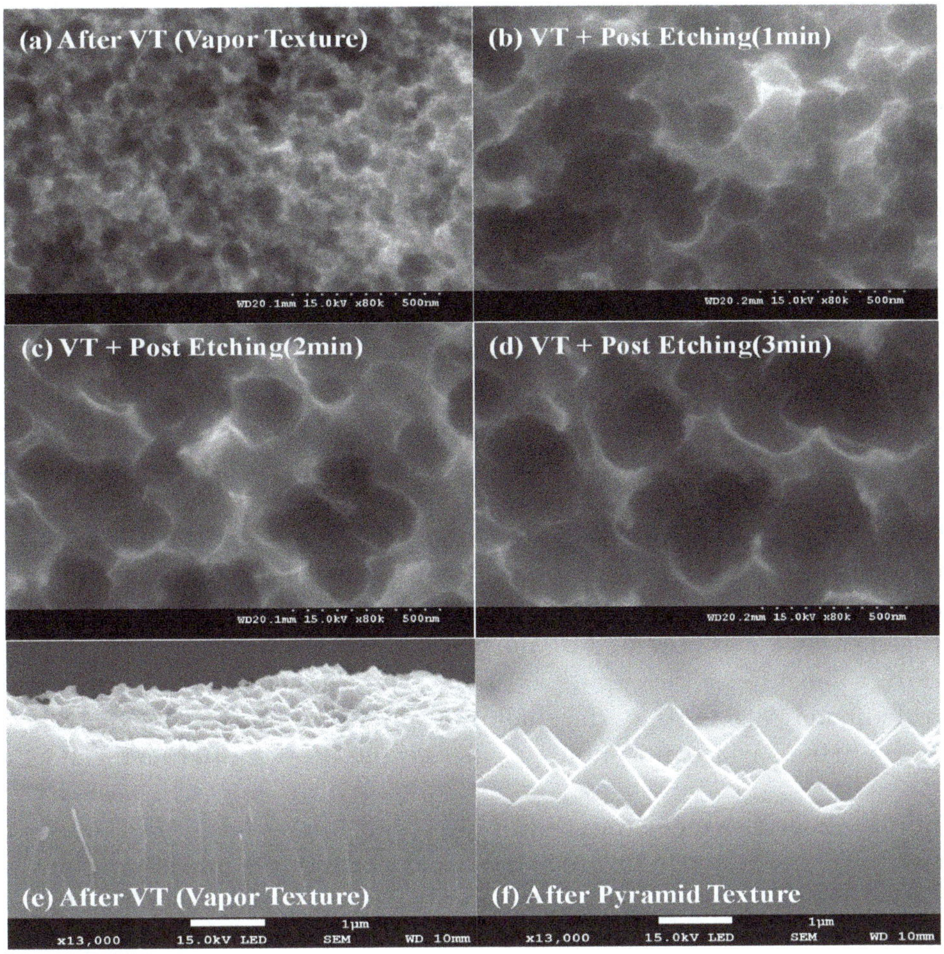

Figure 2. Changes in surface morphology according to nanostructure of vapor texture (**a**) and after post-etching for 1 (**b**), 2 (**c**), and 3 min (**d**); cross-section of nanostructure from the vapored-texture process (**e**); cross-section of micro-pyramid structure from the pyramid textured process (**f**).

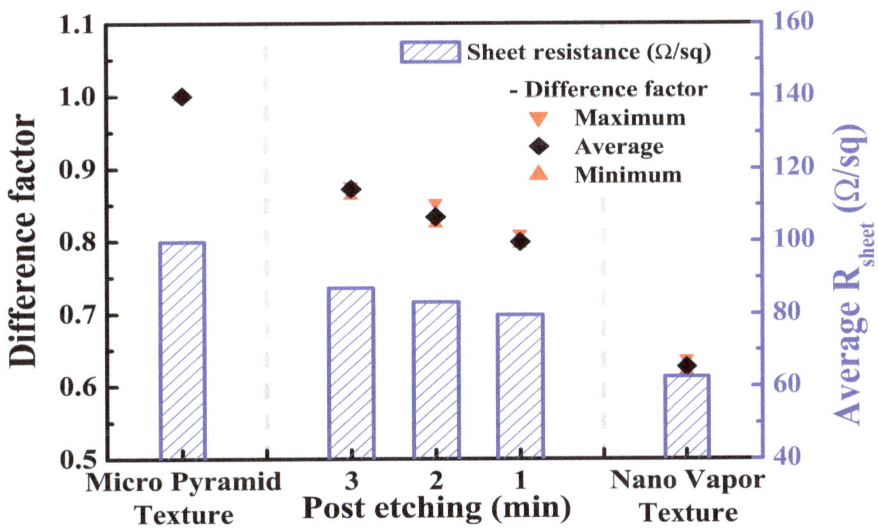

Figure 3. Change in emitter sheet resistance after phosphorus doping process, according to the variation in surface morphology.

3.2. Characteristics of Phosphorus Doping According to Nanosurface Morphology

The samples, with approximately 50-nm size nanostructure on one side and micro-pyramid on the other side, were analyzed with the difference factor index of surface morphology through a doping process about the variation of temperature and time from a high sheet resistance to low sheet resistance. Figure 4 showed the results of the difference factor index of surface morphology by the variation in sheet resistance. For the deep emitter doping conditions, during 30 min at 860 °C, the average sheet resistances obtained for the pyramid-structured and nanostructured surfaces were 40 and 39 Ω/sq, respectively. In this time, the difference factor index for the 50 nm nanostructured- and micro-pyramid-structured surfaces was 0.97, which was attributed to almost equal sheet resistance characteristics for both surfaces. Figure 4 shows the results, and it is observed that as the sheet resistance increased by the doping process condition from a low sheet resistance, the difference factor index became smaller and the SE characteristic was increased. The best difference factor index region of the SE characteristic was when the doping condition with the sheet resistance for the pyramid structure surface was approximately 100 to 150 Ω/sq, and for the nanostructured surface, the range was about 74 to 122 Ω/sq with the difference factor converged to the level of 0.76. This result was compared with the difference factor index of 0.64 shown in Figure 3. The nano-vapor texture was predicted to result from some large-sized nanostructure. Further, sheet resistance of more than 150 Ω/sq on the pyramid-structured surface was obtained by the doping process condition, and the difference factor index increased for a sheet resistance up to 360 Ω/sq. The experimental results have shown a sheet resistance of approximately 360 Ω/sq for the shallow emitter, and limited change occurred in the sheet resistance according to the surface morphology, whereas the deep emitter had a sheet resistance of 40 Ω/sq.

The variation in sheet resistance for the doping condition between the nanostructures and the micro-pyramid structure (as shown in Figure 3) indicated a reduced to maximum sheet resistance in the range of about 100 to 150 Ω/sq by the size of the nanostructure. It can be seen that the SE characteristic was maximized in this doping region. Shown in Figure 4, the sheet resistance of the "A" sample as the shallow emitter and the "B" sample as the general emitter were analyzed by SIMS to investigate the factors of change in sheet resistance of the nano- and micro-surface structure according to the doping condition.

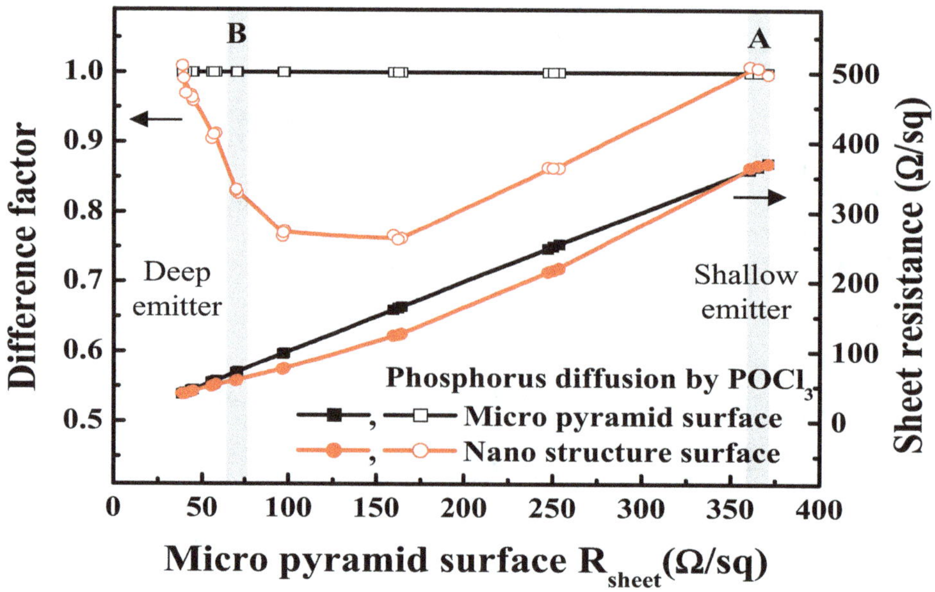

Figure 4. Change in emitter sheet resistance of micro-pyramid structure and nanostructure, according to the doping process condition.

The SIMS result in Figure 5 indicated that the analysis of the surface morphology of the sample can be measured by the change in silicon (Si) intensity and phosphorus (P) concentration according to the sputter time. On the surface of the nanostructures shown in Figure 5a,c, the signal of the initial Si was low, and it was detected as 3×10^3 cps level. After that, it increased rapidly and then stabilized at 6×10^3 cps level over 300 s, and an intensity of 7×10^3 cps level was detected over 550 s. On the surface of the micro-pyramid structure shown in Figure 5b,d, the low signal of the initial Si was higher, and it was detected at the level of 6×10^3 and 4×10^3 cps compared with the nanosurface. It can be observed that it is relatively easier to detect Si than the nanostructure surface. In other words, the doped silicon surface of the nanostructure morphology was coated with many dopant materials, more than the micro-pyramid surface.

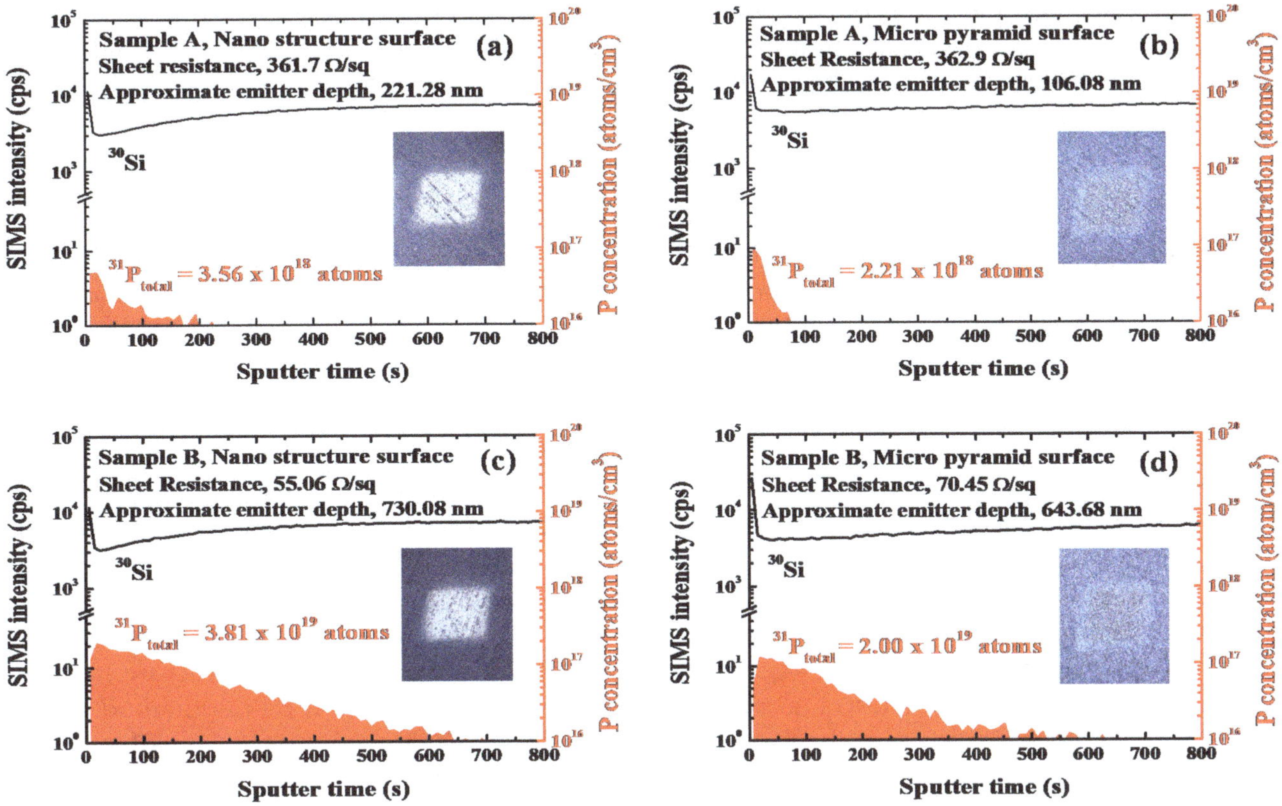

Figure 5. Secondary ion mass spectroscope (SIMS) analysis of surface morphology according to the doping process condition, (**a**) nanostructure surface in shallow-doping condition A, (**b**) micro-pyramid structure surface in shallow-doping condition A, (**c**) nanostructure surface in traditional-doping condition B, (**d**) micro-pyramid structure surface in traditional-doping condition B.

The results of the analysis of the P concentration by depth to SIMS analysis did not show accurate values due to the shape of the surface structure [27]. However, the results of that detected by SIMS are valid in the comparative analysis of the total amount of P in the pyramid structure and the nanostructure. In Figure 5a, for the "A" sample of nanostructure morphology, the total amount of P detection concentration is 3.56×10^{18} atoms/cm^3 at a sheet resistance of 361.7 Ω/sq. Regarding the micro-pyramid morphology of the "A" sample, the total amount of P detection concentration was 2.21×10^{18} atoms/cm^3 and the sheet resistance was 362.9 Ω/sq, as shown in Figure 5b. In the similar sheet resistance, approximately 360 Ω/sq by shallow doping was the highest total amount of detected concentration of P for the nanostructure morphology compared with the micro-pyramid structure. This result is due to the fact that many dopant materials were inactive on the large surface area of the nanostructure surface by the shallow doping process [28].

In Figure 5c, for the "B" sample with the SE characteristic of nanostructure morphology, the total amount of P detection concentration was 3.81×10^{19} atoms/cm^3 at a sheet resistance of 55.06 Ω/sq. For the micro-pyramid surface morphology of "B" samples, the total amount of P detection concentration was 2.00×10^{19} atoms/cm^3 and sheet resistance 70.45 Ω/sq, as shown in Figure 5d. Even under the doping conditions of "B" samples with the SE characteristic, the total amount of P doped with the surface characteristics of the nanostructure was higher than that of the pyramid structure. In deep doping conditions, the emitter layer became deeper, and the emitter from the deeper part of the nanostructure was fused [29,30]. As a result, for the deeper emitter doping conditions, it had a similar sheet resistance of approximately 40 Ω/sq and a difference factor index of approximately 1 for the nanostructured surface and the micro-textured surface, as shown in Figure 4.

3.3. Mechanism of SE Formation by Nano–Micro-Morphology

The results of sheet resistance analysis in Figures 3 and 4 clearly shows that the best SE structure had a smaller size of the nanostructure, and the doping process condition for the sheet resistance region on the pyramid structure was 100–150 Ω/sq.

Figure 6 shows the SE mechanism in which nanostructures and micro-pyramid structures were formed through the doping process. The local nanostructures for SEs were formed at the texture processing stage with pyramidal structures. The formed textured surface structure had two surface structures simultaneously, as shown in Figure 6a. To produce emitter layer formation, the phosphorus silicate glass (PSG) layer was formed on the surface of the silicon by a pre-deposition process of $POCl_3$ doping in a quartz tube furnace, as shown in Figure 6b. The report of Catherine et al. analyzes in detail the thickness of the formed PSG layer and the depth of the emitter formed in the peaks, valleys, and flank of the pyramid structure during the doping process [31]. It can be seen from the literature that the PSG layer in the valley position "A" is formed relatively thick in the pyramid structure compared to the peak and flank as shown in Figure 6b. In the nanostructure, it can be understood that the thick PSG layer formed more than the pyramid structure thick PSG layer because the valley position "A" had more frequency than the pyramid structure with respect to the same area. As a result, the doped silicon surface of the nanostructure morphology was coated with many dopant materials more than the surface of the micro-pyramid. Dastgheib-Shirazi et al. reported an analysis of the depth variation of the thermally diffused emitter according to the thickness of the PSG layer formed in the doping process [32]. It can be seen from the literature that the thick PSG layer formed a higher surface concentration of phosphorus, forming a deeper emitter layer through the thermal diffusion of the doping process, as shown in Figure 6c. Moreover, it can be seen from the literature that deeper emitters for the micro-pyramid structure were formed in the peak "D" than in the valleys "B" and flanks "C" [31]. In the nanostructure region, it can be understood that the sheet resistance was lower than that in the micro-pyramid structure formed by a higher frequency structure valley "A" of a thicker PSG layer and the deep diffused layer of peak "D" through the emitter doping process. In conclusion, the surface morphology of nanostructure and micro-pyramid structure was through the conventional thermal diffusion emitter doping process becoming the novel SE structure as shown in Figure 7.

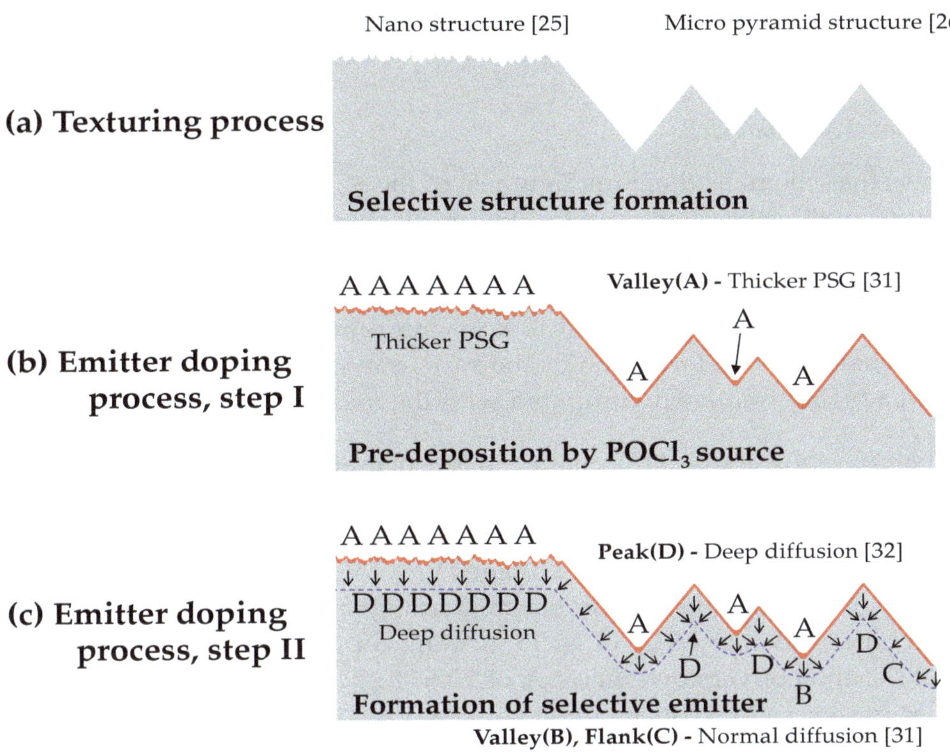

Figure 6. Formation mechanism of the novel selective emitter (SE) according to surface morphology, (**a**) selective structure formation on texturing process [25,26], (**b**) phosphorus pre-deposition on doping process [31], (**c**) phosphorus diffusion on doping process [31,32].

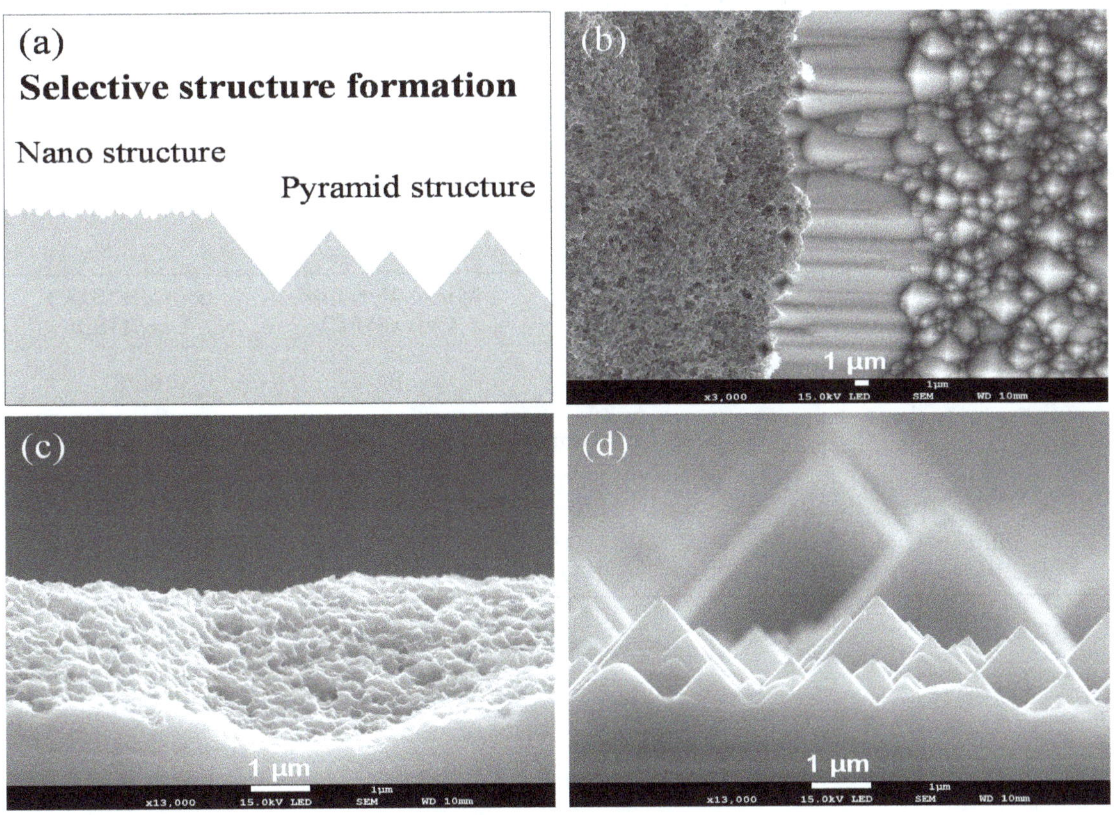

Figure 7. SEM image of the novel SE surface using a nanostructure and a micro-pyramid structure, (**a**) schematic diagram of the novel SE structure, (**b**) top view of the novel SE surface in 3 K magnification,

(**c**) cross-section view of the nanostructure side in 13 K magnification, (**d**) cross-section view of the micro-pyramid structure side in 13 K magnification. Included in all is scale bar of 1 μm.

3.4. Fabrication of Novel SE Solar Cell.

Figure 7 shows the schematic diagram Figure 7a of the SE using a nano-micro-pyramid structure and the SEM analysis of the fabricated samples. Figure 7b shows the surface of the novel SE structure sample analyzed through SEM at 3 K magnification on which the nanostructure and micro-pyramid structure were formed together. Figure 7c,d show the enlarged SEM result of the cross-section of the nanostructure and micro-pyramid structure at 13 K magnification. In the nanostructure of Figure 7c, as described in the mechanism of SE formation by the surface morphology of Figure 6, the valley and the peak observed had a higher frequency compared with the micro-pyramid structure in Figure 7d.

Table 1 and Figure 8 show the best J–V characteristics of the solar cells fabricated with nanostructure cells (NSC), micro-pyramid structure cells (MPC), and the novel SE structure cells (SEC), respectively. The J–V curve results from the NSC emitter had a sheet resistance of 62 Ω/sq, short-circuit current density (J_{SC}) of 39.81 mA/cm^2, open-circuit voltage (V_{OC}) of 638 mV, fill factor (FF) of 79.82%, and conversion efficiency of 20.27%. The J–V curve results from the MPC emitter had sheet resistance of 82 Ω/sq, short-circuit current density (J_{SC}) of 40.35 mA/cm^2, open-circuit voltage (V_{OC}) of 647 mV, fill factor (FF) of 79.00%, and conversion efficiency of 20.61%. The NSC showed a short-circuit current density (J_{SC}) of 0.54 mA/cm^2 and an open-circuit voltage (V_{OC}) of 9 mV, which was degraded compared to that of the MPC. As shown in this result, it was difficult to emitter passivate the nanostructure surface compared to the micro-pyramid surface [29]. However, the NSC showed 0.82% improvement in the fill factor (FF) compared to the MPC. The improved fill factor (FF) in the NSC can be seen in the literature as an improved series resistance by electrode contact with the sheet resistance of the front emitter [33]. Despite improvements in the fill factor (FF) of the NSC, the degradation of efficiency characteristics due to short circuit current density (J_{SC}) and open-circuit voltage (V_{OC}) was due to the low sheet resistance characteristics and higher recombination characteristics [29].

Table 1. Best and average solar cell characteristics according to surface structure with a comparison of sheet resistance.

Light J–V Analysis	**Nanostructure Cell (NSC)**		**Micro-Pyramid Cell (MPC)**		**SE Cell (SEC) Low/High**	
Sheet Resistance of Emitter (Ω/sq)	62		82		62/82	
Cell remark (9 samples fabricated)	Best	Average	Best	Average	Best	Average
J_{SC} (mA/cm^2)	39.81	39.07	40.35	40.24	40.79	40.63
V_{OC} (V)	0.638	0.637	0.647	0.643	0.643	0.642
Fill Factor (%)	79.82	80.18	79.00	78.61	79.60	79.29
Efficiency (%)	20.27	19.94	20.61	20.35	20.88	20.67

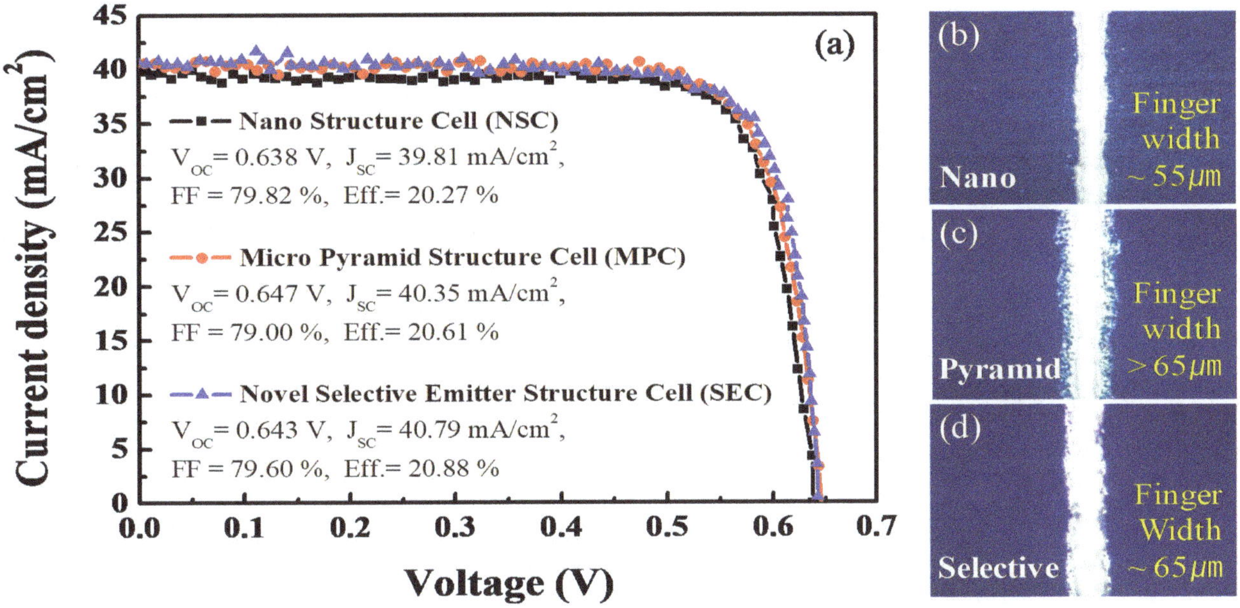

Figure 8. Light density–voltage (J–V) characteristic analysis of crystalline silicon (c-Si) solar cells according to surface morphology and printed front Ag electrode shape, (**a**) result of light J–V, (**b**) optical microscope (OM) image of printed front Ag electrode shape on nanostructured solar cell surface, (**c**) OM image of printed front Ag electrode shape on micro-pyramid-structured solar cell surface, (**d**) OM image of printed front Ag electrode shape on the novel SE-structured solar cell surface.

The J–V curve results for the novel SE structure cell (SEC) using nanostructure and micro-pyramid structure had a current density (J_{SC}) of 40.79 mA/cm^2, open-circuit voltage (V_{OC}) of 643 mV, fill factor (FF) of 79.60%, and conversion efficiency of 20.88%. The improvement in the SEC efficiency of the selective surface structure emitter was analyzed by an FF increase (0.6%) and a J_{SC} increase (0.44 mA/cm^2) with respect to those of the MPC. The improvement in the fill factor (FF) was also observed in the internal quantum efficiency (IQE) results of Figure 9 due to the low contact resistance of the nanostructure electrode contacts.

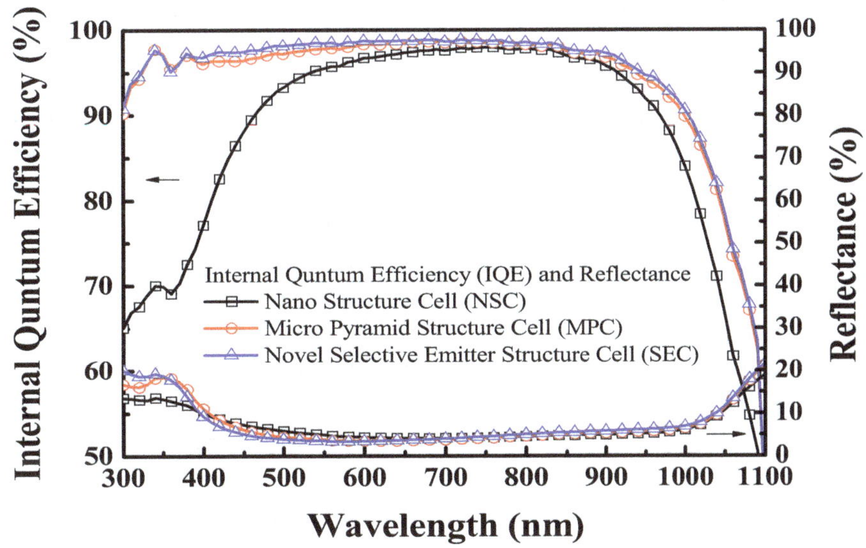

Figure 9. Result of reflectance and internal quantum efficiency (IQE) of c-Si solar cells according to surface morphology.

In the 300 to 400 nm region observed as the recombination characteristics of the surface emitter, similar analytical results as the micro-pyramid structure were shown, but improved IQE of average

0.56% and maximum 1.1% in the wavelength range of 400 to 1100 nm were obtained. The improvement in the IQE shows the characteristic of decreasing the series resistance, which also affects the improvement in the fill factor (FF) [34]. The improvement in the short circuit current density (J_{SC}) is confirmed by the improvement in the shading loss of the screen-printing electrode according to the surface structure of Figure 8b–d. The screen-printed electrode lines on the nanostructure surface showed a clean edge electrode shape of about 55 μm shown in Figure 8b. In the micro-pyramid structure of Figure 8c, the screen-printed electrode line width was approximately 65 μm or more by electrode spreading. It can be seen from the literature that the short-circuit current density was lowered due to the shading loss by the electrode spreading caused by the action of the Ag particle size of the screen-printing electrode and the micro-pyramid structure size [35]. As shown in Figure 8d, the printed electrode line width of the selective surface structure was approximately 65 μm and a clean edge electrode shape was observed. The shape of the clean edge electrode by the selective surface structure minimized the shading loss and could obtain improved characteristics of short circuit current density (J_{SC}) compared to a micro-pyramid structure.

In this study, we fabricated a new SE solar cell using the final nanostructure and micro-pyramid structure. This was performed by controlling the difference in sheet resistance in the doping process together with the pyramid structure formed in the texturing process using the nanostructure formed before the texturing process. As the size of the nanostructure became smaller, the emitter sheet resistance of the doped sample became smaller. The difference in sheet resistance between the nanostructure and the pyramid structure was found to be the largest in the range of the emitter doping condition of 100 to 150 Ω/sq at the micro-pyramid structure. These experimental facts have been confirmed by the SIMS analysis and the literature, which show that the nanostructure had more frequent valleys and peaks than that of the micro-pyramid structure. As a result of the characteristics of the solar cell having the novel SE using the nanostructure, the efficiency of 20.88% was improved by about 0.27% compared to the conventional pyramid structure with the efficiency of 20.61%. This improvement was obtained by improving the fill factor (FF) through the contact of the electrode with the low sheet resistance surface on the nanostructure. It also obtained an improvement in the short circuit current density (J_{SC}) by minimizing the shadow loss of the electrode printed on the nanostructure surface.

4. Conclusions

In this study, a novel SE technique that uses selective nanosurface morphologies for a heavy emitter area was proposed before the pyramid texturing process [3–24]. The SE technology, which used different surface morphologies, exhibited different sheet resistances despite the nano/micro-structure regions being doped with a single doping process. It was also possible to control the difference in sheet resistance during the doping process by changing the temperature, pre-deposition time, and drive-in time. The result of the uniformly doped sheet resistance on the standard micro-pyramidal surface was about 82 Ω/sq, and the doping results of the selectively formed nano/micro-surface were 62 and 82 Ω/sq, respectively. As a result, doped-on selectively formed nano/micro-surfaces SE cells showed a J_{SC} increase (0.44 mA/cm^2) and an FF increase (0.6%) compared to the homogeneously doped cells on the micro-pyramid surface, which resulted in about 0.27% enhanced conversion efficiency.

The novel SE technology that uses nano/micro-structures with similar form to the selective epitaxial layer growth SE technology blocked the risk of solar cell manufacturing processes such as contamination, surface damage, reflectance loss, and over-alignment loss. At each different height location, the nanostructure and the micro-pyramid structure were formed on the silicon surface. Therefore, alignment loss could be minimized by easily aligning the front printed electrode to the high-level location region on the nanosurface.

Author Contributions: Conceptualization, M.J.; methodology, M.J.; validation, Y.H.C., Y.K., and E.-C.C.; formal analysis, M.J., J.P., and Y.H.C.; investigation, M.J.; data curation, M.J., J.P., and Y.H.C.; writing—original draft preparation, M.J. and Y.K.; writing—review and editing, M.J., Y.K., and E.-C.C.; supervision, D.L., E.-C.C., and J.Y.;

project administration, E.-C.C.; funding acquisition, D.L, E.-C.C., and J.Y. All authors have read and agreed to the published version of the manuscript.

References

1. Green, M.A. The path to 25% silicon solar cell efficiency: History of silicon cell evolution. *Prog. Photovolt. Res. Appl.* **2009**, *17*, 183–189. [CrossRef]
2. Swanson, R.M. A vision for crystalline silicon photovoltaics. *Prog. Photovolt. Res. Appl.* **2006**, *14*, 443–453. [CrossRef]
3. ITRPV 2020. *International Technology Roadmap for Photovoltaic (ITRPV) Eleventh Edition—2019 Results*; ITRPV: Frankfurt, Germany, 2020.
4. Gassenbauer, Y.; Ramspeck, K.; Bethmann, B.; Dressler, K.; Moschner, J.D.; Fiedler, M.; Brouwer, E.; Drößler, R.; Lenck, N.; Heyer, F.; et al. Rear-Surface Passivation Technology for Crystalline Silicon Solar Sells: A Versatile Process for Mass Production. *IEEE J. Photovolt.* **2013**, *3*, 125–130. [CrossRef]
5. Green, M.A. The passivated emitter and rear cell (PERC): From conception to mass production. *Sol. Energy Mater. Sol. Cells* **2015**, *143*, 190–197. [CrossRef]
6. Chunduri, S.K.; Schmela, M. High Efficiency Cell Technologies 2019 From PERC to Passicated Contacts and HJT. *Taiyang News*, 2019.
7. Min, B.; Wagner, H.; Müller, M.; Neuhaus, H.; Brendel, R.; Altermatt, P.P. Incremental efficiency improvements of mass-produced PERC cells up to 24%, predicted solely with continuous development of existing technologies and wafer materials. In Proceedings of the 31st European Photovoltaic Solar Energy Conference and Exhibition, Hamburg, Germany, 14–18 September 2015; pp. 473–476.
8. Hahn, G. Status of selective emitter technology. In Proceedings of the 25th European Photovoltaic Solar Energy Conference and Exhibition and 5th World Conference on photovoltaic Energy Conversion, Munich, Germany, 6–10 September 2010; pp. 1091–1096.
9. Rahman, M.Z. Status of selective emitters for p-type c-Si solar cells. *Opt. Photonics J.* **2012**, *2*, 129–134. [CrossRef]
10. Lv, Y.; Zhuang, Y.F.; Wang, W.J.; Wei, W.W.; Sheng, J.; Zhang, S.; Shen, W.Z. Towards high-efficiency industrial p-type mono-like Si PERC solar cells. *Sol. Energy Mater. Sol. Cells* **2020**, *204*, 110202. [CrossRef]
11. Tonini, D.; Borrosso, C.; Cellere, G.; Furin, V.; Galiazzo, M.; Kumar, P.; Tanner, D.; Voltan, A. Efficiency gain in c-Si cells through selective emitter and double printing. *Energy Procedia* **2011**, *8*, 598–606. [CrossRef]
12. Horzel, J.; Szlufcik, J.; Nijs, J.; Mertens, R. A simple processing sequence for selective emitters [Si solar cells]. In Proceedings of the Conference Record of the Twenty Sixth IEEE Photovoltaic Specialists Conference, Anaheim, CA, USA, 29 September–3 October 1997; pp. 139–142.
13. Antoniadis, H. Silicon ink high efficiency solar cells. In Proceedings of the 34th IEEE Photovoltaic Specialists Conference, Philadelphia, PA, USA, 7–12 June 2009; pp. 000650–000654.
14. Zhong, S.; Shen, W.; Liu, F.; Li, X. Mass production of high efficiency selective emitter crystalline silicon solar cells employing phosphorus ink technology. *Sol. Energy Mater. Sol. Cells* **2013**, *117*, 483–488. [CrossRef]
15. Lauermann, T.; Book, F.; Dastgheib-Shirazi, A.; Hahn, G.; Haverkamp, H.; Bleidiessel, R.; Fleuster, M. The optimal choice of the doping levels in an inline selective emitter design for screen printed multicrystalline silicon solar cells. In Proceedings of the 24th European Photovoltaic Solar Energy Conference and Exhibition, Hamburg, Germany, 21–24 September 2009; pp. 1795–1797.
16. Basu, P.K.; Cunnusamy, J.; Sarangi, D.; Boreland, M.B. Novel selective emitter process using non-acidic etch-back for inline-diffused silicon wafer solar cells. *Renew. Energy* **2014**, *66*, 69–77. [CrossRef]
17. Book, F.; Raabe, B.; Hahn, G. Two diffusion step selective emitter: Comparison of mask opening by laser or etching paste. In Proceedings of the 23rd European Photovoltaic Solar Energy Conference and Exhibition, Valencia, Spain, 1–5 September 2008; pp. 1546–1549.
18. Engelhardt, J.; Kromer, H.; Hahn, G.; Terheiden, B. Laser doping from as-deposited CVD layers for high-efficiency crystalline silicon solar cells. In Proceedings of the Silicon PV 2019, the 9th International Conference on Crystalline Silicon Photovoltaics, Leuven, Belgium, 8–10 April 2019; Volume 2147, p. 070002.
19. Kim, M.; Kim, D.; Kim, D.; Kang, Y. Influence of laser damage on the performance of selective emitter solar cell fabricated using laser doping process. *Sol. Energy Mater. Sol. Cells* **2015**, *132*, 215–220. [CrossRef]

20. Raabe, B.; Haverkamp, H.; Book, F.; Dastgheib-Shirazi, A.; Moll, R.; Hahn, G. Monocrystalline silicon: Future cell concepts. In Proceedings of the 22nd European Photovoltaic Solar Energy Conference and Exhibition, Milano, Italy, 3–7 September 2007; pp. 1024–1029.
21. Pal, B.; Ray, S.; Gangopadhyay, U.; Ray, P.P. Novel technique for fabrication of n-type crystalline silicon selective emitter for solar cell processing. *Mater. Res. Express* **2019**, *6*, 075523. [CrossRef]
22. Low, R.; Gupta, A.; Bateman, N.; Ramappa, D.; Sullivan, P.; Skinner, W.; Mullin, J.; Peters, S.; Weiss-Wallrath, H. High efficiency selective emitter enabled through patterned ion implantation. In Proceedings of the 35th IEEE Photovoltaic Specialists Conference, Honolulu, HI, USA, 20–25 June 2010; pp. 001440–001445.
23. Dubé, C.E.; Tsefrekas, B.; Buzby, D.; Tavares, R.; Zhang, W.; Gupta, A.; Low, R.J.; Skinner, W.; Mullin, J. High efficiency selective emitter cells using patterned ion implantation. *Energy Procedia* **2011**, *8*, 706–711. [CrossRef]
24. Payo, M.R.; Li, Y.; Russell, R.; Singh, S.; Filipek, I.K.; Duerinckx, F.; Szlufcik, J.; Poortmans, J. Efficiency gain in plated bifacial n-type PERT cells by means of a selective emitter approach using selective epitaxy. *Sol. Energy Mater. Sol. Cells* **2020**, *204*, 110173. [CrossRef]
25. Ju, M.; Balaji, N.; Lee, Y.; Park, C.; Song, K.; Choi, J.; Yi, J. Novel vapor texturing method for EFG silicon solar cell applications. *Sol. Energy Mater. Sol. Cells* **2012**, *107*, 366–372. [CrossRef]
26. Ju, M.; Balaji, N.; Park, C.; Nguyen, H.T.T.; Cui, J.; Oh, D.; Jeon, M.; Kang, J.; Shim, G.; Yi, J. The effect of small pyramid texturing on the enhanced passivation and efficiency of single c-Si solar cells. *RSC Adv.* **2016**, *6*, 49831–49838. [CrossRef]
27. Komatsu, Y.; Harata, D.; Schuring, E.W.; Vlooswijk, A.H.G.; Katori, S.; Fujita, S.; Venema, P.R.; Cesar, I. Calibration of electrochemical capacitance-voltage method on pyramid texture surface using scanning electron microscopy. *Energy Procedia* **2013**, *38*, 94–100. [CrossRef]
28. Li, H.; Ma, F.; Hameiri, Z.; Wenham, S.; Abbott, M. On elimination of inactive phosphorus in industrial $POCl_3$ diffused emitters for high efficiency silicon solar cells. *Sol. Energy Mater. Sol. Cells* **2017**, *171*, 213–221. [CrossRef]
29. Kafle, B.; Schön, J.; Fleischmann, C.; Werner, S.; Wolf, A.; Clochard, L.; Duffy, E.; Hofmann, M.; Rentsch, J. On the emitter formation in nanotextured silicon solar cells to achieve improved electrical performances. *Sol. Energy Mater. Sol. Cells* **2016**, *152*, 94–102. [CrossRef]
30. Es, F.; Baytemir, G.; Kulakci, M.; Turan, R. Metal-assisted nano-textured solar cells with SiO_2/Si_3N_4 passivation. *Sol. Energy Mater. Sol. Cells* **2017**, *160*, 269–274. [CrossRef]
31. Voyer, C.; Buettner, T.; Bock, R.; Biro, D.; Preu, R. Microscopic homogeneity of emitters formed on textured silicon using in-line diffusion and phosphoric acid as the dopant source. *Sol. Energy Mater. Sol. Cells* **2009**, *93*, 932–935. [CrossRef]
32. Dastgheib-Shirazi, A.; Steyer, M.; Micard, G.; Wagner, H.; Altermatt, P.P.; Hahn, G. Relationships between diffusion parameters and phosphorus precipitation during the $POCl_3$ diffusion process. *Energy Procedia* **2013**, *38*, 254–262. [CrossRef]
33. Shanmugam, V.; Cunnusamy, J.; Khanna, A.; Basu, P.K.; Zhang, Y.; Chen, C.; Stassen, A.F.; Boreland, M.B.; Mueller, T.; Hoex, B.; et al. Electrical and microstructural analysis of contact formation on lightly doped phosphorus emitters using thick-film Ag screen printing pastes. *IEEE J. Photovolt.* **2013**, *4*, 168–174. [CrossRef]
34. Yang, Y.; Seyedmohammadi, S.; Kumar, U.; Gnizak, D.; Graddy, E.; Shaikh, A. Screen printable silver paste for silicon solar cells with high sheet resistance emitters. *Energy Procedia* **2011**, *8*, 607–613. [CrossRef]
35. Ju, M.; Mallem, K.; Dutta, S.; Balaji, N.; Oh, D.; Cho, E.; Cho, Y.H.; Kim, Y.; Yi, J. Influence of small size pyramid texturing on contact shading loss and performance analysis of Ag-screen printed mono crystalline silicon solar cells. *Mater. Sci. Semicond. Process.* **2018**, *85*, 68–75. [CrossRef]

16

Efficient CdTe Nanocrystal/TiO_2 Hetero-Junction Solar Cells with Open Circuit Voltage Breaking 0.8 V by Incorporating a Thin Layer of CdS Nanocrystal

Xianglin Mei [1], Bin Wu [1], Xiuzhen Guo [1], Xiaolin Liu [1], Zhitao Rong [1], Songwei Liu [1], Yanru Chen [1], Donghuan Qin [1,2,*], Wei Xu [1,2,*], Lintao Hou [3,*] and Bingchang Chen [1]

1. School of Materials Science and Engineering, South China University of Technology, Guangzhou 510640, China; mxianglin@sina.com (X.M.); damien_wu@foxmail.com (B.W.); XzGuo_19@163.com (X.G.); LiuLin9708@163.com (X.L.); rzt1512388848@163.com (Z.R.); majestyv@sina.com (S.L.); caro_ccyr@163.com (Y.C.); 17728100969@163.com (B.C.)
2. Institute of Polymer Optoelectronic Materials & Devices, State Key Laboratory of Luminescent Materials & Devices, South China University of Technology, Guangzhou 510640, China
3. Guangdong Provincial Key Laboratory of Optical Fiber Sensing and Communications, Guangzhou Key Laboratory of Vacuum Coating Technologies and New Energy Materials, Siyuan Laboratory, Department of Physics, Jinan University, Guangzhou 510632, China

* Correspondence: qindh@scut.edu.cn (D.Q.); xuwei@scut.edu.cn (W.X.); thlt@jnu.edu.cn (L.H.)

Abstract: Nanocrystal solar cells (NCs) allow for large scale solution processing under ambient conditions, permitting a promising approach for low-cost photovoltaic products. Although an up to 10% power conversion efficiency (PCE) has been realized with the development of device fabrication technologies, the open circuit voltage (V_{oc}) of CdTe NC solar cells has stagnated below 0.7 V, which is significantly lower than most CdTe thin film solar cells fabricated by vacuum technology (around 0.8 V~0.9 V). To further improve the NC solar cells' performance, an enhancement in the V_{oc} towards 0.8–1.0 V is urgently required. Given the unique processing technologies and physical properties in CdTe NC, the design of an optimized band alignment and improved junction quality are important issues to obtain efficient solar cells coupled with high V_{oc}. In this work, an efficient method was developed to improve the performance and V_{oc} of solution-processed CdTe nanocrystal/TiO_2 hetero-junction solar cells. A thin layer of solution-processed CdS NC film (~5 nm) as introduced into CdTe NC/TiO_2 to construct hetero-junction solar cells with an optimized band alignment and *p-n* junction quality, which resulted in a low dark current density and reduced carrier recombination. As a result, devices with improved performance (5.16% compared to 2.63% for the control device) and a V_{oc} as high as 0.83 V were obtained; this V_{oc} value is a record for a solution-processed CdTe NC solar cell.

Keywords: nanocrystal; CdTe; TiO_2; CdS; solar cells; solution processed

---

1. Introduction

Since the first reported solution-processed CdTe nanocrystal solar cells (NCs) in 2005, they have been rapidly developed due to their potential for next-generation photovoltaic products (including NCs, quantum dots, polymers, Sb_2Se_3, and perovskite solar cells) at low cost, low material consumption, and simple fabricating techniques [1–8]. During the past decade, intensive research has been focused on preparing high-quality CdTe NC films to improve the performance of NC solar cells [9–11]. Advances in CdTe NC thin film treatment and device architecture have led to a significant increase in the performance of solar cells from 2.9% in 2005 to ~7% in 2011 [12]. Efficient CdTe NC solar cells are prepared by using a planar *p-n* hetero-junction configuration. In this device structure, carriers are mainly generated in the CdTe NC film and electrons are injected from the conducting band of the CdTe NC to an n type partner (such as CdSe, CdS, ZnO, or TiO_2 et al.), while the hole travels to the back contact of the device. Nowadays, solution-processed CdTe NC solar cells mainly suffer from a low open circuit voltage (V_{oc}): most CdTe NC solar cells have a V_{oc} between 0.5 V and 0.7 V [13], while these values are 0.8 V~0.9 V for CdTe thin film solar cells that have been prepared by the close space sublimation (CSS) method [14], which limits further improvement in performance. The loss in potential for CdTe NC solar cells is defined as $E_{loss} = E_g - eV_{oc}$, where E_g is the bandgap of the CdTe NC thin film (~1.45 eV). The value of E_{loss} is greater than 0.7 eV for the CdTe NC solar cells, while this value is below 0.4 eV for most perovskite or III–V group semiconductor solar cells [15,16]. According to the Shockley–Queisser constraint, the minimum E_{loss} is about 0.3 eV for CdTe NC solar cells and the maximum theoretical V_{oc} with a bandgap of 1.45 eV is 1.15 eV [17]. The E_{loss} for CdTe NC solar cells can be mainly attributed to the recombination existing in the *p-n* junction and the back contact, given that the CdTe NC film has been prepared at optimized conditions. To obtain low resistance ohmic contacts to CdTe thin films, a heavily doped region at the surface of the CdTe should be formed before back contact formation via wet etching (using a bromine/methanol or phosphoric/nitric treatment) [18–20]. Unfortunately, the wet etching will result in the NC thin film being removed from the substrate or device shunt, which was confirmed by Panthani et al. [21].

Another way to make good ohmic contacts to the CdTe NC film is by using metal/p^+ semiconductor/metal oxide/organic hole transport materials with a high work function as a back contact. Occasionally, Au is selected as the back contact for CdTe NC solar cells, and a V_{oc} of 0.65 V can be obtained [22] due to the low work function (5.1 eV for Au~5.5 eV for CdTe). Recently, Kurley et al. demonstrated ohmic contacts could be realized by inserting transparent ZnTe:Cu, etched CdTe:Cu, or a Te buffer layer between the CdTe and ITO (Indium Tin Oxide) [23]. Unfortunately, although as high as 8.6% (without light soaking/current treatment) of the PCE coupled with a high fill factor (~60%) and J_{sc} were attained in this case, the V_{oc} was below 0.7 V, which limits the device's performance for further improvement. *p*-doping spiro-OMeTAD [24] or P3KT [25] have also been employed as hole transport materials for the CdTe NC thin film's back contact, and a high efficiency (~6%) was obtained in optimized NC solar cells. Most recently, a novel crosslinkable conjugated polymer poly(diphenylsilane-co-4-vinyl-triphenylamine) (Si-TPA) with high work function (5.38 eV) was introduced successfully into solution-processed CdTe/CdSe (or CdS) NC solar cells with an inverted structure of (ITO/ZnO/CdSe/CdTe/Si-TPA/Au); a PCE as high as 8.34% was obtained due to the decreased carrier recombination and dipole effects [26]. Another important issue for increasing the V_{oc} of CdTe NC solar cells is preparing a high-quality *p-n* junction and optimizing the band alignment of the whole device. As the size of a CdTe NC is in the range of 1~0 nm, the *n*-type partner is expected to have a similar size to obtain a homogeneous interface. In our previous work, we found that using solution-processed CdS NC or CdSe NC to replace the widely used CBD–CdS (chemical bath deposition CdS) as an *n*-type partner for CdTe NC solar cells improved V_{oc} and performance due to the high junction quality and reduced carrier recombination in the *p*/*n* junction [27,28]. Most recently, we found that a higher V_{oc} (0.66–0.74 V) could be obtained in CdTe NC/TiO_2 heterojunction solar cells by using Sb doped TiO_2 as the buffer layer due to the improved band alignment. However, the large differences in crystal type (solution-processed TiO_2 has an anatase structure [29] while CdTe NC has a zinc blende structure [30]) and lattice constant (0.948 nm for TiO_2 and 0.648 nm for CdTe) resulted in low junction quality. On the contrary, when compared to TiO_2, CdS had a lower lattice mismatch with CdTe and a high-quality hetero-junction is expected by incorporating a CdS thin film, which suppresses the leakage current due to the reduced defect density. In this paper, we developed an efficient method to simultaneously enhance the V_{oc} and PCE of solution-processed CdTe NC/TiO_2 solar cells by inserting a thin layer of CdS NC between the CdTe and TiO_2 film. The CdS NC possesses a similar size and structure as that of CdTe NC, which can efficiently decrease the lattice mismatch between CdTe and TiO_2; in addition, CdS has suitable energy levels, which are well matched with CdTe, therefore decreasing the energy loss and improving the V_{oc} of the device. The incorporation of a CdS NC thin film optimizes the band alignment of the CdTe/TiO_2 junction and reduces the interface recombination. Compared to the control device (with the structure FTO(SnO_2:F)/TiO_2/CdTe/Au), all of the devices with a CdS interlayer showed a significantly higher V_{oc} (0.72–0.83 V). A V_{oc} as high as 0.83 V was obtained with an optimal thickness of the CdS NC film (3.74 nm), which is a record for a solution-processed CdTe NC solar cell. When further optimizing the device fabrication conditions, we achieved CdTe/TiO_2 NC solar cells that exhibited a J_{sc} of 17.38 mA/cm^2, a V_{oc} of 0.73 V, an FF (fill factor) of 40.67%, and a high PCE of 5.16%. This PCE value was almost two times higher than the control device (with a PCE of 2.65%). As a simple fabrication process, we believe that this design holds potential for efficient CdTe NC solar cells with a PCE of up to 10%.

2. Experiment Procedure

A TiO_2 sol-gel precursor was synthesized via a convenient method according to our previous work [31]. In a typical process, 4.25 mL titanium *n*-butoxide, 3.75 mL ethanolamine, and 25 mL ethyl alcohol were mixed and gently stirred in a 50 mL beaker for 2 h to form a transparent sol-gel. Next, 5 mL of acetic acid in 5 mL of deionized water was gradually dropped into the mixture and continuously stirred for 24 h. Finally, the mixture was transferred to the fume hood to accelerate the condensation procedure. When the total volume of the mixture decreased to 15 mL, it was taken out for the fabrication of TiO_2 thin film. The synthesis of the CdS NC and CdTe NC solutions was conducted following methods published previously [23,28]. Transmission electron microscope (TEM) images of the CdS and CdTe NC are presented in Figure S1a,b. The CdS NC showed a spherical morphology while the CdTe NC showed a rod-shaped structure. The transmission spectrum of FTO/TiO_2/CdS with different thicknesses of CdS is shown in Figure S2. It is evident that the introduction of a thin layer CdS NC film had little impact on the transmission of FTO/ TiO_2 (less than 10% decrease when compared to the NC device without CdS NC film), which is prospective for increasing the spectrum response in short wavelengths.

Solar cells with the configuration of FTO/TiO_2/CdS/CdTe/Au were prepared by a simple solution process under ambient conditions, as shown in Figure 1. A TiO_2 film with a thickness of 40 nm was prepared by depositing a Ti^{2+} precursor onto the FTO substrate and spin-casted at 2500 rpm for 15 s, then the substrate was annealed at 500 °C for 1 h to eliminate any organic solvent and form a compact TiO_2 thin film. Several drops of the CdS NC solution with different concentrations (5 mg/mL, 10 mg/mL, 15 mg/mL, and 20 mg/mL) were then deposited onto the FTO/TiO_2 and spin-casted at 3000 rpm for 20 s. Following this, the substrate was transferred to a hot plate and annealed at 150 °C for 10 minutes, then transferred to another hot plate and annealed at 380 °C for 30 min. One wash with isopropanol was used to remove any impurities. The CdTe NCs were then deposited layer by layer onto the FTO/TiO_2/CdS substrate with a process described previously in [23]. Finally, several drops of saturated $CdCl_2$/methanol were put onto the FTO/TiO_2/CdS/CdTe substrate and spin-casted at 1100 rpm for 20 s, then transferred onto a hot plate at 330–420 °C for 15 min. Sixty nanometers of Au was deposited via thermal evaporation through a shadow mask with an active area of 0.16 cm^2 to make the electrode contact.

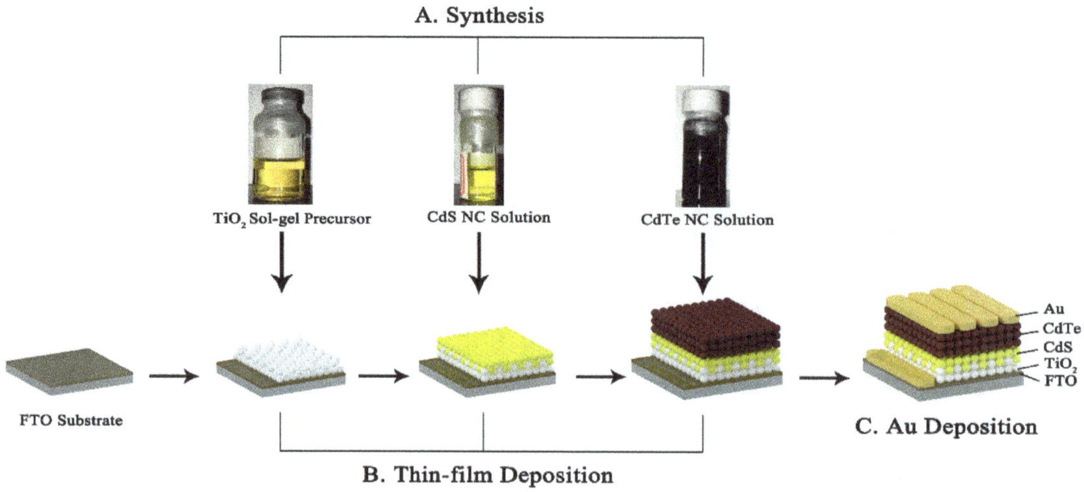

Figure 1. A schematic of the fabrication process of the NC solar cells.

The PCE of the NC solar cells were investigated under an illumination of 100 mW cm^{-2} with an air mass 1.5 (AM 1.5) solar simulator (Oriel model 91192), while the *J-V* characteristics were measured with a Keithley 2400. The *C-V* (Capacity-Voltage) measurements were taken with an Autolab PGSTAT-30 equipped with an impedance analyzer module. The external quantum efficiency (EQE) of the NC solar cell was measured using Solar Cell Scan 100 (Zolix, Beijing, China). Atomic force microscopy (AFM) images were obtained using a NanoScope NS3A system (Veeko, CA, USA). Transient photovoltage measurements (TPV) were taken by using the OmniFluo system (Zolix, Beijing, China).

3. Results and Discussion

The cross-section scanning electron microscope (SEM) image of the optimal CdTe/TiO_2 NC heterojunction solar cells is shown in Figure 2a. A high-temperature prepared TiO_2 thin film that was compatible with the FTO substrate was selected as a buffer layer for electron collecting. Zinc-blende CdS NC has a similar structure and size as CdTe NC and was deposited on the TiO_2 film. A gold electrode was deposited onto the CdTe NC film to collect photo-generated holes. The introduction of the CdS NC film was anticipated to decrease the lattice mismatch and interface defects between the CdTe and TiO_2. From the energy dispersive spectrum (EDS, Figure S3, supporting information), the emergence of an S element implied that the CdS had been introduced into the NC solar cells. The XRD pattern of the FTO/TiO_2/CdTe and FTO/TiO_2/CdS/CdTe thin films is presented in Figure S4; peaks for the zinc blende CdS were found when the CdS NC film was introduced. The band alignment of the FTO, TiO_2, CdS, CdTe, and Au is presented in Figure 2b. In this device architecture, light passes through the FTO, TiO_2, then CdS, and is absorbed by the CdTe NC active layer. The photon-generated carriers are separated by the built-in field of CdTe/TiO_2. Electrons are injected from the conducting band of CdTe to CdS then TiO_2, and collected by the FTO electrode, while the hole transfers from the valence band of CdTe to the gold electrode. To investigate the morphology changes of the TiO_2 thin film after the deposition of the CdS thin layer, atomic force microscopy (AFM) was used to characterize the surface images of FTO/TiO_2/CdS with different thicknesses of CdS NC. As shown in Figure 2c–f, a smooth surface was observed in the case of FTO/TiO_2/CdS with the thin CdS NC film (0.78 nm, Figure 2d). The TiO_2 film was totally covered with the CdS NC film when the CdS NC thickness was increased to 3.74 nm (Figure 2e). When the thickness of the CdS NC film reached 9.51 nm, although the TiO_2 was totally covered by the CdS NC film, the surface was very undulating. It was noted that the root mean squares were 3.01 nm, 4.00 nm, 13.80 nm, and 13.90 nm for a CdS NC thickness increase from 0 to 9.51 nm, respectively. A smooth CdS NC surface is essential to enhance the physical contact between CdS and CdTe and decrease interfacial recombination, leading to improved device performance.

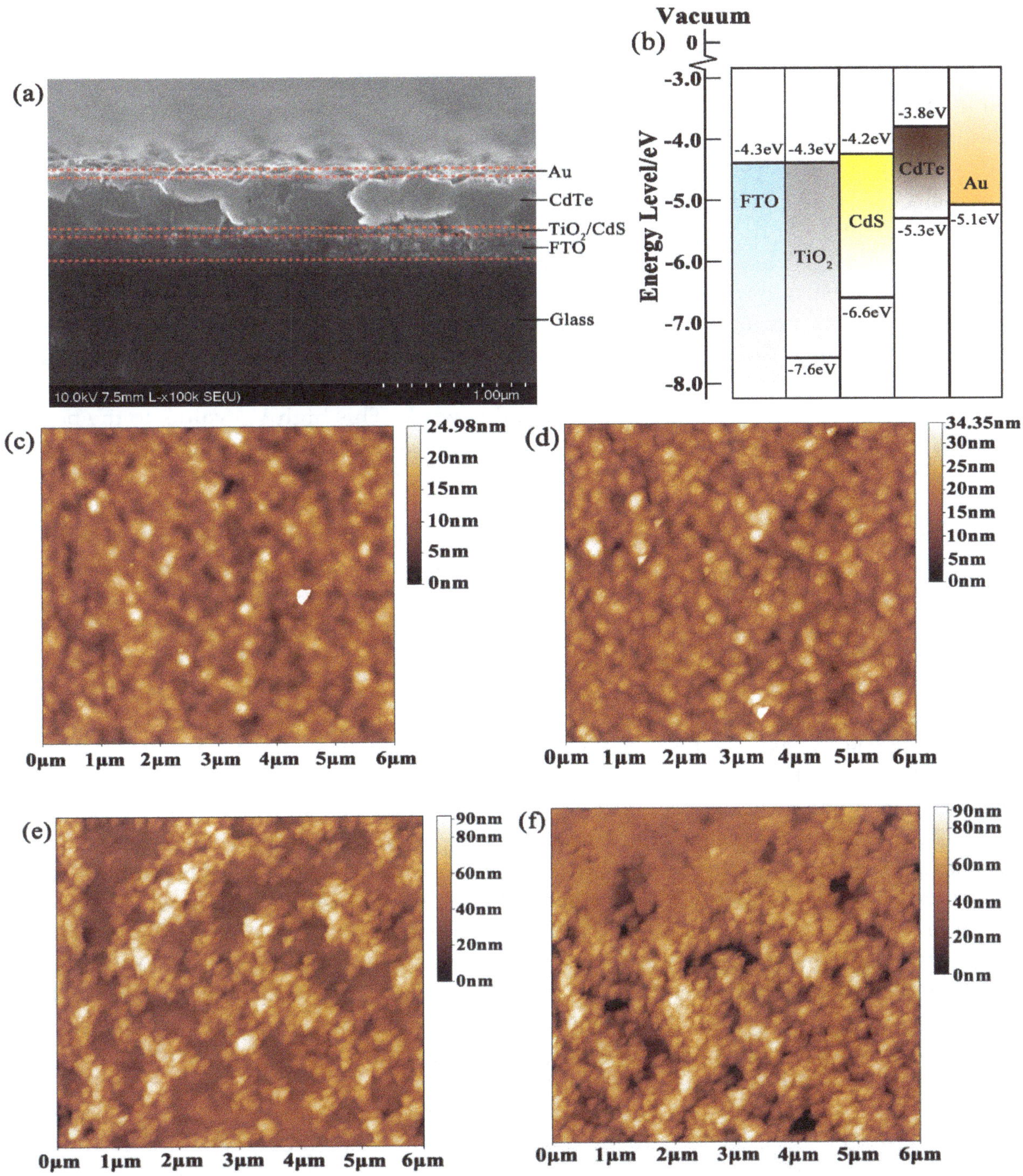

Figure 2. (**a**) Cross-section SEM image of the NC solar cells; (**b**) Energy levels of FTO, TiO_2, CdS, CdTe, and Au; (**c**) AFM images of FTO/TiO_2 without CdS; AFM images of FTO/TiO_2 with CdS; (**d**) 0.78 nm CdS NC; (**e**) 3.74 nm CdS NC; and (**f**) 9.51 nm CdS NC.

To decrease the interface defects between CdTe and TiO_2, a thin layer of CdS NC was deposited onto the TiO_2 film with different thicknesses via a solution process. It was reported in our previous works that an optimal annealing temperature for the CdTe NC/TiO_2 heterojunction was around 400 °C [32]. Figure 3a presents the current density vs. voltage (J-V) curves of the devices with CdS (3.74 nm) under air mass 1.5 G (AM 1.5 G) illumination, and the detailed parameters are summarized in Table 1. The NC solar cell with a CdS NC interlayer showed a V_{oc} of 0.83 V, a J_{sc} of 16.02 mA/cm^2, and a fill factor (FF) of 30.46%, resulting in a PCE of 4.05%; NC solar cells without the CdS NC interlayer only showed a V_{oc} of 0.69 V, a J_{sc} of 12.32 mA/cm^2, and an FF of 31.17%, leading to a PCE of 2.65%. Therefore, the PCE observed from the NC solar cells with a CdS NC interlayer showed a 52.8% improvement when compared to devices without the CdS NC interlayer. The parallel resistance (R_{sh}) of the NC solar cells was found to be slightly improved after inserting the CdS NC film, which implies a decreasing carrier recombination for CdS NC devices (Table 1). From the external quantum efficiency (EQE) spectrum (Figure 3b), one can see that the CdS NC interlayer device had a higher photon-to-electron conversion efficiency over the whole wavelength; when they were integrated, current densities of 15.99 mA/cm^2 and 12.30 mA/cm^2 were predicted, respectively, which were consistent with our *J-V* curves (Figure 3a). It is interesting that the NC devices with a CdS NC interlayer had a drastically improved V_{oc} (0.83 V for the CdS NC device, 0.69 V for the control device), demonstrating the advantage of the CdS NC interlayer. Figure 3c shows the V_{oc} of efficient CdTe NC solar cells with the different device structures (the device parameters are summarized in Table 2) that have been reported in recent years. Most devices showed a V_{oc} below 0.7 V, which was significantly lower than the devices fabricated in this work. This high V_{oc} value, to the best of our knowledge, is the highest V_{oc} reported for solution-processed CdTe NC solar cells with different structures. The V_{oc} obtained in this work was 13–40% higher than that of the conventional CdTe–ZnO NCs solar cells and ~18% higher than that of the inverted CdTe–TiO_2 NCs solar cells previously reported. The annealing temperature and thickness of the CdS NC film evidently has an influence on the junction quality of the NC solar cells. To investigate the annealing temperature on the performance of the devices, all devices with a 3.74 nm CdS NC interlayer were fabricated at the same conditions except for the final annealing procedure. As shown in Figure 3d (the *J-V* curves are presented in Supporting Information Figure S2, while the parameters are summarized in Table 1), the PCE increased linearly with an annealing temperature from 330 °C to 400 °C, then dropped when the annealing temperature was further increased to 420 °C. It was noted that all the devices showed a V_{oc} up to 0.7 V, and devices annealed at 390 °C/400 °C showed the highest V_{oc}, surpassing 0.8 V (0.82 V/0.83 V). It is well known that a TiO_2 thin film prepared by the decomposition of a Ti^{2+} precursor shows a porous structure, which is of benefit for separating the hole/electron pair in the case of dye sensitization solar cells [33]. However, a planar heterojunction is expected for thin film solar cells as a reduced interface area. We anticipated that the incorporation of CdS NC on top of a TiO_2 thin film would fill the hole of the TiO_2 film and permit the formation of a smooth and compact CdTe NC film on top of it. Furthermore, when compared to TiO_2, the CdS NC had a similar size and structure to that of CdTe NC, and therefore a high junction quality was attained in this case due to decreased defects and reduced nonradiative recombination in the interface. On the other hand, due to the low band offset between CdTe and CdS, a high V_{oc} was expected once the junction quality was improved (improving annealing temperature resulted in a higher junction quality). Further increases of the annealing temperature up to 400 °C may result in the oxidation of CdTe or pin-holes in the CdTe NC thin film, and therefore low device performances will be obtained in this case. It was also found that with increases in annealing temperature from 330 °C to 400 °C, the R_s decreased from 142.7 $\Omega\cdot cm^{-2}$ to ~100 $\Omega\cdot cm^{-2}$, while R_{sh} decreased from 400 $\Omega\cdot cm^{-2}$ to ~150 $\Omega\cdot cm^{-2}$. We speculated that with the increase in annealing temperature, the NC may grow larger, therefore resulting in a low R_s. However, as the annealing is conducted under ambient conditions, the surface of CdTe may oxidize, forming CdO at high temperatures, which could increase the series resistance (R_s) of the NC solar cells. On the other hand, aggressive $CdCl_2$ treatment at higher temperatures may lead to the formation of some pinholes, which will decrease the R_{sh} of the NC solar cells. In CdS thickness experiments, the thickness varied from 0.78 to 9.51 nm, whereas for TiO_2, the CdTe was fixed at 40 nm and 400 nm with the same structure (FTO/TiO_2/CdS/CdTe/Au). The PCEs with different CdS NC thicknesses are presented in Figure 3e (the *J-V* curves for different CdS NC thicknesses under light are provided in Supporting Information Figure S5, while the detailed photovoltaic parameters are summarized in Table 1). It was evident that the PCEs of the NC solar cells with different thicknesses of CdS were higher than those without a CdS NC interlayer. The PCEs of the NC solar cells increased with a CdS NC from 0 to 2.23 nm, then degraded when the CdS NC thickness exceeded 2.23 nm. The best device was obtained in the case of a 2.23 nm CdS NC interlayer, which showed the following of merits: a J_{sc} of 17.38 mA/cm^2, a V_{oc} of 0.73 V, an FF of 40.67%, and a PCE of 5.16%. The best PCE value was almost one time higher than the control device. It was noted that, although a high PCE was obtained in the 2.23 nm CdS NC device, the V_{oc} was significantly lower than that of the 3.74 nm CdS NC device. We anticipated that the built-in field was weak for devices with a too-thin CdS thickness due to the inadequate coverage of the TiO_2 film. The V_{oc} value was proportional to the built-in field in the NC solar cells. With an increase in the CdS NC thickness, the built-in field of the NC solar cells increased and a high V_{oc} was expected in this case, which conformed to our experiment results. However, with the increase in the CdS NC film thickness, the R_s of the NC solar cells increased, which may affect the J_{sc} and FF of the NC solar cells. Furthermore, the junction quality of CdTe/CdS/TiO_2 also had significant effects on the FF and the efficiency of the NC solar cells. One of the major issues for solution-processed NC solar cells is device stability. We examined the stability of NC solar cells without a CdS NC interlayer under ambient operating conditions. In this device configuration, the stability of the NC solar cells is mainly related to the CdTe/TiO_2 interface, the CdTe NC active layer, and the back contact. A device with a CdS NC interlayer maintained 96.7% of its PCE after being placed at ambient conditions (Figure 3f) for 30 days. In contrast, the control device only maintained 79.2% of its initial efficiency. We speculated that the introduction of a CdS NC interlayer restrains the diffusion of defects on the surface of the CdTe NC, therefore improving the stability of the NC solar cells.

Table 1. Summary of the photovoltaic parameters of the NC solar cells prepared under different conditions.

Annealing Temperature (°C)	CdS Layer Thickness (nm)	V_{oc} (V)	J_{sc} (mA/cm^2)	FF (%)	PCE (%)	R_s ($\Omega\cdot cm^{-2}$)	R_{sh} ($\Omega\cdot cm^{-2}$)
400	0	0.69	12.32	31.17	2.65	96.7	149.1
330	3.74	0.71	6.88	29.07	1.42	142.7	408.0
350	3.74	0.71	12.67	26.01	2.34	96.1	101.4
380	3.74	0.75	14.66	30.56	3.36	101.7	228.4
390	3.74	0.82	15.61	30.62	3.92	103.0	135.8
400	3.74	0.83	16.02	30.46	4.05	108.8	163.4
420	3.74	0.71	9.11	24.89	1.61	148.4	157.8
400	0.78	0.73	14.56	31.24	3.32	93.0	103.9
400	2.23	0.73	17.38	40.67	5.16	51.9	268.3
400	9.51	0.72	11.78	20.28	1.72	126.8	54.4

Table 2. Summary of the V_{oc} obtained in efficient CdTe NC solar cells in the literature.

Device Architecture	V_{oc} (V)	J_{sc} (mA/cm^2)	FF (%)	AM 1.5G Efficiency (%)	Ref.
ITO/CdTe/CdSe/Ca/Al	0.45	13.2	49	2.9	[1]
ITO/CdTe/Al	0.50	4.1	51	1.1	[9]
ITO/CdTe/ZnO/Al	0.59	20.7	56	6.9	[12]
ITO/CdTe/In:ZnO/Al	0.68	25.8	71	12.3	[21]
ITO/CdTe/ZnO/Al	0.69	25.5	64.7	11.3	[5]
ITO/ZnO/CdSe/CdTe/Au	0.65	15.28	58.5	5.81	[22]
ITO/TiO_2/CdTe/spiro-OMeTAD/Au	0.71	15.82	45.2	5.16	[25]
ITO/ZnO/CdSe/CdSe:CdTe/CdTe/Au	0.60	21.06	49.5	6.25	[23]
ITO/$(N_2H_5)_2CdTe_2$/CdTe/ZnO:In/Al	0.73	24.6	71	12.7	[24]
FTO/ZnO/Sb:TiO_2/CdTe/Au	0.74	11.16	30.13	2.49	[32]
ITO/ZnO/CdS/CdTe/Si-TPA/Au	0.67	20.58	52.76	7.27	[27]
FTO/TiO_2/CdS/CdTe/Au	0.83	16.02	30.5	4.05	This Work

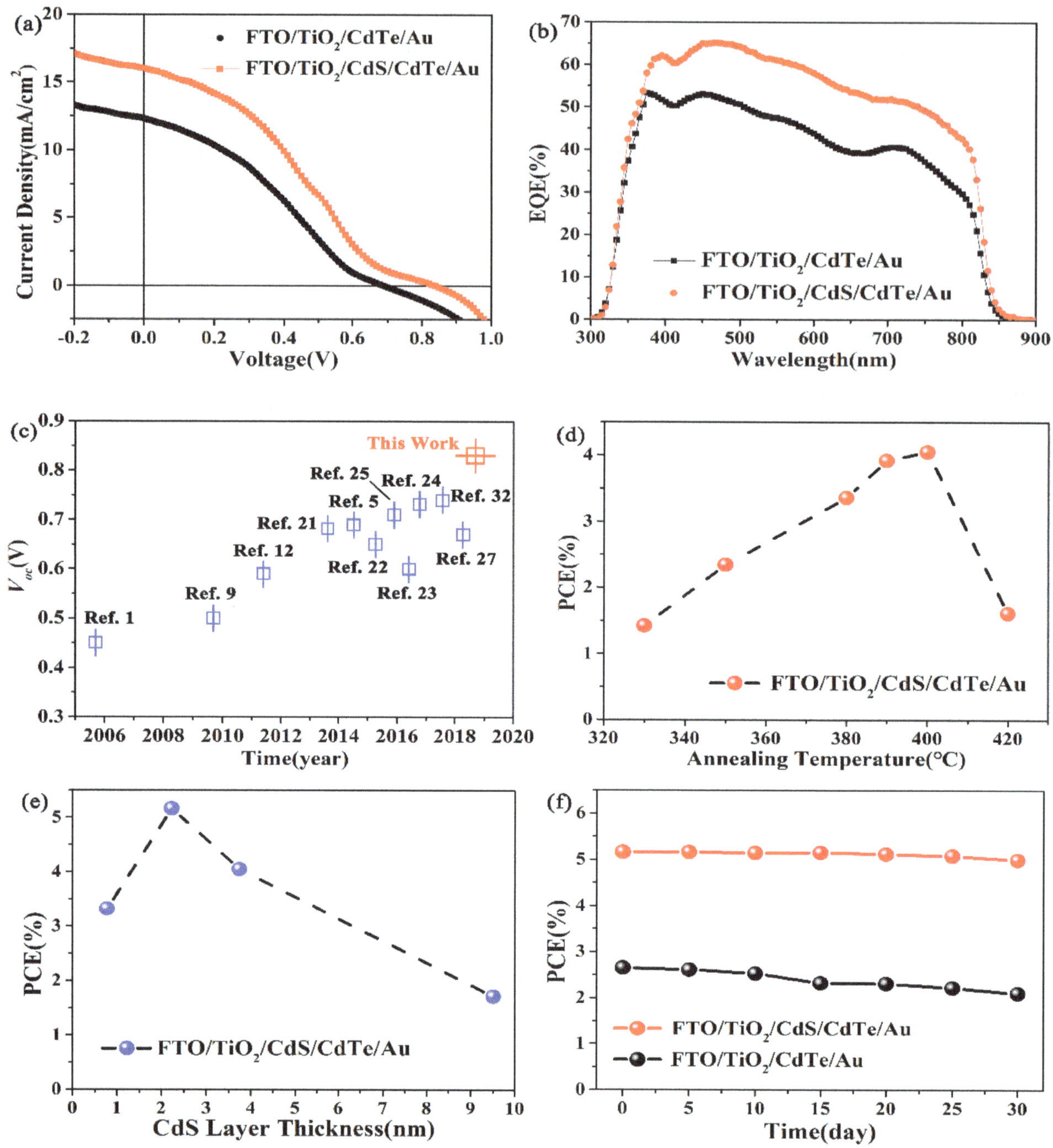

Figure 3. (**a**) *J-V* characteristics of the NC solar cells with/without a CdS (3.74 nm) NC interlayer in a structure of FTO/TiO_2/CdS (with/without)/CdTe/Au. The *J-V* curves were measured under 100 mW·cm^{-2} AM 1.5 G illumination, which were corrected by a calibrated Si solar cell. Corresponding (**b**) external quantum efficiency (EQE) spectrum; (**c**) Summary of the V_{OC} of efficient CdTe NC solar cells reported in the literature; NC solar cells with (**d**) different annealing temperatures and (**e**) different thicknesses of CdS NC film; and (**f**) The stabilized PCEs of NC solar cells with/without a CdS NC interlayer.

To gain more insight into the performance improvement in NC solar cells with a CdS NC interlayer, we characterized the *J-V* curves under dark. As shown in Figure 4a, the current at the reversed bias from a device with a CdS NC interlayer was almost one order lower than that from a device without a CdS NC interlayer. The low leakage current implied that the CdS NC interlayer could decrease the CdTe/TiO_2 interface defects and carrier recombination, resulting in a significant improvement in device performance. The built-in potential (V_{bi}) of the NC solar cells was mainly determined by the *p*-*n* junction between the CdTe and *n*-type partner. Compared to TiO_2, CdS had a lower lattice mismatch with CdTe and a lower band offset; therefore, a higher V_{oc} was expected in devices with a CdS NC interlayer, which agreed well to our experimental results. Capacitance–voltage curves (measured at a constant frequency of 1000 Hz) were carried out to investigate the built-in field of NC solar cells with/without a CdS NC interlayer. The C^{-2} with voltage (V) plotted is shown in Figure 4b. According to the Mott–Schottky equation [34],

$$C^{-2} = \frac{2}{A^2 q \varepsilon_0 \varepsilon N_A}(V_{bi} - V) \quad (1)$$

where A is the active area (0.16 cm^2); ε is the relative dielectric constant of CdTe (10.6); ε_0 is the vacuum permittivity; and N_A is the net acceptor concentration. The V_{bi} was extracted at a forward bias from the interception of the fitted line with the x axis. A higher V_{bi} (0.82 V) for devices with a CdS NC interlayer was observed, while this value was 0.72 V for devices without a CdS NC interlayer, which agreed well with their *J*-*V* curves, as shown in Figure 3a. The N_A of NC solar cells calculated from the above formula was ~$10^{16}/cm^3$. Altering the annealing temperature or using a different structure did not have a significant effect on the N_a value. To further investigate the effects of the CdS NC interlayer on the recombination process of the NC solar cells, the transient photovoltage (TPV) was used to measure the charge recombination in the NC solar cells with/without a CdS NC interlayer. In the case of TPV measurement, a steady state equilibrium was obtained when the NC solar cells were placed under a white light bias and an additional number of charges were generated by applying another weak laser pulse. As shown in Figure 4c, the charge recombination was characterized by tracking the transient voltage associated with the perturbations in charge population. The charge recombination times for NC devices with/without a CdS NC interlayer were 2.96 μs and 1.26 μs, respectively, which implied a lower charge recombination rate in the CdS NC interlayer device when compared to devices without a CdS NC interlayer. The roll-over of *J*-*V* at high annealing temperatures (400 °C or above, Figure 3a, Figures S4 and S6), which was also found in our previous work, was also noteworthy [32]. The roll-over mainly originated from the non-ohmic contact between the CdTe and Au, which can be mainly attributed to the large resistance present in the surface of the CdTe NC thin film. We speculated that CdO formed on the surface of the CdTe NC film at high annealing temperatures under ambient conditions. As CdO is an n-type semiconductor material, the device showed a *n*(TiO_2)–*p*(CdTe)–*n*(CdO) structure, so a *J*-*V* curve with roll-over was very likely to be obtained in this case. Pin-holes that formed in some parts of the NC thin film (due to the large inner stress in NC thin film at high temperature) or the diffusion of $CdCl_2$ across the whole NC thin film may also result in device shunt at high annealing temperatures. In this case, a low FF is likely to be obtained. In order to improve the contact quality, we also fabricated an NC device with a MoO_x/Au back contact or devices that were ozone etched before the Au electrode was deposited. The *J*-*V* curves for the devices with the structure FTO/TiO_2/CdS/CdTe/MoO_x (5 nm)/Au and the devices with different ozone etching times are presented in the Supporting Information Figure S7a,b, while the photovoltaic parameters are summarized in Table S1. Unfortunately, all of these attempts may result in devices shunting, or the degradation of the device performance. Further work should be carried out to eliminate the roll over to improve the performance of the NC solar cells.

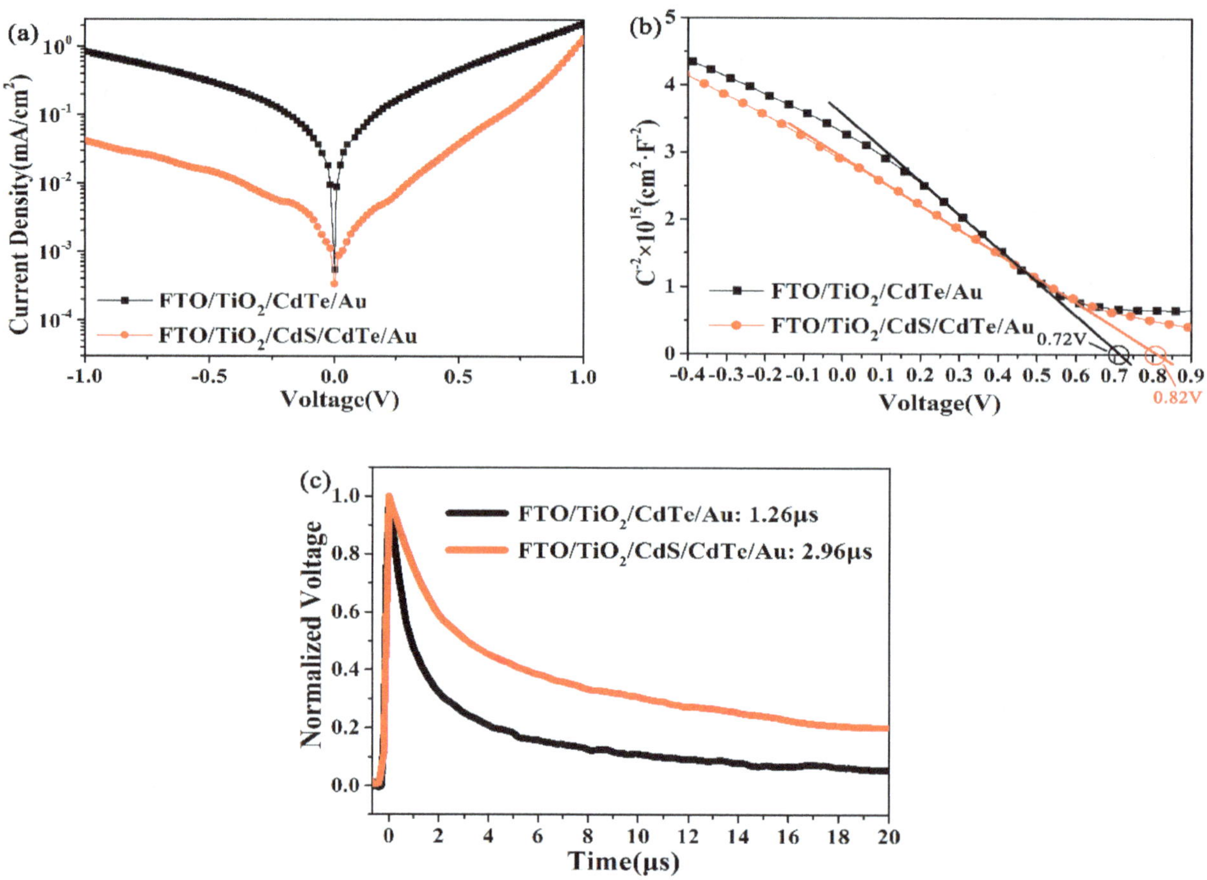

Figure 4. (**a**) *J*-*V* curves of the NC solar cells with/without a CdS thin film under dark; (**b**) Mott–Schottky curves in dark conditions measured at a constant frequency of 1000 Hz for the NC solar cell device with/without a CdS interlayer; and (**c**) Transient photovoltage measurements of the NC solar cells with/without a CdS interlayer.

4. Conclusions

In conclusion, we fabricated efficient CdTe NC/TiO_2 heterojunction solar cells through a simple layer by layer sintering solution process. The introduction of a thin layer of CdS NC between the CdTe and TiO_2 resulted in optimized band alignment and reduced the interface defects. Compared to the control devices, drastic improvements in V_{oc} and PCE were observed for the devices with a CdS NC interlayer. A V_{oc} as high as 0.83 V was attained by optimizing the thickness of the CdS NC, which was the highest record for solution-processed CdTe NC solar cells. After carefully optimizing the fabrication parameters, we obtained a device with a PCE of 5.16%, showing a 94.7% increase when compared to the control device. Our work here provides a new way to improve the V_{oc} and performance of CdTe NC solar cells.

Supplementary Materials
Figure S1: TEM images of as prepared (a) CdS and (b) CdTe nanocrystal, Figure S2: Transmission spectrum of FTO/TiO_2/CdS with different thickness of CdS NC film, Figure S3: EDS obtained on the cross-section of CdTe NC solar cells with configuration of FTO/TiO_2/CdS/CdTe/Au, Figure S4: *J*-*V* characteristic of NC solar cells with different annealing temperatures (all devices with 3.74 nm CdS interlayer), Figure S5: XRD pattern of FTO/TiO_2 and FTO/TiO_2/CdS, Figure S6: *J*-*V* characteristic of NC solar cells with different thicknesses of CdS NC film (all devices annealing at 400°C), Figure S7: (a) *J*-*V* curves for NC solar cells with/without MoOx buffer layer (b) *J*-*V* curves for NCs solar cells with different ozone etching times, Table S1: Summarized photovoltaic parameters from Figure S4.

Author Contributions: D.Q. and X.M. conceived and designed the experiments; X.M., B.W., and X.L. performed the experiments; X.M. and B.C. analyzed the data; X.G., Z.R., S.L., and Y.C. contributed reagents/materials/analysis tools; and X.M., D.Q., W.X. and L.H. wrote the paper.

Acknowledgments: We thank the financial support of the National Natural Science Foundation of China (Nos. 91333206, 61274062, and 61774077), the Guangzhou Science and Technology Planning Project (No. 201804010295), the National Undergraduate Innovative and Entrepreneurial Training Program (No. 201810561016), and the Fundamental Research Funds for the Central Universities.

References

1. Gur, I.; Fromer, N.A.; Geier, M.L.; Alivisators, A.P. Air-Stable All-Inorganic Nanocrystal Solar Cells Processed from Solution. *Science* **2005**, *310*, 462–465. [CrossRef] [PubMed]
2. Kim, T.; Firdaus, Y.; Kirmani, A.R.; Liang, R.Z.; Hu, H.; Liu, M.; Labban, A.E.; Hoogland, S.; Beaujuge, P.M.; Sargent, E.H.; et al. Hybrid Tandem Quantum Dot/Organic Solar Cells with Enhanced Photocurrent and Efficiency via Ink and Interlayer Engineering. *ACS Energy Lett.* **2018**, *3*, 1307–1314. [CrossRef]
3. Wen, X.; Chen, C.; Lu, S.; Li, K.; Kondrotas, R.; Zhao, Y.; Chen, W.; Gao, L.; Wang, C.; Zhang, J.; et al. Vapor transport deposition of antimony selenide thin film solar cells with 7.6% efficiency. *Nat. Commun.* **2018**, *9*, 2179. [CrossRef] [PubMed]
4. Li, D.B.; Yin, X.; Grice, C.R.; Guan, L.; Song, Z.; Wang, C.; Chen, C.; Li, K.; Cimaroli, A.J.; Awni, R.A.; et al. Stable and efficient CdS/Sb_2Se_3 solar cells prepared by scalable close space sublimation. *Nano Energy* **2018**, *49*, 346–353. [CrossRef]
5. Crisp, R.W.; Panthani, M.G.; Rance, W.L.; Duenow, J.N.; Parilla, P.A.; Callahan, R.; Dabney, M.S.; Berry, J.J.; Talapin, D.V.; Luther, J.M. Nanocrystal Grain Growth and Device Architectures for High-Efficiency CdTe Ink-Based Photovoltaics. *ACS NANO* **2014**, *8*, 9063–9072. [CrossRef] [PubMed]
6. Chen, Z.; Du, X.; Zeng, Q.; Yang, B. Recent development and understanding of polymer-nanocrystal hybrid solar cells. *Mater. Chem. Front.* **2017**, *1*, 1502–1513. [CrossRef]
7. Zhao, Y.; Tan, H.; Yuan, H.; Yang, Z.; Fan, J.Z.; Kim, J.; Voznyy, O.; Gong, X.; Quan, L.N.; Tan, C.S.; et al. Perovskite seeding growth of formamidinium-leadiodide-based perovskites for efficient and stable solar cells. *Nat. Commun.* **2018**, *9*, 1607. [CrossRef] [PubMed]
8. He, Z.; Zhong, C.; Su, S.; Xu, M.; Wu, H.; Cao, Y. Enhanced power-conversion efficiency in polymer solar cells using an inverted device structure. *Nat. Photon.* **2012**, *6*, 591–595. [CrossRef]
9. Anderson, I.E.; Breeze, A.J.; Olson, J.D.; Yang, L.; Sahoo, Y.; Carter, S.A. All-inorganic spin-cast nanoparticle solar cells with nonselective electrodes. *Appl. Phys. Lett.* **2009**, *94*, 063101. [CrossRef]
10. Townsend, T.K.; Foos, E.E. Fully solution processed all inorganic nanocrystal solar cells. *Phys. Chem. Chem. Phys.* **2014**, *16*, 16458. [CrossRef] [PubMed]
11. Tian, Y.; Zhang, Y.; Lin, Y.; Gao, K.; Zhang, Y.; Liu, K.; Yang, Q.; Zhou, X.; Qin, D.; Wu, H.; et al. Solution-processed efficient CdTe nanocrystal/CBD-CdS hetero-junction solar cells with ZnO interlayer. *J. Nanopart. Res.* **2013**, *15*, 2053. [CrossRef]
12. Jasieniak, J.; MacDonald, B.I.; Watkins, S.E.; Mulvaney, P. Solution-Processed Sintered Nanocrystal Solar Cells via Layer-by-Layer Assembly. *Nano Lett.* **2011**, *11*, 2856–2864. [CrossRef] [PubMed]
13. Xue, H.; Wu, R.; Xie, Y.; Tan, Q.; Qin, D.; Wu, H.; Huang, W. Recent Progress on Solution-Processed CdTe Nanocrystals Solar Cells. *Appl. Sci.* **2016**, *6*, 197. [CrossRef]
14. Kumar, S.G.; Rao, K.S.R.K. Physics and chemistry of CdTe/CdS thin film heterojunction photovoltaic devices: fundamental and critical aspects. *Energy Environ. Sci.* **2014**, *7*, 45. [CrossRef]
15. Tan, H.; Jain, A.; Voznyy, O.; Lan, X.; Arquer, F.P.G.D.; Fan, J.Z.; Quintero-Bermudez, R.; Yuan, M.; Zhang, B.; Zhao, Y.; et al. Efficient and stable solution-processed planar perovskite solar cells via contact passivation. *Science* **2017**, *355*, 722–726. [CrossRef] [PubMed]
16. Green, M.A.; Hishikawa, Y.; Dunlop, E.D.; Levi, D.H.; Hohl-Ebinger, J.; Ho-Baillie, A.W.Y. Solar cell efficiency tables (version 51). *Prog. Photovolt. Res. Appl.* **2018**, *26*, 3–12. [CrossRef]
17. Grätzel, M. The light and shade of perovskite solar cells. *Nat. Mater.* **2014**, *13*, 838–842. [CrossRef] [PubMed]
18. Kotina, I.M.; Tukhkonen, L.M.; Patsekina, G.V.; Shchukarev, A.V.; Gusinskii, G.M. Study of CdTe etching process in alcoholic solutions of bromine. *Semicond. Sci. Technol.* **1998**, *13*, 890–894. [CrossRef]
19. Seymour, F.H.; Kaydanov, V.; Ohno, T.R. Cu and $CdCl_2$ influence on defects detected in CdTe solar cells with admittance spectroscopy. *Appl. Phys. Lett.* **2005**, *87*, 153507. [CrossRef]
20. Rose, D.H.; Hasoon, F.S.; Dhere, R.G.; Albin, D.S.; Ribelin, R.M.; Li, X.S.; Mahathongdy, Y.; Gessert, T.A.;

Sheldon, P. Fabrication Procedures and Process Sensitivities for CdS/CdTe Solar Cells. *Prog. Photovolt. Res. Appl.* **1999**, *7*, 331–340. [CrossRef]

21. Panthani, M.G.; Kurley, J.M.; Crisp, R.W.; Dietz, T.C.; Ezzyat, T.; Luther, J.M.; Talapin, D.V. High Efficiency Solution Processed Sintered CdTe Nanocrystal Solar Cells: The Role of Interfaces. *Nano Lett.* **2014**, *14*, 670–675. [CrossRef] [PubMed]
22. Liu, H.; Tian, Y.; Zhang, Y.; Gao, K.; Lu, K.; Wu, R.; Qin, D.; Wu, H.; Peng, Z.; Hou, L.; et al. Solution processed CdTe/CdSe nanocrystal solar cells with more than 5.5% efficiency by using an inverted device structure. *J. Mater. Chem. C.* **2015**, *3*, 4227–4234. [CrossRef]
23. Xie, Y.; Tan, Q.; Zhang, Z.; Lu, K.; Li, M.; Xu, W.; Qin, D.; Zhang, Y.; Hou, L.; Wu, H. Improving performance in CdTe/CdSe nanocrystals solar cells by using bulk nano-heterojunctions. *J. Mater. Chem. C.* **2016**, *4*, 6483. [CrossRef]
24. Kurley, J.M.; Panthani, M.G.; Crisp, R.W.; Nanayakkara, S.U.; Pach, G.F.; Reese, M.O.; Hudson, M.H.; Dolzhnikov, D.S.; Tanygin, V.; Luther, J.M.; et al. Transparent Ohmic Contacts for Solution-Processed, Ultrathin CdTe Solar Cells. *ACS Energy Lett.* **2017**, *2*, 270–278. [CrossRef]
25. Du, X.; Chen, Z.; Liu, F.; Zeng, Q.; Jin, G.; Li, F.; Yao, D.; Yang, B. Improvement in Open-Circuit Voltage of Thin Film Solar Cells from Aqueous Nanocrystals by Interface Engineering. *ACS Appl. Mater. Interfaces* **2016**, *8*, 900–907. [CrossRef] [PubMed]
26. Zeng, Q.; Hu, L.; Cui, J.; Feng, T.; Du, X.; Jin, G.; Liu, F.; Ji, T.; Li, F.; Zhang, H.; et al. High-Efficiency Aqueous-Processed Polymer/CdTe Nanocrystals Planar Heterojunction Solar Cells with Optimized Band Alignment and Reduced Interfacial Charge Recombination. *ACS Appl. Mater. Interfaces* **2017**, *9*, 31345–31351. [CrossRef] [PubMed]
27. Guo, X.; Tan, Q.; Liu, S.; Qin, D.; Mo, Y.; Hou, L.; Liu, A.; Wu, H.; Ma, Y. High-efficiency solution-processed CdTe nanocrystal solar cells incorporating a novel crosslinkable conjugated polymer as the hole transport layer. *Nano Energy* **2018**, *46*, 150–157. [CrossRef]
28. Liu, S.; Liu, W.; Heng, J.; Zhou, W.; Chen, Y.; Wen, S.; Qin, D.; Hou, L.; Wang, D.; Xu, H. Solution-Processed Efficient Nanocrystal Solar Cells Based on CdTe and CdS Nanocrystals. *Coatings* **2018**, *8*, 26. [CrossRef]
29. Wen, S.; Li, M.; Yang, J.; Mei, X.; Wu, B.; Liu, X.; Heng, J.; Qin, D.; Hou, L.; Xu, W.; et al. Rationally Controlled Synthesis of $CdSe_xTe_{1-x}$ Alloy Nanocrystals and Their Application in Efficient Graded Bandgap Solar Cells. *Nanomaterials* **2017**, *7*, 380. [CrossRef] [PubMed]
30. Liu, H.; Tang, J.; Kramer, I.J.; Debnath, R.; Koleilat, G.I.; Wang, X.; Fisher, A.; Li, R.; Brzozowski, L.; Levina, L.; et al. Electron Acceptor Materials Engineering in Colloidal Quantum Dot Solar Cells. *Adv. Mater.* **2011**, *23*, 3832–3837. [CrossRef] [PubMed]
31. Yang, Y.; Zhao, B.; Gao, Y.; Liu, H.; Tian, Y.; Qin, D.; Wu, H.; Huang, W.; Hou, L. Novel Hybrid Ligands for Passivating PbS Colloidal Quantum Dots to Enhance the Performance of Solar Cells. *Nano-Micro Lett.* **2015**, *7*, 325–331. [CrossRef]
32. Li, M.; Liu, X.; Wen, S.; Liu, S.; Heng, J.; Qin, D.; Hou, L.; Wu, H.; Xu, W.; Huang, W. CdTe Nanocrystal Hetero-Junction Solar Cells with High Open Circuit Voltage Based on Sb-doped TiO_2 Electron Acceptor Materials. *Nanomaterials* **2017**, *7*, 101. [CrossRef] [PubMed]
33. Galliano, S.; Bella, F.; Gerbaldi, C.; Falco, M.; Viscardi, G.; Grätzel, M.; Barolo, C. Photoanode/Electrolyte Interface Stability in Aqueous Dye-Sensitized Solar Cells. *Energy Technol.* **2017**, *5*, 300–311. [CrossRef]
34. Zhu, J.; Yang, Y.; Gao, Y.; Qin, D.; Wu, H.; Hou, L.; Huang, W. Enhancement of open-circuit voltage and fill factor in CdTe nanocrystals solar cells by using interface materials. *Nanotechnology* **2014**, *25*, 365203. [CrossRef] [PubMed]

17

Efficient Ni/Au Mesh Transparent Electrodes for ITO-Free Planar Perovskite Solar Cells

Dazheng Chen [1,2,*], Gang Fan [1], Hongxiao Zhang [2], Long Zhou [2], Weidong Zhu [1], He Xi [1,3], Hang Dong [1], Shangzheng Pang [1], Xiaoning He [1], Zhenhua Lin [1], Jincheng Zhang [1,2], Chunfu Zhang [1,2,*] and Yue Hao [1]

1 Wide Bandgap Semiconductor Technology Disciplines State Key Laboratory, School of Microelectronics, Xidian University, Xi'an 710071, China
2 Shaanxi Joint Key Laboratory of Graphene, Xidian University, Xi'an 710071, China
3 School of Advanced Materials and Nanotechnology, Xidian University, Xi'an 710071, China
* Correspondence: dzchen@xidian.edu.cn (D.C.); cfzhang@xidian.edu.cn (C.Z.)

Abstract: Indium thin oxide (ITO)-free planar perovskite solar cells (PSCs) were fabricated at a low temperature (150 °C) in this work based on the transparent electrode of photolithography processed nickel/gold (Ni/Au) mesh and the high conductivity polymer, PH1000. Ultrathin Au was introduced to increase the conductivity of metal mesh, and the optimal hexagonal Ni (30 nm)/Au (10 nm) mesh (line width of 5 μm) shows a transmittance close to 80% in the visible light region and a sheet resistance lower than 16.9 Ω/sq. The conductive polymer PH1000 not only smooths the raised surface of the metal mesh but also enhances the charge collection ability of metal mesh. The fabricated PSCs have the typical planar structure (glass/Ni-Au mesh/PH1000/PEDOT:PSS/$MA_yFA_{1-y}PbI_xCl_{3-x}$/PCBM/BCP/Ag) and the champion PSC (0.09 cm^2) obtains a power conversion efficiency (PCE) of 13.88%, negligible current hysteresis, steady current density and PCE outputs, and good process repeatability. Its photovoltaic performance and stability are comparable to the reference PSC based on the ITO electrodes (PCE = 15.70%), which demonstrates that the Ni/Au mesh transparent electrodes are a promising ITO alternative to fabricate efficient PSCs. The relatively lower performance of Ni/Au based PSC results from the relatively slower charge extraction and stronger charge recombination than the ITO based PSC. Further, we tried to fabricate the large area (1 cm^2) device and achieve a PCE over 6% with negligible hysteresis and steady current density and PCE outputs. The improvements of perovskite film quality and interface modification should be an effective approach to further enhance the device performance of Ni/Au based PSCs, and the Ni/Au mesh electrode may find wider applications in PSCs and flexible devices.

Keywords: metal mesh; transparent electrode; photolithography; perovskite solar cell; large-area solar cell

---

1. Introduction

Organic-inorganic hybrid perovskites solar cells (PSCs) have attracted more and more attention due to their advantages of low fabrication cost, light weight, solution processability, tunable light absorption range, bipolar transport properties, large-area manufacturing, and compatibility to both rigid and flexible substrates. Their rapid progress of increased power conversion efficiency (PCEs) from 3.8% to 23.32% and improved stability indicate many potential applications, including in photovoltaic plants, photovoltaic curtains, building integrated photovoltaic materials, wearable electronics devices, and even space power systems [1–10]. And the efficient bottom transparent electrode is primary and crucial to meeting these versatile applications. Nowadays, transparent conductive oxides, such as the indium thin oxide (ITO) is the most commonly used transparent electrode in PSCs. However, ITO prices are becoming more and more expensive because of the rising cost of indium. On the other hand, the ITO shows a relatively large sheet resistance, brittleness, and poor mechanical robustness [11,12]. There will be cracks on ITO surface at a bending radius (<10 mm) against repeated bending, which will further decrease its conductivity and degrade the device performance, making it incompatible with low-cost technology, such as roll-to-roll printing [13].

To overcome the drawbacks of ITO, many alternatives have been developed with good electrical and optical properties, such as graphene [14,15], carbon nanotubes [16], metal nanowires [17,18], transparent conducting oxide nanocrystal [19], conducting polymer [20,21] ultrathin metal films [21–24], and metal-mesh transparent conductive electrodes (TCEs) [25,26]. Among the alternatives, metal meshes have excellent ductility and can be easily fabricated by mature evaporation process, which is more suitable for large-area device production. Metal-mesh TCEs are highly bendable as well, and silver metal-mesh TCEs have been successfully used to fabricate organic solar cells, PSCs, and flexible devices [5,27]. More importantly, the conductivity and transmittance of metal mesh can be adjusted by structure parameters such as the metal-grid pitch, line width, and film thickness. And there are many ways to achieve size-tunable metal-mesh TCEs, such as laser sintering of nanoparticle ink, lithographic patterning, grain boundary lithography, nanoimprint lithography, and photolithography [28–31]. Photolithography is a standard fabrication process in semiconductor devices and integrated circuits, which determines their feature size (≤7 nm) and operation performance. It is believed that the photolithography process provides more precise control of the shape and size of the objects it creates and can create patterns over an entire surface cost-effectively [32]. In addition, the photolithography related processes are low-temperature technology and compatible to large area substrate (18 inches) or flexible substrates. Therefore, photolithography is a promising process to prepare highly consistent metal mesh for transparent electrode and solar cell applications [33–35].

We have proved in our previous work [36] that the adhesion of Ag on a glass substrate is relatively poorer than that of Ni, and Ni mesh processed by photolithography has been successfully used to fabricate PSCs, but the device performance (PCE = 5.74%) should be further improved. In this work, we optimize the thickness and shape of Ni meshes, Au is used to increase the conductivity of metal mesh, the high conductivity polymer PH1000 is employed to smooth the raised surface of metal mesh and enhance the charge collection ability of metal mesh, the optimized two-step $MA_yFA_{1-y}PbI_xCl_{3-x}$ perovskite is used as the light absorber, and the fabricated ITO free PSCs have a structure of Glass/hexagonal Ni (30 nm)-Au (10 nm) mesh/PH1000/PEDOT:PSS/$MA_yFA_{1-y}PbI_xCl_{3-x}$/PCBM/BCP/Ag. The champion PSC (0.09 cm^2) obtains a PCE of 13.88%, negligible current hysteresis, steady current density and PCE outputs, good repeatability and storing stability. Further, we tried to fabricate the large area (1 cm^2) devices under the same process, and achieve a PCE over 6% with negligible hysteresis and stable steady-state outputs. The comparable performance to the ITO based reference PSC demonstrates that the Ni/Au mesh transparent electrodes are a promising ITO alternative to fabricate efficient PSCs.

2. Materials and Methods

2.1. Materials

All materials and reagents, Methylammonium iodide (MAI, 99.8%, Dyesol, Queanbeyan, Australia), Formamidinium iodide (FAI, 99.8%, Dyesol, Queanbeyan, Australia), Lead iodide (PbI_2, 99.999%, Sigma-Aldrich, Saint Louis, MI, USA), Lead chloride ($PbCl_2$, 99.999%, Sigma-Aldrich, Saint Louis, MI, USA), Phenyl-C_{61}-butyric acid methyl ester (PCBM, 98%, Nano-C, Westwood, MA, USA), Poly(3,4-ethy-lenedioxythiophene) poly(styrenesulfonate) (PEDOT:PSS, Clevios PVP Al 4083, Heraeus, Hanau, Germany), high conductivity polymer PH1000 (Clevios PH1000, Heraeus, Hanau, Germany), N,N'-Dimethylformamide (DMF, 99.8%, Aladdin, Beijing, China), Chlorobenzene (CB, 99.8%, Sigma-Aldrich, Saint Louis, MI, USA), Bathocuproine (BCP, 96%, Sigma-Aldrich, Saint Louis, MI, USA) Isopropanol (IPA, 99.5%, Sigma-Aldrich, Saint Louis, MI, USA), photoresist (AZ6112,

Ruicai, Suzhou, China), and developing solution (2.38%, NMD-3, Ruicai, Suzhou, China), were used as received without further purification.

2.2. Metal Mesh Preparation

Firstly, glass substrates ($2 \times 2.5\ cm^2$) were ultrasonically cleaned in deionized (DI) water, acetone, ethyl alcohol, and DI water for 20 min, respectively. Secondly, the nitrogen gun was used to dry the substrates, and spin-coat the positive photoresist on substrates by two-steps of 500 rpm for 5s and 4000 rpm for 30 s with the acceleration of 4000 rpm/s, followed by baking on a hot plate at 100 °C for 90 s to cure the photoresist. Thirdly, the substrates were naturally cooled and exposed for 2.3 s under a shadow mask and then developed for 60 s in the developing solution. Then, the samples were rinsed in flowing DI water and dried by nitrogen gun. Fourthly, an optical microscope was employed to check the defined patterns and the samples were transferred into the E-beam evaporation system and nickel/gold (Ni/Au) mesh of x nm/10 nm were deposited under a pressure below 5×10^{-4} Pa. After the metal mesh deposition, the samples were immersed in the acetone solution and lifted off by a low power ultrasonic bath for several ten seconds. The completed metal-mesh show a line width of 5 μm, and the active areas of 0.09 mm^2 and 1 cm^2.

2.3. PSCs Preparation and Characterization

At first, the prepared metal-mesh substrates were UV-ozone treated by 15 min, and the high conductive polymer PH 1000 were spin-coated on the metal mesh at 1000 rpm for 60 s and annealed on a hotplate at 150 °C for 15 min. Secondly, the PEDOT:PSS was spin-coated at 6000 rpm for 45 s and 150 °C annealing for 15 min. Thirdly, the $MA_yFA_{1-y}PbI_xCl_{3-x}$ precursor solution was prepared by mixing PbI_2, $PbCl_2$ that was dissolved in DMF solution, and stirred for 2 h at 75 °C, the solution was then spin coated onto the PEDOT/PSS layer at 3000 rpm for 45 s; Mixing FAI, MAI in the solvent of DMF, stirring at room temperature until completely dissolved, then spin coated it onto the $PbI_2/PbCl_2$ layer at 3000 rpm for 45 s; after thermally annealing at 100 °C for 10 min the perovskite layer was formed. Fourthly, the PCBM (20 mg/mL in CB) was spin-coated on the perovskite layer at 2000 rpm for 40s, and the BCP (0.5 mg/mL in IPA) was spin-coated on the PCBM at 6000 rpm for 45s. Finally, the Ag (100 nm) electrode was thermally evaporated under a shadow mask and the fabricated devices (seeing Figures S1 and S2 in Supporting Information) have an active area of 0.09 cm^2 and 1 cm^2. As a reference, the ITO based PSCs, ITO/PEDOT:PSS/Perovskite/PCBM/BCP/Ag, were also fabricated under the same process conditions.

The current density-voltage (J-V) characteristics were measured with a source measurement unit of Keithley 2400 and simulated AM1.5 G sun light (100 mW/cm^2, SEN-EI Electric. Co. Ltd, XES-300T1, Osaka, Japan). The incident photo-to-current conversion efficiency (IPCE) was measured by the quantum efficiency measurement system (SCS10-X150, Zolix instrument. Co. Ltd, Beijing, ChinaZolix Instrument. Co. Ltd). The four-point-probe system was utilized to measure the sheet resistance of electrodes. The UV–visible spectrophotometer (Perkin-Elmer Lambda 950, Waltham, MA, USA) was used to characterize the transmittance spectra of different samples. The film morphology was characterized by a JSM-7800F extreme-resolution analytical field emission scanning electron microscope (SEM) (JEOL Ltd., Tokyo, Japan) and atomic force microscopy (AFM) (Agilent 5500, Santa Clara, CA, USA). Electrochemical impedance spectroscopy (EIS) measurements were performed on an electrochemical workstation (CHI600E, Shanghai Chenhua, Shanghai, China) with a 10 mV amplitude perturbation and frequencies between 100 Hz and 1 MHz. M-S plots were recorded on the same system under AC excitation amplitude of 30 mV at a frequency of 5 kHz. Transient photocurrent (TPC) measurement was performed with a system excited by a 532 nm (1000 Hz, 3.2 ns) pulse laser. Transient photovoltage (TPV) measurement was performed with the same system excited by a 405 nm (50 Hz, 20 ms) pulse laser. A digital oscilloscope (Tektronix, D4105, Beaverton, OR, USA) was used to record the photocurrent or photovoltage decay process with a sampling resistor of 50 Ω or 1 MΩ, respectively. All the measurements were performed under ambient atmosphere at room temperature.

3. Results and Discussion

The geometry of ITO-free planar PSCs with an inverted structure based on a metal mesh transparent electrode is shown in Figure 1. Here, the two-step solution processed $MA_yFA_{1-y}PbI_xCl_{3-x}$ perovskite layer is chosen as the light absorber; the PEDOT:PSS and PCBM act as the hole and electron transport layers (HTL and ETL), respectively; the BCP is further used to modify the electron collection at the PCBM/Ag interface. In particular, to obtain optimal metal mesh with high transmittance and low resistance, the photolithography process was chosen to precisely define the line width and space between metal lines. Two curial interfaces related to the metal mesh were designed to improve the properties of Ni/Au transparent electrode, then the corresponding performance of ITO-free PSCs with small and large active areas are discussed as follows.

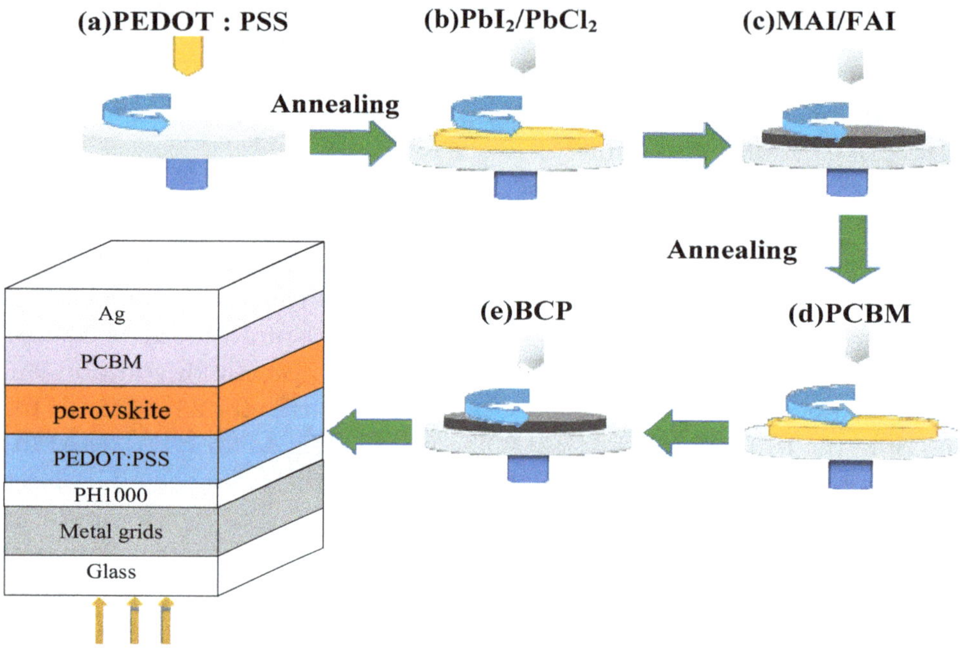

Figure 1. Process of indium thin oxide (ITO)-free perovskite solar cells (PSCs) based on a metal mesh transparent electrode.

3.1. Optimization of Ni/Au Mesh Transparent Electrode

There are two curial interfaces introduced by the metal mesh, including the metal/substrate and metal/HTL interfaces. First, the good adhesion of metal mesh on substrate is necessary to fabricate the PSCs. Although the silver material possesses the lowest resistivity, it has been found in our previous work [36] that the adhesion of Ag mesh on glass substrates is relatively weak, and the mesh is easily damaged in the lift-off process, thus the Ni mesh is chosen to fabricate PSCs. However, the relatively low conductivity of Ni limits the improvement of device performance. In this work, an ultrathin Au is introduced to enhance the conductivity of Ni, and the shape and thickness of Ni/Au mesh has been further optimized. On the other hand, as the PEDOT:PSS HTLs (about 10 nm) is prepared by solution coating, the film quality is highly related to the smoothness of the underlying layer. Considering the thickness of the Ni/Au metal mesh has exceeded 50 nm, the high conductivity PHI000 (about 100 nm) is deposited to smooth the raised surface of the metal mesh, and simultaneously enhance the hole collection ability of Ni/Au electrode. It is noted that the PH1000 only causes a little loss of transmittance in the visible region [5], therefore the Ni/Au/PH1000 electrode is selected in the ITO-free PSCs.

Figure 2a provides the transmittance of spectra of Ni/Au mesh (square, hexagon) and ITO on glass, and the glass substrate. It can be seen that the glass shows the highest transmittance at over 90% in 300 nm to 850 nm, the commercial ITO coated glass substrate possesses an average transmittance over

80%, with the oscillation behavior related to the optical interference at the ITO/glass interface. For the Ni/Au meshes, although the highest transmittance is lower than the ITO sample, a relatively higher transmittance is obtained in the 300–370 nm and 420–500 nm regions. Their average transmittance has also exceeded 80% and the hexagon mesh shows a slightly low transmittance compared to the square mesh. Consequently, the optical performance of the Ni/Au mesh can meet the requirement of the transparent electrode. For the electrical property, the four-point-probe method is used to measure their sheet resistances. The corresponding values are also in Figure 2, showing that the hexagon mesh (22.6 Ω/sq) has lower resistance than the square mesh (30.7 Ω/sq), but larger resistance than that of the ITO (10 Ω/sq). From Figure S3, it is clear that the PSC with the Ni/Au square mesh electrode shows better performance compared to the device with a pure Ni electrode (58 Ω/sq); and the PSC with a hexagon Ni/Au electrode performs better than the device with a square Ni/Au electrode. Their improved PCEs mainly come from the increased J_{SC} values, which is in line with the results of sheet resistance. Thus, the hexagon Ni/Au mesh should be more suitable for fabricating ITO-free PSCs, and its thickness is further optimized. Here, the thickness of Au is fixed on 10 nm, and the thicknesses of Ni vary by 10 nm, 20 nm, 30 nm, and 40 nm. Table 1 shows the corresponding sheet resistances of the Ni/Au meshes. When the thickness of Ni increases from 10 to 40 nm, the sheet resistance values are 33.8 Ω/sq, 22.6 Ω/sq,16.9 Ω/sq and 13.6 Ω/sq, respectively. It is clear that the thicker the metal is, the lower the sheet resistance is. However, too thick a metal will lead to a worse lift-off effect, such as uneven edges of the metal wire and partial fracturing of the grid, which will induce a weak conductivity in the electrode. Meanwhile, if the thickness is less than 10 nm, the adhesion between Ni and glass substrate becomes very poor. Also, with metal deposited by electron beam evaporation it is difficult to form a homogeneous film at a thickness lower than 10 nm, as that will lead to a weaker conductivity [37]. Therefore, the thickness of Ni is a trade-off, which should be further determined by the device performance. As shown in Figure 2b and Table 1, all the devices show similar V_{OC} and FF values, and the J_{SC} dominants the overall PCEs. When the thickness of Ni is 10 nm or 40 nm, a lower PCE of about 10% is limited by the relatively poor J_{SC} values of 15.43 mA/cm^2 and 16.16 mA/cm^2. While if the 20 nm or 30 nm Ni is used, a J_{SC} exceeding 20 mA/cm^2 can be achieved for PSCs, and the PSCs with30 nm Ni obtain superior PCE of 13.72%. As a result, the optimal metal mesh transparent electrode should be the hexagon Ni (30 nm)/Au (10 nm) mesh electrode, and the corresponding device performance is further investigated.

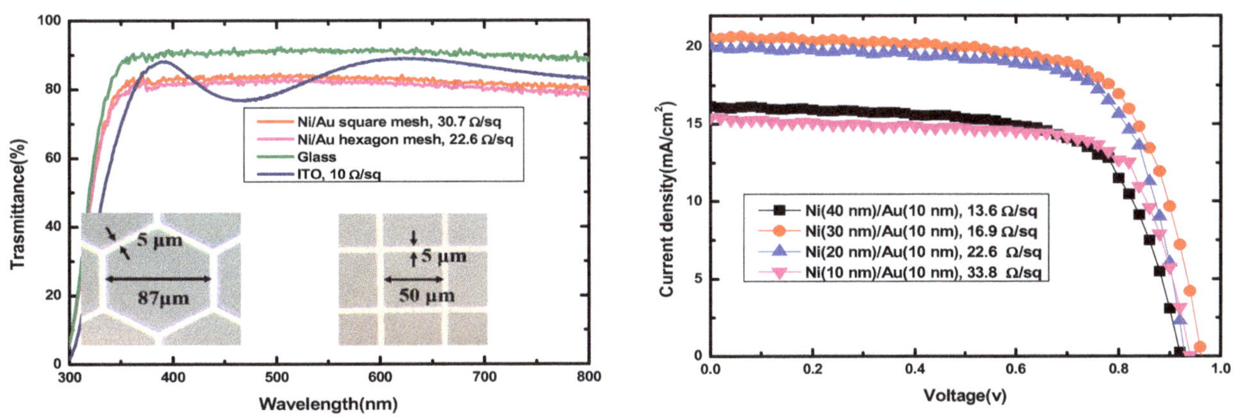

(a) (b)

Figure 2. (**a**) Transmittance spectra of Ni (20 nm)/Au (10 nm) mesh (square, hexagon) and ITO on glass, and the glass substrate; (**b**) density-voltage (J-V) curves for PSCs with hexagon Ni(x nm)/Au (10 nm) (x = 10, 20, 30, 40 nm) metal mesh electrodes. The sheet resistances of Ni/Au and ITO electrode are also displayed in Figure 2.

Table 1. Photovoltaic parameters of PSCs with various Ni(x nm)/Au (10 nm) meshes.

Ni Thickness (nm)	Ni/Au Resistance (Ω/sq)	J_{SC} (mA/cm^2)	V_{OC} (V)	FF (%)	PCE (%)
40	13.6	16.16	0.90	68.76	10.13
30	16.9	20.67	0.96	69.09	13.72
20	22.6	20.04	0.92	71.22	13.14
10	33.8	15.43	0.94	71.68	10.43

3.2. *Performance of PSCs with Optimal Ni/Au Metal Mesh*

Figure 3a presents the J-V curves of the champion PSC with the electrode and reference PSC with an ITO electrode. It can be seen that the PSC based on Ni/Au mesh obtains a champion PCE of 13.88%, V_{OC} of 0.94 V, J_{SC} of 21.14 mA/cm^2, and FF of 69.75% at reverse voltage scan direction; while the forward scanned PCE is 13.39% with V_{OC} = 0.93 V, J_{SC} = 21.04 mA/cm^2, and FF = 68.42%(Figure 3c). This negligible current hysteresis in the PSC based on Ni/Au mesh electrode is strongly related to the effects of PCBM passivation and BCP interface medication, which has been proved in our previous works [3,4]. Further, the steady-state outputs of current density and PCE at the maximum power point voltage of PSC based on Ni/Au mesh are shown in Figure 3d, the nearly unchanged current density and PCE outputs during 160 s illustrate the good operation stability of PSC. At the same time, the reference ITO based PSC shows a PCE of 15.70%, V_{OC} of 0.96 V, J_{SC} of 21.60 mA/cm^2, and FF of 75.80%. Meanwhile, as shown in Figure S4, the lower leakage current and better rectification characteristics are in line with the relatively higher performance of PSC based on the ITO electrode. It is clear that the PSC based on Ni/Au mesh can obtain comparable PCE to that of the ITO based PSC, which demonstrates that the hexagon Ni/Au mesh is a promising ITO alternative. On the other hand, compared to ITO based PSC, the low V_{OC} and FF of Ni/Au based PSC may be explained by the relatively weak charge transport and collections ability. As J-V curves under AM 1.5 G illumination reveal the photovoltaic performance of PSC in the entire absorption range, to further study the photovoltaic conversion at single incident light wavelengths, the IPCE spectra are measured and the results are shown in Figure 3b. The integrated J_{SC} values of 19.29 mA/cm^2 (Ni/Au-PSC) and 19.95 mA/cm^2 (ITO-PSC) agree well with the measured J_{SC} in J-V measurements, which manifest the dependability of the J-V curve measurement. However, it should be noted that the IPCEs of ITO based PSC are higher than that of Ni/Au based PSC at each single wavelength from 300 nm to 800 nm. As the overall optical transmittance of Ni/Au and ITO are close, the perovskite film quality and the charge transport properties may account for the relatively low performance of the Ni/Au based PSC.

First, as the same preparation method of perovskite film has been evaluated in our previous work [3], in the present work the primary concern is the morphology of perovskite films. Figure 4a,c and Figure 4b,d show the SEM and AFM images of perovskite film based on Ni/Au mesh and ITO electrodes. It has been observed that both the perovskite films are compact and smooth, there is no significant difference in the grain sizes and root-mean-square roughness (RMS) values. From this it can be understood that the PH1000 buffer layer has smoothened the raised surface of Ni/Au mesh, which ensures a similar film morphology of perovskite on PEDOT:PSS/PH1000/Ni/Au/Glass and PEDOT:PSS/ITO/Glass samples. Here, we extracted the series resistance (R_s) and shunt resistance (R_{sh}) from the J-V curves by the parameter extraction method in our previous work [38]. As show in Figure S5, the experimental data are well reproduced by the fitting curves and the corresponding R_s, R_{sh}, ideality factor, saturation current are listed in Table S1. It is clear that the PSC based on Ni/Au mesh show a R_S of 1.2 Ω·cm^2 and R_{SH} of 5.548 kΩ·cm^2, while that of the ITO based PSC are 1.0 Ω·cm^2 and 6.398 kΩ·cm^2, combining the larger ideality factor and saturation current values, thus the more efficient carrier transport contributes to the better performance of ITO based PSC. To further study the charge transport properties, transient photocurrent (TPC) and transient photovoltage (TPV) measurements are carried out. The TPC can reflect extraction and transport property of carriers, and the TPV provides insight to carrier recombination property in a solar cell. As shown in Figure 5a, the ITO based PSC has a relatively faster photocurrent decay (0.46 μs) compared with the Ni/Au based PSC (1.96 μs), suggesting that the extraction and transport of carriers are more efficient in ITO based PSC. Meanwhile, the photovoltage decay processes are displayed in Figure 5b, the fitted charge recombination lifetimes are 1389.04 μs and 1183.34 μs for PSC based on ITO and the Ni/Au mesh, respectively. What is more, EIS measurements were carried out to evaluate the carriers' recombination in the corresponding PSCs. The corresponding Nyquist plots, the equivalent circuit diagram, and fitted results are shown in Figure S6 and Table S2. It can be found that, the PSC based on Ni/Au mesh achieved a relatively smaller R_{rec} (carrier recombination resistance) and shorter electron lifetime. Simultaneously, the sheet resistance of Ni (30 nm)/Au (10 nm) mesh (16.9 Ω/sq) is higher than that of ITO. Combining the results of AFM, SEM, TPC, TPV, and conductivity, the relatively low performance of PSC based on Ni/Au mesh can be understood.

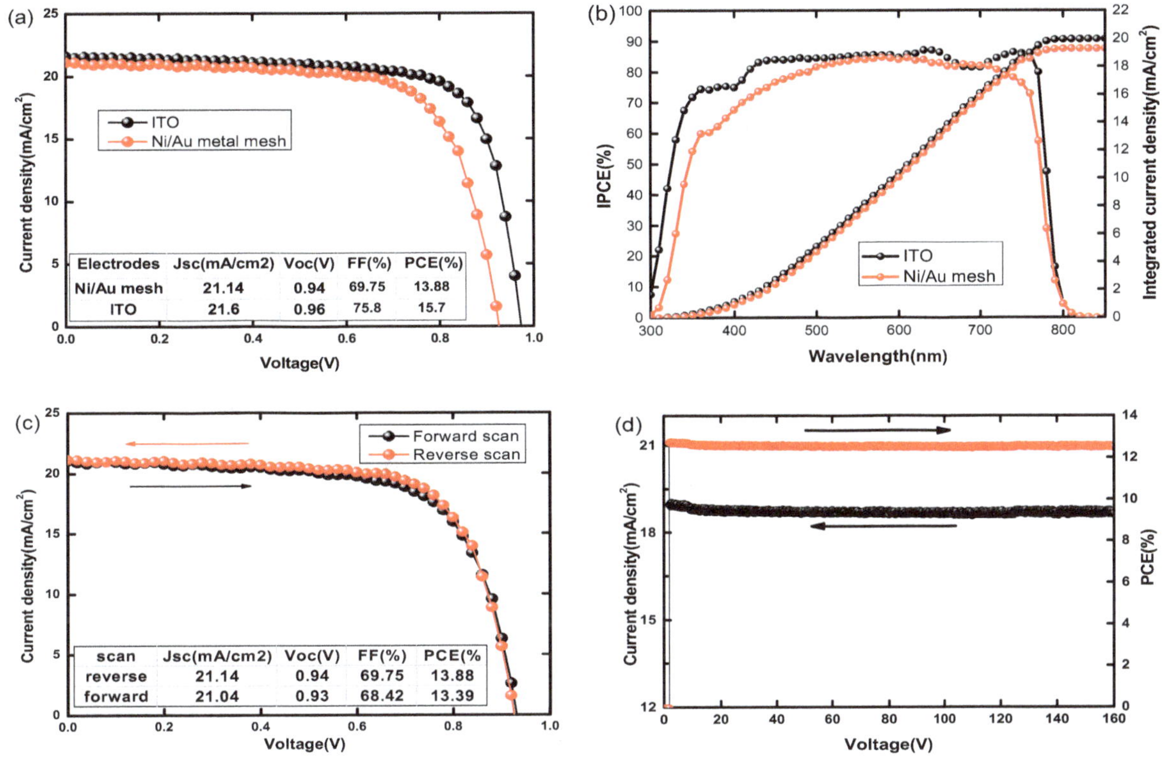

Figure 3. (**a**) J-V curves and (**b**) incident photo-to-current conversion efficiency (IPCE) spectra of PSCs based on Ni/Au mesh and PSCs based ITO measured under 100 mW/cm^2 AM 1.5 G illumination; (**c**) J-V curves in forward and reverse scans and (**d**) steady output current density and PCE of PSCs based on Ni/Au mesh and the maximum power point (Vmax = 0.67 V).

In addition, the statistical results of photovoltaic parameters and the store stability of Ni/Au based PSCs are shown in Figures 6 and 7, the device number in the statistical analysis is 15. From Figure 6, it can be seen that most of the V_{OC} values concentrate in the range from 0.84 V to 0.98 V and eight of them exceed 0.92 V; the J_{SC} ranges from 14 mA/cm^2 to 24 mA/cm^2, with eight of them higher than 19 mA/cm^2; the FF from 62% to 76%, with nine of them larger than 69%; the PCE from 9% to 14% with eight of them exceeding 12%. All the photovoltaic parameters exhibit Gaussian distribution, which suggests the good reproducibility of Ni/Au based PSCs. Here, their FF and V_{OC} values should be further improved to obtain high performance PSCs. Besides the optimization of Ni/Au mesh electrode, more efficient charge transport layers [3,39], and interface passivation [1,40] are feasible approaches to achieve this goal. Furthermore, from Figure 7, the PSCs based on Ni/Au mesh and ITO present a similar trend of PCE degradation. After being stored in N_2 for 240 h, the Ni/Au based PSCs keep 58% of their initial PCE values, and the ITO based devices maintain 60% of their initial PCE values. Therefore, the PSCs based on Ni/Au mesh possess comparable photovoltaic performance and stability to the reference ITO based PSCs, and the Ni/Au mesh is one of the promising ITO alternatives to fabricate PSCs. Nowadays, owing to the improvement of perovskite quality and interface modification, the PCE of planar ITO based PSCs has exceeded 23% [1]. It is believed that there is plenty of room for performance improvement of PSCs based on metal mesh transparent electrodes, which will be further investigated in our future works.

Figure 4. SEM and AFM images of perovskite film based on (**a**,**c**) Ni/Au mesh and (**b**,**d**) ITO electrodes. The unit of RMS values are nm.

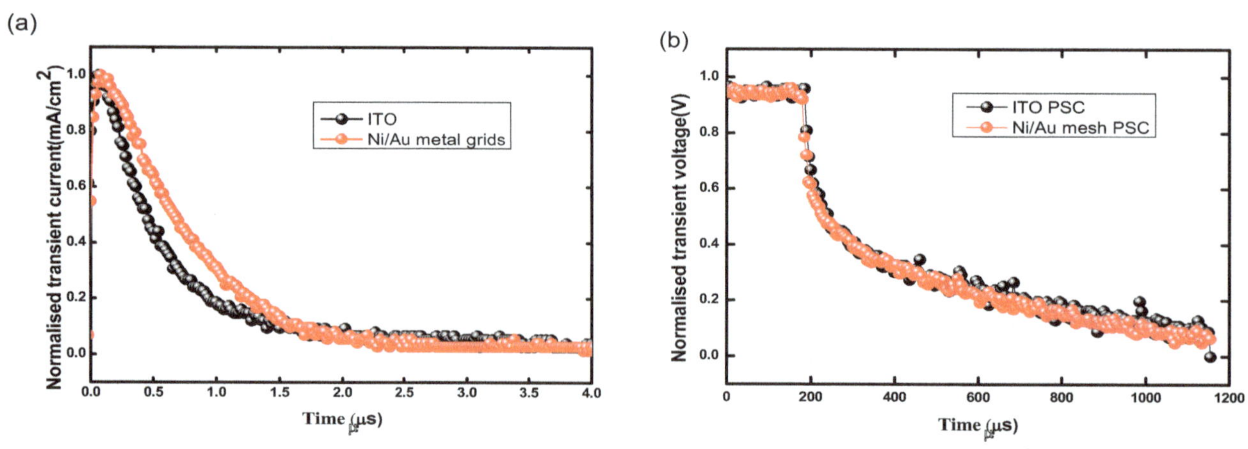

Figure 5. (**a**) Transient photocurrent (TPC) and (**b**) transient photovoltage (TPV) decay curves of the PSCs based on Ni/Au mesh and ITO electrodes.

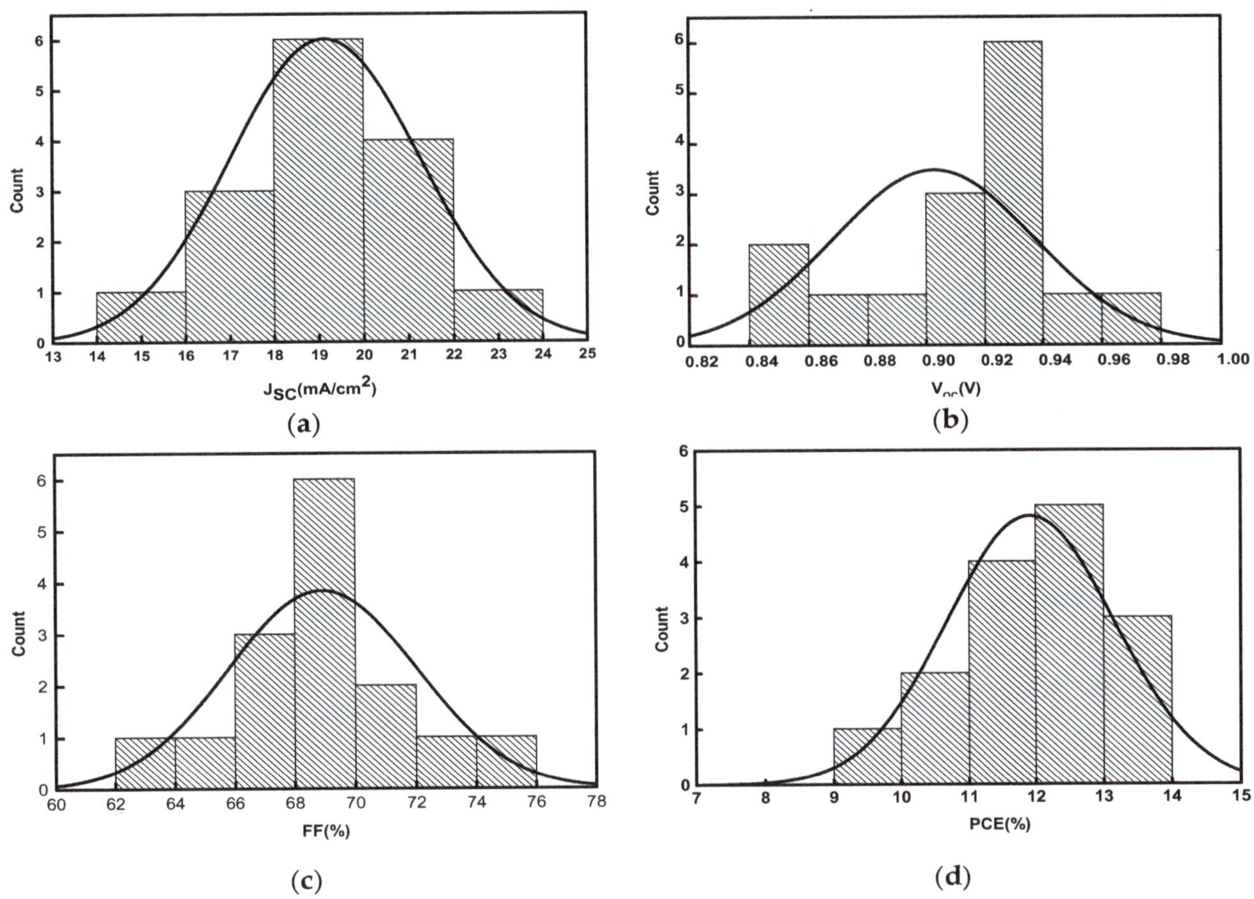

Figure 6. Histograms of photovoltaic parameters. (**a**) J_{SC}, (**b**) V_{OC}, (**c**) FF, and (**d**) PCE for PSC based on Ni/Au mesh electrode. The histograms shown are the photovoltaic parameters for 15 devices.

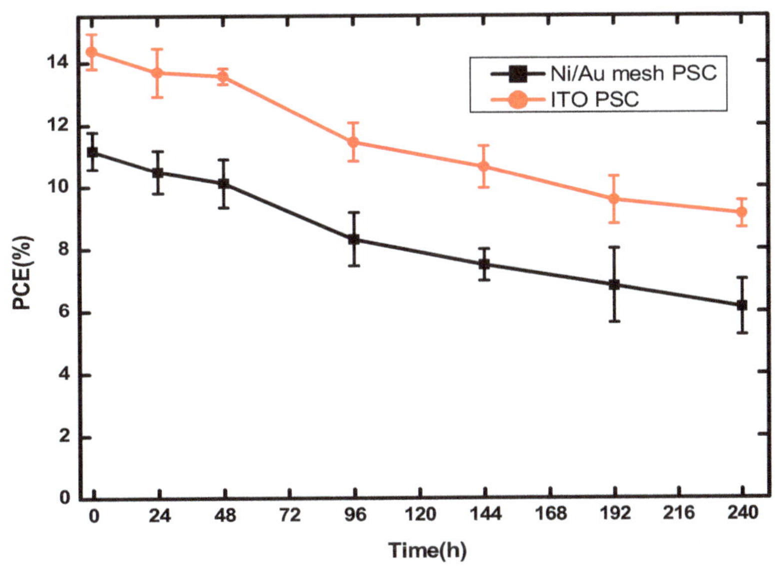

Figure 7. Storing stability of Ni/Au mesh and ITO based PSCs without encapsulation in N_2 for 240 h.

What is more, the large-area (1 cm^2) Ni/Au mesh-based PSCs are fabricated under the same process conditions as the small-area (0.09 cm^2) PSCs. Figure 8a,b shows the J-V curve, photovoltaic parameters, and steady outputs of large-area Ni/Au based PSCs and ITO based reference PSCs. The Ni/Au based PSC obtains a PCE of 6.01%, with V_{OC} = 1.04 V, J_{SC} = 11.28 mA/cm^2, and FF = 51.28% at the reverse voltage scan direction, as well the steady PCE and current density outputs under continuous AM 1.5 G illumination during 160 s. The forward scanned PCE = 5.93%, V_{OC} = 1.05 V, J_{SC} = 10.81 mA/cm^2, and FF = 52.27%, thus the current hysteresis effect is negligible. While the ITO based PSC shows a higher PCE of 9.01% with V_{OC} = 1.02 V, J_{SC} = 16.24 mA/cm^2, and FF = 55.59%, it should be noted that the J_{SC} and FF of large area PSCs are obviously lower than the small area PSCs (Figure 3), which is due to relatively poor charge extraction and strong charge recombination related to the more defects and traps in the larger-area perovskite layer. This can be partly proved by the TPC and TPV results. As shown in Figure 8c,d, the large-area PSC displays a slower photocurrent decay (3.59 μs) compared to the small-area cell (1.96 μs), suggesting inefficient charge transport and extraction in the large-area cell. Meanwhile, a faster photovoltage decay (644.49 μs) than the small-area one (1183.34 μs) reveals the relatively strong charge recombination in the large-area cell. Therefore, the perovskite film quality and interface modification are crucial to further improving the performance of large-area devices. Considering the drawbacks of ITO, the meal mesh based large-area PSCs may find wider applications in the near future.

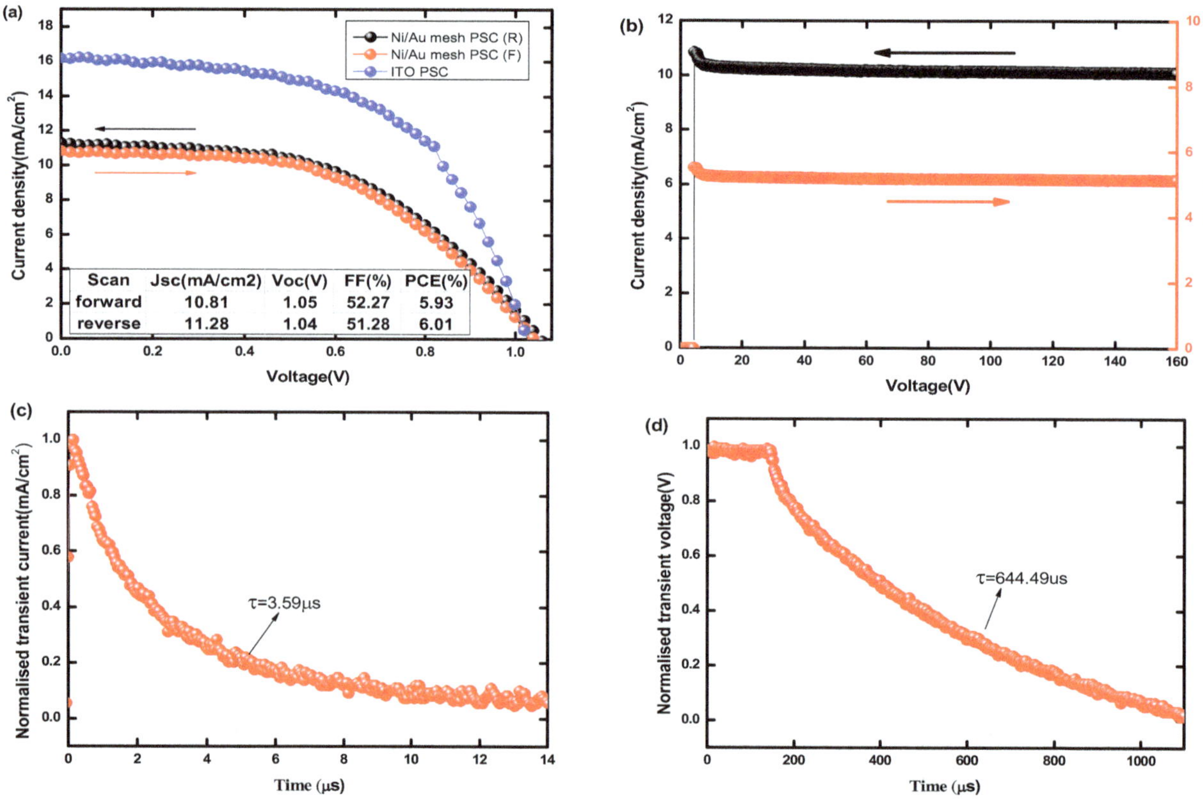

Figure 8. (**a**) J-V curves in forward and reverse scans for large-area PSCs based metal mesh, (**b**) steady output characteristics curve, (**c**) TPC and (**d**) TPV decay curves.

4. Conclusions

In summary, we carefully designed and deposited hexagon Ni/Au metal mesh transparent electrodes with high transmittance and low resistance by photolithography and the e-beam evaporation process. To be an efficient electrode for PSCs, the Ni was used to improve the adhesion of metal mesh to glass substrate and the Au was used to increase the conductivity of the metal mesh. The conductive polymer PH1000 not only smooths the raised surface of metal mesh but also enhances the charge collection ability of metal mesh. The optimal hexagonal Ni (30 nm)/Au (10 nm)/PH1000 electrodes were employed to fabricate ITO-free PSCs. The champion PSC (0.09 cm^2), with a typical planar structure, obtained a PCE of 13.88%, negligible current hysteresis, steady current density and PCE outputs, good repeatability and storing stability. The comparable performance to the ITO based reference PSC demonstrates that the Ni/Au mesh transparent electrodes are a promising ITO alternative to fabricate efficient PSCs. Further, we tried to fabricate the large area (1 cm^2) devices under the same low-temperature process, and achieved a PCE over 6% with negligible hysteresis and stable steady outputs. And the perovskite film quality and interface modification are crucial to further improving the performance of large-area devices. Considering the drawbacks of ITO, the meal mesh-based PSCs may find wider applications in the near future.

Supplementary Materials

Figure S1: Photos of ITO-free PSCs (0.09 cm^2) based on Ni/Au mesh electrode (a) light incident surface (b) backlight surface, Figure S2: Photos of large-area ITO-free PSCs (1 cm^2) based on Ni/Au mesh electrode (a) light incident surface (b) backlight surface., Figure S3: J-V curves of ITO-free PSCs (0.09cm^2) based on Ni (30 nm) square, Ni(20 nm)/Au(10 nm) square, and Ni(20 nm)/Au(10 nm) hexagon mesh electrodes. The sheet resistance of pure Ni and Ni/Au meshes are about 58 Ω/sq, 30.7 Ω/sq, and 22.6 Ω/sq, thus the PSC with Ni/Au hexagon electrode obtains higher Jsc, FF, and PCE values, Figure S4: Semi-log plots of dark JV curves for PSC based on Ni/Au mesh and ITO electrodes, Figure S5: JV curves under AM 1.5G illumination for PSC based on Ni/Au mesh and ITO electrodes. Here the symbols represent the experimental data and the solid lines indicate the fitting curves, Figure S6: Nyquist curves of PSCs based on Ni/Au mesh and ITO electrodes. The insert is the equivalent circuit for the fittings, where R_s represents the series resistive elements related to connections and devices, R_{rec} the carrier recombination resistance and, C_{PE} the constant phase element.; Table S1: R_s, R_{sh}, saturation current, and ideality factor extracted from JV curves under AM 1.5G illumination, Table S2: EIS parameters extracted from Nyquist curves. The electron lifetime is the reciprocal of the frequency of the maximum point of the semi-circular response.

Author Contributions: D.C. and C.Z. conceived the idea, designed the experiment and guided the experiment. G.F., H.Z. and L.Z. conducted most of the electrode preparation, device fabrication, data collection and wrote the manuscript; D.C. and C.Z. revised the manuscript; W.Z., H.X. helped the device measurement, X.H. and Z.L. helped the film characterization, H.D. and S.P. helped the data analysis. J.Z. and Y.H. supervised the group. All authors read and approved the manuscript.

References

1. Jiang, Q.; Zhao, Y.; Zhang, X.; Yang, X.; Chen, Y.; Chu, Z.; Ye, Q.; Li, X.; Yin, Z.; You, J. Surface passivation of perovskite film for efficient solar cells. *Nat. Photonics* **2019**, in press. [CrossRef]
2. Green, M.A.; Hishikawa, Y.; Warta, W.; Dunlop, E.D.; Levi, D.H.; Hohl-Ebinger, J.; Ho-Baillie, A.W.H. Solar cell efficiency tables (version 50). *Prog. Photovolt. Res. Appl.* **2017**, *25*, 668–676. [CrossRef]
3. Liu, Z.; Chang, J.; Lin, Z.; Zhou, L.; Yang, Z.; Chen, D.; Zhang, C.; Liu, S.; Hao, Y. High-performance planar perovskite solar cells using low temperature, solution–combustion-based nickel oxide hole transporting layer with efficiency exceeding 20%. *Adv. Energy Mater.* **2018**, *8*, 1703432. [CrossRef]
4. Pang, S.; Chen, D.; Zhang, C.; Chang, J.; Lin, Z.; Yang, H.; Sun, X.; Mo, J.; Xi, H.; Han, G.; et al. Efficient bifacial semitransparent perovskite solar cells with silver thin film electrode. *Sol. Energy Mater. Sol. Cells* **2017**, *170*, 278–286. [CrossRef]
5. Li, Y.; Meng, L.; Yang, Y.; Xu, G.; Hong, Z.; Chen, Q.; You, J.; Li, G.; Yang, Y.; Li, Y. High-efficiency robust perovskite solar cells on ultrathin flexible substrates. *Nat. Commun.* **2016**, *7*, 10214. [CrossRef] [PubMed]
6. Li, L.; Zhang, S.; Yang, Z.; Berthold, E.E.S.; Chen, W. Recent advances of flexible perovskite solar cells. *J. Energy Chem.* **2018**, *27*, 673–689. [CrossRef]
7. Fu, F.; Feurer, T.; Jäger, T.; Avancini, E.; Bissig, B.; Yoon, S.; Buecheler, S.; Tiwari, A.N. Low-temperature-processed efficient semi-transparent planar perovskite solar cells for bifacial and tandem applications. *Nat. Commun.* **2015**, *6*, 1–9. [CrossRef]
8. McMeekin, D.P.; Sadoughi, G.; Rehman, W.; Eperon, G.E.; Saliba, M.; Hörantner, M.T.; Haghighirad, A.; Sakai, N.; Korte, L.; Rech, B.; et al. A mixed-cation lead mixed-halide perovskite absorber for tandem solar cells. *Science* **2016**, *351*, 3–8. [CrossRef]
9. Cardinaletti, I.; Vangerven, T.; Nagels, S.; Cornelissen, R.; Schreurs, D.; Hruby, J.; Vodnik, J.; Devisscher, D.; Kesters, J.; D'Haen, J.; et al. Organic and perovskite solar cells for space applications. *Sol. Energy Mater. Sol. Cells* **2018**, *182*, 121–127. [CrossRef]
10. Heo, J.H.; Lee, D.S.; Shin, D.H.; Im, S.H. Recent advancements in and perspectives on flexible hybrid perovskite solar cells. *J. Mater. Chem. A* **2019**, *7*, 888–900. [CrossRef]

11. Loh, K.P.; Tong, S.W.; Wu, J. Graphene and graphene-like molecules: Prospects in solar cells. *J. Am. Chem. Soc.* **2016**, *138*, 1095–1102. [CrossRef] [PubMed]
12. Tang, H.; Jiang, Y.; Tang, C.W.; Kwok, H.S. Grid optimization of large-area OLED lighting panel electrodes. *J. Disp. Technol.* **2016**, *12*, 605–609. [CrossRef]
13. Dou, B.; Whitaker, J.B.; Bruening, K.; Moore, D.T.; Wheeler, L.M.; Ryter, J.; Breslin, N.J.; Berry, J.J.; Garner, S.M.; Barnes, F.S.; et al. Roll-to-roll printing of perovskite solar cells. *ACS Energy Lett.* **2018**, *3*, 2558–2565. [CrossRef]
14. Han, T.H.; Lee, Y.; Choi, M.R.; Woo, S.H.; Bae, S.H.; Hong, B.H.; Ahn, J.H.; Lee, T.W. Extremely efficient flexible organic light-emitting diodes with modified graphene anode. *Nat. Photonics* **2012**, *6*, 105–110. [CrossRef]
15. Kang, J.; Jang, Y.; Kim, Y.; Cho, S.H.; Suhr, J.; Hong, B.H.; Choi, J.B.; Byun, D. An Ag-grid/graphene hybrid structure for large-scale, transparent, flexible heaters. *Nanoscale* **2015**, *7*, 6567–6573. [CrossRef] [PubMed]
16. Chan, J.; Venugopal, A.; Pirkle, A.; McDonnell, S.; Hinojos, D.; Magnuson, C.W.; Ruoff, R.S.; Colombo, L.; Wallace, R.M.; Vogel, E.M. Reducing extrinsic performance-limiting factors in graphene grown by chemical vapor deposition. *ACS Nano* **2012**, *6*, 3224–3229. [CrossRef] [PubMed]
17. Ye, S.; Rathmell, A.R.; Chen, Z.; Stewart, I.E.; Wiley, B.J. Metal nanowire networks: The next generation of transparent conductors. *Adv. Mater.* **2014**, *26*, 6670–6687. [CrossRef]
18. Wu, J.; Que, X.; Hu, Q.; Luo, D.; Liu, T.; Liu, F.; Russell, T.P.; Zhu, R.; Gong, Q. Multi-length scaled silver nanowire grid for application in efficient organic solar cells. *Adv. Funct. Mater.* **2016**, *26*, 4822–4828. [CrossRef]
19. Song, J.; Kulinich, S.A.; Li, J.; Liu, Y.; Zeng, H. A general one-pot strategy for the synthesis of high-performance transparent-conducting-oxide nanocrystal inks for all-solution-processed devices. *Angew. Chem. Int. Ed.* **2015**, *127*, 472–476. [CrossRef]
20. Xiao, Y.; Han, G.; Wu, J.; Lin, J.Y. Efficient bifacial perovskite solar cell based on a highly transparent poly(3,4-Ethylenedioxythiophene) as the P-type hole-transporting material. *J. Power Sour.* **2016**, *306*, 171–177. [CrossRef]
21. Sun, K.; Li, P.; Xia, Y.; Chang, J.; Ouyang, J. Transparent conductive oxide-free perovskite solar cells with PEDOT:PSS as transparent electrode. *ACS Appl. Mater. Interfaces* **2015**, *7*, 15314–15320. [CrossRef] [PubMed]
22. Abachi, T.; Cattin, L.; Louarn, G.; Lare, Y.; Bou, A.; Makha, M.; Torchio, P.; Fleury, M.; Morsli, M.; Addou, M.; et al. Highly flexible, conductive and transparent MoO_3/Ag/MoO_3 multilayer electrode for organic photovoltaic cells. *Thin Solid Films* **2013**, *545*, 438–444. [CrossRef]
23. Pang, S.; Li, X.; Dong, H.; Chen, D.; Zhu, W.; Chang, J.; Lin, Z.; Xi, H.; Zhang, J.; Zhang, C.; et al. Efficient bifacial semitransparent perovskite solar cells using Ag/V 2 O 5 as transparent anodes. *ACS Appl. Mater. Interfaces* **2018**, *10*, 12731–12739. [CrossRef] [PubMed]
24. Chang, C.-Y.; Chang, Y.-C.; Huang, W.-K.; Liao, W.-C.; Wang, H.; Yeh, C.; Tsai, B.-C.; Huang, Y.-C.; Tsao, C.-S. Achieving high efficiency and improved stability in large-area ito-free perovskite solar cells with thiol-functionalized self-assembled monolayers. *J. Mater. Chem. A* **2016**, *4*, 7903–7913. [CrossRef]
25. Mao, L.; Chen, Q.; Li, Y.; Li, Y.; Cai, J.; Su, W.; Bai, S.; Jin, Y.; Ma, C.Q.; Cui, Z.; et al. Flexible silver grid/PEDOT: PSS hybrid electrodes for large area inverted polymer solar cells. *Nano Energy* **2014**, *10*, 259–267. [CrossRef]
26. Sam, F.L.M.; Mills, C.A.; Rozanski, L.J.; Silva, S.R.P. Thin film hexagonal gold grids as transparent conducting electrodes in organic light emitting diodes. *Laser Photonics Rev.* **2014**, *8*, 172–179. [CrossRef]
27. Kim, W.; Kim, S.; Kang, I.; Jung, M.S.; Kim, S.J.; Kim, J.K.; Cho, S.M.; Kim, J.H.; Park, J.H. Hybrid silver mesh electrode for ITO-free flexible polymer solar cells with good mechanical stability. *ChemSusChem* **2016**, *9*, 1042–1049. [CrossRef] [PubMed]
28. Suh, Y.D.; Kwon, J.; Lee, J.; Lee, H.; Jeong, S.; Kim, D.; Cho, H.; Yeo, J.; Ko, S.H. Maskless fabrication of highly robust, flexible transparent Cu conductor by random crack network assisted Cu nanoparticle patterning and laser sintering. *Adv. Electron. Mater.* **2016**, *7*, 5024–5031. [CrossRef]
29. Kim, W.K.; Lee, S.; Hee Lee, D.; Hee Park, I.; Seong Bae, J.; Woo Lee, T.; Kim, J.Y.; Hun Park, J.; Chan Cho, Y.; Ryong Cho, C.; et al. Cu mesh for flexible transparent conductive electrodes. *Sci. Rep.* **2015**, *5*, 10715–10722. [CrossRef]
30. Guo, C.F.; Sun, T.; Liu, Q.; Suo, Z.; Ren, Z. Highly stretchable and transparent nanomesh electrodes made by grain boundary lithography. *Nat. Commun.* **2014**, *5*, 3121. [CrossRef]

31. Galagan, Y.; Rubingh, J.E.J.; Andriessen, R.; Fan, C.C.; Blom, P.W.; Veenstra, S.C.; Kroon, J.M. ITO-free flexible organic solar cells with printed current collecting grids. *Sol. Energy Mater. Sol. Cells* **2011**, *17*, 349–354. [CrossRef]
32. Available online: https://en.wikipedia.org/wiki/Photolithography (accessed on 8 June 2019).
33. Abdollahi Nejand, B.; Nazari, P.; Gharibzadeh, S.; Ahmadi, V.; Moshaii, A. All-inorganic large-area low-cost and durable flexible perovskite solar cells using copper foil as a substrate. *Chem. Commun.* **2017**, *53*, 747–750. [CrossRef] [PubMed]
34. Chen, L.; Xie, X.; Liu, Z.; Lee, E.C. A transparent poly(3,4-ethylenedioxylenethiophene):poly(styrene sulfonate) cathode for low temperature processed, metal-oxide free perovskite solar cells. *J. Mater. Chem. A* **2017**, *5*, 6974–6980. [CrossRef]
35. Troughton, J.; Bryant, D.; Wojciechowski, K.; Carnie, M.J.; Snaith, H.; Worsley, D.A.; Watson, T.M. Highly efficient, flexible, indium-free perovskite solar cells employing metallic substrates. *J. Mater. Chem. A* **2015**, *3*, 9141–9145. [CrossRef]
36. Zhang, H.X.; Chen, D.Z.; Zhang, C.F. ITO-free perovskite solar cells using photolithography processed metal grids as transparent anodes. In Proceedings of the 2016 13th IEEE International Conference on Solid-State and Integrated Circuit Technology (ICSICT 2016), Hangzhou, China, 25–28 October 2016; pp. 1026–1028.
37. Lai, W.C.; Lin, K.W.; Wang, Y.T.; Chiang, T.Y.; Chen, P.; Guo, T.F. Oxidized Ni/Au transparent electrode in efficient $CH_3NH_3PbI_3$ perovskite/fullerene planar heterojunction hybrid solar cells. *Adv. Mater.* **2016**, *28*, 3290–3297. [CrossRef] [PubMed]
38. Zhang, C.; Zhang, J.; Hao, Y.; Lin, Z.; Zhu, C. A simple and efficient solar cell parameter extraction method from a single current-voltage curve. *J. Appl. Phys.* **2011**, *110*, 064504. [CrossRef]
39. Macdonald, T.J.; Batmunkh, M.; Lin, C.-T.; Kim, J.; Tune, D.D.; Amboz, F.; Li, X.; Xu, S.; Sol, C.; Papakonstantinou, I.; et al. Origin of performance enhancement in TiO_2-carbon nanotube composite perovskite solar cells. *Small Methods* **2019**, 1900164, Early View. [CrossRef]
40. Chen, D.; Pang, S.; Zhou, L.; Li, X.; Su, A.; Zhu, W.; Chang, J.; Zhang, J.; Zhang, C.; Hao, Y. Efficient TeO_2/Ag transparent top electrode for 20%-efficiency bifacial perovskite solar cells with a bifaciality factor exceeding 80%. *J. Mater. Chem. A* **2019**. accepted manuscript. [CrossRef]

Permissions

All chapters in this book were first published in MDPI; hereby published with permission under the Creative Commons Attribution License or equivalent. Every chapter published in this book has been scrutinized by our experts. Their significance has been extensively debated. The topics covered herein carry significant findings which will fuel the growth of the discipline. They may even be implemented as practical applications or may be referred to as a beginning point for another development.

The contributors of this book come from diverse backgrounds, making this book a truly international effort. This book will bring forth new frontiers with its revolutionizing research information and detailed analysis of the nascent developments around the world.

We would like to thank all the contributing authors for lending their expertise to make the book truly unique. They have played a crucial role in the development of this book. Without their invaluable contributions this book wouldn't have been possible. They have made vital efforts to compile up to date information on the varied aspects of this subject to make this book a valuable addition to the collection of many professionals and students.

This book was conceptualized with the vision of imparting up-to-date information and advanced data in this field. To ensure the same, a matchless editorial board was set up. Every individual on the board went through rigorous rounds of assessment to prove their worth. After which they invested a large part of their time researching and compiling the most relevant data for our readers.

The editorial board has been involved in producing this book since its inception. They have spent rigorous hours researching and exploring the diverse topics which have resulted in the successful publishing of this book. They have passed on their knowledge of decades through this book. To expedite this challenging task, the publisher supported the team at every step. A small team of assistant editors was also appointed to further simplify the editing procedure and attain best results for the readers.

Apart from the editorial board, the designing team has also invested a significant amount of their time in understanding the subject and creating the most relevant covers. They scrutinized every image to scout for the most suitable representation of the subject and create an appropriate cover for the book.

The publishing team has been an ardent support to the editorial, designing and production team. Their endless efforts to recruit the best for this project, has resulted in the accomplishment of this book. They are a veteran in the field of academics and their pool of knowledge is as vast as their experience in printing. Their expertise and guidance has proved useful at every step. Their uncompromising quality standards have made this book an exceptional effort. Their encouragement from time to time has been an inspiration for everyone.

The publisher and the editorial board hope that this book will prove to be a valuable piece of knowledge for researchers, students, practitioners and scholars across the globe.

List of Contributors

Guixia Yang, Yuanlong Pang, Yuqing Yang, Jianyong Liu and Shuming Peng
Institute of Nuclear Physics and Chemistry, China Academy of Engineering Physics, Mianyang 621900, China

Gang Chen
State Key Laboratory of Wide-Bandgap Semiconductor Power Electronics, Nanjing 210000, China

Ming Jiang, Xiaotao Zu, Liang Qiao and Haiyan Xiao
School of Physical, University of Electronic Science and Technology of China, Chengdu 610054, China

Xuan Fang
School of Science, Changchun University of Science and Technology, 7089 Wei-Xing Road, Changchun 130022, China

Hongbin Zhao
State Key Laboratory of Advanced Materials for Smart Sensing, General Research Institute for Nonferrous Metals, Beijing 100088, China

Kai Wang, Haoran Chen, Tingting Niu, Shan Wang and Xiao Guo
Key Laboratory of Flexible Electronics (KLOFE) & Institute of Advanced Materials (IAM), Jiangsu National Synergetic Innovation Center for Advanced Materials (SICAM), Nanjing Tech University (Nanjing Tech), Nanjing 211816, China

Hong Wang
Zhongshan Institute of Modern Industrial Technology, South China University of Technology, Zhongshan 528437, China

Duy Phong Pham
School of Information and Communication Engineering, Sungkyunkwan University, Suwon 16419, Korea

Junsin Yi
School of Information and Communication Engineering, Sungkyunkwan University, Suwon 16419, Korea
College of Information and Communication Engineering, Sungkyunkwan University, Seoul 16419, Korea

Eun-Chel Cho
School of Information and Communication Engineering, Sungkyunkwan University, Suwon 16419, Korea
College of Information and Communication Engineering, Sungkyunkwan University, Seoul 16419, Korea

Youngkuk Kim
School of Information and Communication Engineering, Sungkyunkwan University, Suwon 16419, Korea
College of Information and Communication Engineering, Sungkyunkwan University, Seoul 16419, Korea

Sunhwa Lee
School of Information and Communication Engineering, Sungkyunkwan University, Suwon 16419, Korea
College of Information and Communication Engineering, Sungkyunkwan University, Suwon 16419, Korea

Jinjoo Park
Division of Energy and Optical Technology Convergence, Cheongju University, Cheongju 28503, Korea
Major of Energy and Applied Chemistry, Division of Energy & Optical Technology Convergence, Cheongju University 298, Daeseong-ro, Cheongwon-gu, Cheongju-si, Chungcheongbuk-do, Cheongju 28503, Korea

Calum McDonald and Vladimir Svrcek
Research Center for Photovoltaics, National Institute of Advanced Industrial Science and Technology (AIST), Tsukuba, Ibaraki 305-8568, Japan

Chengsheng Ni
College of Resources and Environment, Southwest University, Beibei, Chongqing 400715, China

Paul Maguire and Davide Mariotti
School of Engineering, Ulster University, Newtownabbey BT37 0QB, UK

Paul Connor and John T. S. Irvine
School of Chemistry, University of St Andrews, North Haugh, St Andrews KY16 9AJ, UK

Filip Ambroz, Joanna L. Donnelly, Jonathan D. Wilden, Thomas J. Macdonald and Ivan P. Parkin
Department of Chemistry, University College London 20 Gordon St., London WC1H 0AJ, UK

Cheolmin Park and Nagarajan Balaji
Department of Energy Science, Sungkyunkwan University, Suwon 16419, Korea

Sungyoon Chung
College of Information and Communication Engineering, Sungkyunkwan University, Suwon 16419, Korea

Shihyun Ahn
Department of Electrical and Computer Engineering, University of Seoul, 163, Seoulsiripdae-ro, Dongdaemun-gu, Seoul 02504, Korea

Hang Dong, Shangzheng Pang, Yi Zhang, Weidong Zhu, Jingjing Chang and Yue Hao
State Key Discipline Laboratory of Wide Band Gap Semiconductor Technology, School of Microelectronics, Xidian University, 2 South Taibai Road, Xi'an 710071, China

Radhouane Chtourou
Laboratory of Nanomaterials and Systems for Renewable Energies (LaNSER), Research and Technology Center of Energy, Techno-Park Borj-Cedria, Bp 95, 2050 Hammam-Lif, Tunis, Tunisia

Pablo Aitor Postigo
Instituto de Micro y Nanotecnología, IMN-CNM, CSIC (CEI UAM+CSIC) Isaac Newton, 8, E-28760 Tres Cantos, Madrid, Spain

Mohamed Ali Zaïbi
Laboratory of Nanomaterials and Systems for Renewable Energies (LaNSER), Research and Technology Center of Energy, Techno-Park Borj-Cedria, Bp 95, 2050 Hammam-Lif, Tunis, Tunisia
Tunis University—National High School of Engineering of Tunis, 5 Av Taha Hussein 1008 Tunis, Tunisia

Mariem Naffeti
Laboratory of Nanomaterials and Systems for Renewable Energies (LaNSER), Research and Technology Center of Energy, Techno-Park Borj-Cedria, Bp 95, 2050 Hammam-Lif, Tunis, Tunisia
Instituto de Micro y Nanotecnología, IMN-CNM, CSIC (CEI UAM+CSIC) Isaac Newton, 8, E-28760 Tres Cantos, Madrid, Spain
Tunis University—National High School of Engineering of Tunis, 5 Av Taha Hussein 1008 Tunis, Tunisia

Dan Craciunescu and Silvian Fara
Department of Physics, Faculty of Applied Sciences, Polytechnic University of Bucharest, 060042 Bucharest, Romania

Laurentiu Fara
Department of Physics, Faculty of Applied Sciences, Polytechnic University of Bucharest, 060042 Bucharest, Romania
Academy of Romanian Scientists, 050091 Bucharest, Romania

Irinela Chilibon, Dan Savastru, Cristina Vasiliu and Laurentiu Baschir
National Institute of Research and Development for Optoelectronics (INOE-2000), 077125 Bucharest, Romania

Ørnulf Nordseth
Institute for Energy Technology (IFE), NO-2027 Kjeller, Norway

Raj Kumar and Edouard Monakhov
Department of Physics/Center for Materials Science and Nanotechnology (SMN), University of Oslo, Blindern, N-0316 Oslo, Norway

James P. Connolly
GeePs (Group of electrical engineering—Paris), UMR CNRS 8507, CentraleSupélec, Univ. Paris-Sud, Université Paris-Saclay, Sorbonne Université, 11 rue Joliot-Curie, 91192 Plateau de Moulon, Gif-sur-Yvette CEDEX, France

Hsuan-Ta Wu and Yu-Ting Cheng
Department of Electrical Engineering, National Cheng Kung University, Tainan 70101, Taiwan

Ching-Chich Leu
Department of Chemical and Materials Engineering, National University of Kaohsiung, Kaohsiung 81148, Taiwan

Shih-Hsiung Wu
Green Energy and Environment Research Laboratories, Industrial Technology Research Institute, Hsinchu 31040, Taiwan

Chuan-Feng Shih
Department of Electrical Engineering, National Cheng Kung University, Tainan 70101, Taiwan
Hierarchical Green-Energy Materials (Hi-GEM) Research Center, National Cheng Kung University, Tainan 70101, Taiwan

Jia-Ren Wu, Diksha Thakur, Shou-En Chiang and Anjali Chandel
Department of Physics, Chung Yuan Christian University, Taoyuan 32023, Taiwan

Jyh-Shyang Wang, Kuan-Cheng Chiu and Sheng Hsiung Chang
Center for Nano Technology, Chung Yuan Christian University, Taoyuan 32023, Taiwan

Kwan Hong Min
Photovoltaics Laboratory, Korea Institute of Energy Research, Daejeon 34129, Korea
Department of Materials Science and Engineering, Korea University, Seoul 02841, Korea

Min Gu Kang, Hee-eun Song, Sungeun Park and Sang Hee Lee
Photovoltaics Laboratory, Korea Institute of Energy Research, Daejeon 34129, Korea

Donghwan Kim
Department of Materials Science and Engineering, Korea University, Seoul 02841, Korea

Taejun Kim
Hyundai Energy Solutions, PV R&D Center, Eumseong, Cheongju 27711, Korea

Yoonmook Kang and Hae-Seok Lee
Department of Energy Environment Policy and Technology, Green School, Graduate School of Korea Energy and Environment, Korea University, Seoul 02841, Korea

Hongye Chen, Min Li, Xiaoyan Wen, Yingping Yang, Daping He and Haifei Lu
School of Science, Wuhan University of Technology, Wuhan 430070, China

Wallace C. H. Choy
Department of Electrical and Electronic Engineering, The University of Hong Kong, Pok Fu Lam Road, Hong Kong SAR, China

Madeshwaran Sekkarapatti Ramasamy, Ka Yeon Ryu, Ju Won Lim, Asia Bibi, Hannah Kwon, Ji-Eun Lee, Dong Ha Kim and Kyungkon Kim
Deprtment of Chemistry and Nanoscience, Ewha Womans University, 52 Ewhayeodae-gil, Seodaemun-gu, Seoul 03760, Korea

Minkyu Ju, Jeongeun Park and Donggun Lim
School of Electronic, Electrical Engineering, Korea National University of Transportation, Chungju 27469, Korea

Young Hyun Cho
School of Information and Communication Engineering, Sungkyunkwan University, Suwon 16419, Korea

Xianglin Mei, Bin Wu, Xiuzhen Guo, Xiaolin Liu, Zhitao Rong, Songwei Liu, Yanru Chen and Bingchang Chen
School of Materials Science and Engineering, South China University of Technology, Guangzhou 510640, China

Donghuan Qin and Wei Xu
School of Materials Science and Engineering, South China University of Technology, Guangzhou 510640, China
Institute of Polymer Optoelectronic Materials & Devices, State Key Laboratory of Luminescent Materials & Devices, South China University of Technology, Guangzhou 510640, China

Lintao Hou
Guangdong Provincial Key Laboratory of Optical Fiber Sensing and Communications, Guangzhou Key Laboratory of Vacuum Coating Technologies and New Energy Materials, Siyuan Laboratory, Department of Physics, Jinan University, Guangzhou 510632, China

Dazheng Chen, Jincheng Zhang and Chunfu Zhang
Wide Bandgap Semiconductor Technology Disciplines State Key Laboratory, School of Microelectronics, Xidian University, Xi'an 710071, China
Shaanxi Joint Key Laboratory of Graphene, Xidian University, Xi'an 710071, China

Gang Fan, Xiaoning He and Zhenhua Lin
Wide Bandgap Semiconductor Technology Disciplines State Key Laboratory, School of Microelectronics, Xidian University, Xi'an 710071, China

Hongxiao Zhang and Long Zhou
Shaanxi Joint Key Laboratory of Graphene, Xidian University, Xi'an 710071, China

He Xi
Wide Bandgap Semiconductor Technology Disciplines State Key Laboratory, School of Microelectronics, Xidian University, Xi'an 710071, China
School of Advanced Materials and Nanotechnology, Xidian University, Xi'an 710071, China

Sanchari Chowdhury, Jinsu Park, Jaemin Kim, Sehyeon Kim and Younghyun Cho
College of Information and Communication Engineering, Sungkyunkwan University, Seoul 16419, Korea

Index

www.ingramcontent.com/pod-product-compliance
Lightning Source LLC
LaVergne TN
LVHW060524210926
848485LV00005B/1140
* 9 7 8 1 6 4 1 1 6 8 2 6 7 *